Genomic Analysis of Antibiotics Resistance in Pathogens

Editor

Teresa V. Nogueira

MDPI • Basel • Beijing • Wuhan • Barcelona • Belgrade • Manchester • Tokyo • Cluj • Tianjin

Editor
Teresa V. Nogueira
UEISPSA
INIAV–National Institute for
Agricultural and Veterinary
Research
Oeiras
Portugal

Editorial Office
MDPI
St. Alban-Anlage 66
4052 Basel, Switzerland

This is a reprint of articles from the Special Issue published online in the open access journal *Antibiotics* (ISSN 2079-6382) (available at: www.mdpi.com/journal/antibiotics/special_issues/Gene_pathogen).

For citation purposes, cite each article independently as indicated on the article page online and as indicated below:

LastName, A.A.; LastName, B.B.; LastName, C.C. Article Title. *Journal Name* **Year**, *Volume Number*, Page Range.

ISBN 978-3-0365-6524-8 (Hbk)
ISBN 978-3-0365-6523-1 (PDF)

Contents

About the Editor

Teresa V. Nogueira

Teresa Nogueira received a Ph.D. in Physiology and Genetics of Microorganisms (Specialization in Genes, Genomes and Cells) from the University Paris-Sud in France in 2001. She is a researcher at INIAV, National Institute for Agricultural and Veterinary Research, in the Laboratory of Bacteriology and Mycology and at the Center for Ecology, Evolution and Environmental Change of the University of Lisbon, in the team of Evolutionary Ecology of Microorganisms in Portugal.

She is interested in the evolutionary dynamics of antibiotic resistance and bacterial virulence in the human and animal gut, as well as in environmental microbiomes. She also aims to understand the antibiotic resistance and virulence profiles of microbial communities in the environment and in food-producing animals as part of the one health concept through large-scale genomic and metagenomic analyses of microbiomes.

Dr. Teresa Nogueira also teaches bioinformatics in Ph.D. programs and participates in outreach and science communication activities to promote scientific literacy in biological evolution. In this context, she is a member of the scientific associations EvoKE (Evolutionary Knowledge for Everyone) deeply involved in promoting public understanding and acceptance of evolution.

Preface to "Genomic Analysis of Antibiotics Resistance in Pathogens"

The emergence of antibiotic-resistant pathogens currently represents a serious threat to public health and the economy. Due to antibiotic treatments in humans and veterinary medicine, prophylactic use and environmental contamination, bacteria are today more frequently exposed to unnatural doses of antibiotics and their selective effect.

Antibiotic resistance can be encoded on chromosomes, plasmids, or other mobile genetic elements in bacteria. It may also result from mutations that lead to changes in the affinity of antibiotics for their targets or in the ability of antibiotics to act on bacterial growth or death. Exposure of bacteria, bacterial populations, and microbial communities to antibiotics at different concentrations shapes their genomic dynamics, as does the mobilisation and spread of resistance determinants. It is, therefore, essential to understand the dynamics and mobilisation of genes encoding antibiotic resistance, in human, animal, plant, and environmental microbiomes, through genomic and metagenomic approaches and bioinformatics analyses.

This Special Issue gathers research publications on the horizontal transfer of antibiotic-resistance genes, their dissemination and epidemiology, their association with bacterial virulence, between bacterial genotypes and their phenotypes, and other related research topics.

Teresa V. Nogueira
Editor

Editorial

Genomic Analysis of Antibiotics Resistance in Pathogens

Teresa Nogueira [1,2]

1   INIAV—National Institute for Agrarian and Veterinary Research, 2780-157 Oeiras, Portugal;
    teresa.nogueira@iniav.pt
2   cE3c—Center for Ecology, Evolution and Environmental Change & CHANGE—Global Change and
    Sustainability Institute, Faculdade de Ciencias, Universidade de Lisboa, 1749-016 Lisbon, Portugal

Citation: Nogueira, T. Genomic Analysis of Antibiotics Resistance in Pathogens. *Antibiotics* **2022**, *11*, 1013. https://doi.org/10.3390/antibiotics11081013

Received: 13 July 2022
Accepted: 24 July 2022
Published: 28 July 2022

Publisher's Note: MDPI stays neutral with regard to jurisdictional claims in published maps and institutional affiliations.

The emergence of antibiotic-resistant pathogens currently represents a serious threat to public health and the economy worldwide. Due to antibiotic treatments in human and veterinary medicine, pathogenic bacteria are most often exposed to unnatural doses of antibiotics and their selective effect.

Back in 1945, in his Nobel prize lecture, Alexander Fleming said: "The time may come when penicillin can be bought by anyone in the shops. Then there is the danger that the ignorant man may easily underdose himself and by exposing his microbes to non-lethal quantities of the drug make them resistant. Here is a hypothetical illustration. Mr. X. has a sore throat. He buys some penicillin and gives himself, not enough to kill the streptococci but enough to educate them to resist penicillin. He then infects his wife. Mrs. X gets pneumonia and is treated with penicillin. As the streptococci are now resistant to penicillin the treatment fails. Mrs. X dies. Who is primarily responsible for Mrs. X's death? Why Mr. X whose negligent use of penicillin changed the nature of the microbe. Moral: If you use penicillin, use enough" [1].

In fact, after the discovery of penicillin, antibiotics were introduced into clinical practice in humans, and later in veterinary health and agriculture as growth promoters. As a consequence, antibiotics became an environmental contaminant.

Fleming predicted what we know today to be the genomic dynamics of the acquisition of antibiotic resistance and its epidemic nature, eight years before Francis Crick and James Watson determined the structure of the DNA molecule. He also predicted that the widespread, unsupervised use of antibiotics could become a health problem, as he was already aware that antibiotic resistance could be triggered by exposure to sublethal doses of antibiotics.

In bacteria, antibiotic resistance can be encoded on chromosomes, plasmids, or other mobile genetic elements. It can also result from mutations that lead to changes in the affinity of antibiotics for their targets or in the ability of antibiotics to act on bacterial growth or death. It is essential to understand the evolutionary dynamics and mobilization of genes encoding antibiotic resistance in the human, animal, plant, and environmental microbiomes but also the transmission of resistant bacteria between individuals and between humans, animals, and the environment in a One Health approach. These studies can be developed using genomic and metagenomic approaches and bioinformatics analyses. This Special Issue addresses the horizontal transfer of antibiotic resistance genes and their spread, epidemiology, and association with bacterial virulence between bacterial genotypes and their phenotypes.

Although antibiotics are global contaminants, environmental bacteria, commensals, and human and animal pathogens are not expected to be exposed to the same types of antibiotics. Pathogenic bacteria will more often be exposed to therapeutic doses of the antibiotics of medical interest. Darmancier et al. [2] performed a comprehensive bioinformatic study of 16,632 complete bacterial reference genome sequences to discover whether there would be a relationship between bacterial virulence and antibiotic resistance. They found evidence that some categories of virulence and antibiotic resistance genes could be co-selected and that mobile genetic elements, such as integrative and conjugative elements, could play an important role in their co-mobilization. The fact that human pathogens whose therapy requires specific antibiotics have become resistant has emerged as a major obstacle to the treatment of diseases that threaten humans.

This collection also gathers papers on antibiotic resistance of some of the major emerging infections pathogens such as *Mycobacteroides abscessus* [3], *Klebsiella pneumoniae* [4,5], *Salmonella* sp. [6,7], *Francisella tularensis* [8], *Acinetobacter baumannii* [9,10], or *Pseudomonas aeruginosa* [11] and other human pathogens such as *Neisseria gonorrhoea* [12], *Staphylococcus aureus* [13], or *Escherichia coli* [14]. Most of them are high-priority pathogens [15], according to the National Institute of Allergy and Infectious Diseases, and are globally and medically important pathogens.

Ng and Ngeow have published a review on the Tigecycline (a third-generation tetracycline) resistance in the Actinobacteria *Mycobacteroides abscessus* a clinically important human pathogen known to harbor a multidrug-resistance phenotype [3].

Staphylococcus aureus is one of the main etiological agents of skin and wound infections of nosocomial origin, belonging to the phylum Firmicutes. It has been classified as a globally and medically important human and foodborne pathogen [16], and methicillin resistance has been a major challenge to treatment and patient science. Ullah et al. have performed a comparative genomic analysis of a highly virulent methicillin-resistant *S. aureus* (MRSA) of a Pakistan clone [13].

Neisseria gonorrhoeae is a Beta Proteobacteria that is the etiological agent of gonorrhea, a sexually transmitted disease, as well as an eye infection transmitted, for example, during vaginal delivery, which can lead to blindness when left untreated. Gonzalez et al. conducted a study in which they allowed two reference strains of *N. gonorrhoeae* to evolve and then followed the emergence of ciprofloxacin resistance by mutation. They then performed genomic analyses of several of the clones to conclude that there are strain-specific differences in the emergence of ciprofloxacin resistance [12].

The Gamma Proteobacteria phylum encloses many human pathogens. This collection includes a review by Rodrigues and his colleagues who have characterized the resistome of *Acinetobacter baumannii* and described its genome as very plastic and open [10]. Thadtapong et al. have highlighted the epidemic potential of colistin and carbapenem-resistant *A. baumannii* clone carrying a conjugative system and a potential virulence system [9].

Another Gamma Proteobacteria is *Pseudomonas aeruginosa*, which is an opportunistic human pathogen that is often difficult to treat with antibiotics due to its ability to form biofilms but also due to the expression of efflux pumps. Ahmed et al. have studied the impact of titanium dioxide nanoparticles on the quorum-sensing genes controlling biofilm production and efflux pump gene expression [11].

Souder et al. have identified a novel role for two genes *dipA* and *pilD* in *Francisella tularensis* (the etiologic agent of Tularaemia) susceptibility to resazurin [8].

The phylum Gamma-proteobacteria also includes the most common etiologic agents of human gastroenteritis: *Salmonella* sp. and *Escherichia coli*. Hernández-Díaz and colleagues performed a comparative genomic analysis of Typhimurium serotypes of *S. enterica* that led to the identification of 44 genes, 34 plasmids, and 5 point mutations associated with antibiotic resistance, distributed across 220 genomes of ST213 strains [6]. Vázquez et al., on the other hand, performed a genomic analysis of *S. enterica* serotype Infantis to identify *pESI*-like plasmids encoding blaCTX-M-65 [7]. Masoud et al. identified extended-spectrum beta-lactamases (ESBLs), metallo beta-lactamases (MBLs), and plasmid-mediated quinolone resistance in *E. coli* isolated from different clinical specimens in Egypt [14].

Klebsiella pneumoniae is another important human pathogen and one of the etiological agents of pneumonia. Mendes et al. identified Ceftazidime/Avibactam resistance in a patient isolate in Portugal [4]. As the combination resistance to Ceftazidime and Avibactam is new, and as it appears together with resistance to other antibiotics, the authors recommend special epidemiological surveillance.

Altayb et al. studied and isolated a *K. pneumoniae* isolate from a patient with recurrent urinary tract infection, identified as hypermucoviscent, type 2 (K2) capsular polysaccharide, ST14, and multidrug-resistant. The authors performed genomic analysis and demonstrated that it harbors four plasmids, several virulence factors, and antibiotic resistance encoding genes, which are worth studying in more detail [5].

Funding: Fundação para a Ciência e a Tecnologia (FCT) supported T.N. by contract PTDC/BIA-MIC/28824/2017.

Conflicts of Interest: The author declares no conflict of interest.

References

1. The Nobel Prize in Physiology or Medicine 1945. Available online: https://www.nobelprize.org/prizes/medicine/1945/fleming/lecture/ (accessed on 12 July 2022).
2. Darmancier, H.; Domingues, C.P.F.; Rebelo, J.S.; Amaro, A.; Dionísio, F.; Pothier, J.; Serra, O.; Nogueira, T. Are Virulence and Antibiotic Resistance Genes Linked? A Comprehensive Analysis of Bacterial Chromosomes and Plasmids. *Antibiotics* **2022**, *11*, 706. [CrossRef] [PubMed]
3. Ng, H.F.; Ngeow, Y.F. Genetic Determinants of Tigecycline Resistance in Mycobacteroides Abscessus. *Antibiotics* **2022**, *11*, 572. [CrossRef] [PubMed]
4. Mendes, G.; Ramalho, J.F.; Bruschy-Fonseca, A.; Lito, L.; Duarte, A.; Melo-Cristino, J.; Caneiras, C. First Description of Ceftazidime/Avibactam Resistance in a ST13 KPC-70-Producing Klebsiella Pneumoniae Strain from Portugal. *Antibiotics* **2022**, *11*, 167. [CrossRef] [PubMed]
5. Altayb, H.N.; Elbadawi, H.S.; Baothman, O.; Kazmi, I.; Alzahrani, F.A.; Nadeem, M.S.; Hosawi, S.; Chaieb, K. Genomic Analysis of Multidrug-Resistant Hypervirulent (Hypermucoviscous) Klebsiella Pneumoniae Strain Lacking the Hypermucoviscous Regulators (RmpA/RmpA2). *Antibiotics* **2022**, *11*, 596. [CrossRef] [PubMed]
6. Hernández-Díaz, E.A.; Vázquez-Garcidueñas, M.S.; Negrete-Paz, A.M.; Vázquez-Marrufo, G. Comparative Genomic Analysis Discloses Differential Distribution of Antibiotic Resistance Determinants between Worldwide Strains of the Emergent ST213 Genotype of Salmonella Typhimurium. *Antibiotics* **2022**, *11*, 925. [CrossRef]
7. Vázquez, X.; Fernández, J.; Rodríguez-Lozano, J.; Calvo, J.; Rodicio, R.; Rodicio, M.R. Genomic Analysis of Two MDR Isolates of Salmonella Enterica Serovar Infantis from a Spanish Hospital Bearing the BlaCTX-M-65 Gene with or without FosA3 in PESI-like Plasmids. *Antibiotics* **2022**, *11*, 786. [CrossRef] [PubMed]
8. Souder, K.; Beatty, E.J.; McGovern, S.C.; Whaby, M.; Young, E.; Pancake, J.; Weekley, D.; Rice, J.; Primerano, D.A.; Denvir, J.; et al. Role of DipA and PilD in Francisella Tularensis Susceptibility to Resazurin. *Antibiotics* **2021**, *10*, 992. [CrossRef]
9. Thadtapong, N.; Chaturongakul, S.; Soodvilai, S.; Dubbs, P. Colistin and Carbapenem-Resistant Acinetobacter Baumannii Aci46 in Thailand: Genome Analysis and Antibiotic Resistance Profiling. *Antibiotics* **2021**, *10*, 1054. [CrossRef]
10. Rodrigues, D.L.N.; Morais-Rodrigues, F.; Hurtado, R.; dos Santos, R.G.; Costa, D.C.; Barh, D.; Ghosh, P.; Alzahrani, K.J.; Soares, S.C.; Ramos, R.; et al. Pan-Resistome Insights into the Multidrug Resistance of Acinetobacter Baumannii. *Antibiotics* **2021**, *10*, 596. [CrossRef] [PubMed]
11. Ahmed, F.Y.; Aly, U.F.; Abd El-Baky, R.M.; Waly, N.G.F.M. Effect of Titanium Dioxide Nanoparticles on the Expression of Efflux Pump and Quorum-Sensing Genes in MDR Pseudomonas Aeruginosa Isolates. *Antibiotics* **2021**, *10*, 625. [CrossRef] [PubMed]
12. González, N.; Abdellati, S.; De Baetselier, I.; Laumen, J.G.E.; Van Dijck, C.; de Block, T.; Kenyon, C.; Manoharan-Basil, S.S. Alternative Pathways to Ciprofloxacin Resistance in Neisseria Gonorrhoeae: An In Vitro Study of the WHO-P and WHO-F Reference Strains. *Antibiotics* **2022**, *11*, 499. [CrossRef] [PubMed]
13. Ullah, N.; Nasir, S.; Ishaq, Z.; Anwer, F.; Raza, T.; Rahman, M.; Alshammari, A.; Alharbi, M.; Bae, T.; Rahman, A.; et al. Comparative Genomic Analysis of a Panton–Valentine Leukocidin-Positive ST22 Community-Acquired Methicillin-Resistant Staphylococcus Aureus from Pakistan. *Antibiotics* **2022**, *11*, 496. [CrossRef] [PubMed]
14. Masoud, S.M.; Abd El-Baky, R.M.; Aly, S.A.; Ibrahem, R.A. Co-Existence of Certain ESBLs, MBLs and Plasmid Mediated Quinolone Resistance Genes among MDR E. Coli Isolated from Different Clinical Specimens in Egypt. *Antibiotics* **2021**, *10*, 835. [CrossRef] [PubMed]
15. NIAID Emerging Infectious Diseases/Pathogens | NIH: National Institute of Allergy and Infectious Diseases. Available online: https://www.niaid.nih.gov/research/emerging-infectious-diseases-pathogens (accessed on 12 July 2022).
16. Ecker, D.J.; Sampath, R.; Willett, P.; Wyatt, J.R.; Samant, V.; Massire, C.; Hall, T.A.; Hari, K.; McNeil, J.A.; Büchen-Osmond, C.; et al. The Microbial Rosetta Stone Database: A Compilation of Global and Emerging Infectious Microorganisms and Bioterrorist Threat Agents. *BMC Microbiol.* **2005**, *5*, 19. [CrossRef] [PubMed]

MDPI

Article

Comparative Genomic Analysis Discloses Differential Distribution of Antibiotic Resistance Determinants between Worldwide Strains of the Emergent ST213 Genotype of *Salmonella* Typhimurium

Elda Araceli Hernández-Díaz [1,†], Ma. Soledad Vázquez-Garcidueñas [2,†], Andrea Monserrat Negrete-Paz [1] and Gerardo Vázquez-Marrufo [1,*]

[1] Centro Multidisciplinario de Estudios en Biotecnología, Facultad de Medicina Veterinaria y Zootecnia, Universidad Michoacana de San Nicolás de Hidalgo, Km 9.5 Carretera Morelia-Zinapécuaro, Col. La Palma Tarímbaro, Morelia 58893, Michoacán, Mexico; 0617452b@umich.mx (E.A.H.-D.); andrea.negrete@umich.mx (A.M.N.-P.)

[2] División de Estudios de Posgrado, Facultad de Ciencias Médicas y Biológicas "Dr. Ignacio Chávez", Universidad Michoacana de San Nicolás de Hidalgo, Ave. Rafael Carrillo esq. Dr. Salvador González Herrejón, Col. Cuauhtémoc, Morelia 58020, Michoacán, Mexico; soledad.vazquez@umich.mx

[*] Correspondence: gvazquez@umich.mx; Tel./Fax: +52-01-443-2-95-80-29

[†] These authors contributed equally to this work.

Citation: Hernández-Díaz, E.A.; Vázquez-Garcidueñas, M.S.; Negrete-Paz, A.M.; Vázquez-Marrufo, G. Comparative Genomic Analysis Discloses Differential Distribution of Antibiotic Resistance Determinants between Worldwide Strains of the Emergent ST213 Genotype of *Salmonella* Typhimurium. *Antibiotics* **2022**, *11*, 925. https://doi.org/10.3390/antibiotics11070925

Academic Editor: Teresa V. Nogueira

Received: 20 June 2022
Accepted: 7 July 2022
Published: 9 July 2022

Publisher's Note: MDPI stays neutral with regard to jurisdictional claims in published maps and institutional affiliations.

Abstract: *Salmonella enterica* constitutes a global public health concern as one of the main etiological agents of human gastroenteritis. The Typhimurium serotype is frequently isolated from human, animal, food, and environmental samples, with its sequence type 19 (ST19) being the most widely distributed around the world as well as the founder genotype. The replacement of the ST19 genotype with the ST213 genotype that has multiple antibiotic resistance (MAR) in human and food samples was first observed in Mexico. The number of available genomes of ST213 strains in public databases indicates its fast worldwide dispersion, but its public health relevance is unknown. A comparative genomic analysis conducted as part of this research identified the presence of 44 genes, 34 plasmids, and five point mutations associated with antibiotic resistance, distributed across 220 genomes of ST213 strains, indicating the MAR phenotype. In general, the grouping pattern in correspondence to the presence/absence of genes/plasmids that confer antibiotic resistance cluster the genomes according to the geographical origin where the strain was isolated. Genetic determinants of antibiotic resistance group the genomes of North America (Canada, Mexico, USA) strains, and suggest a dispersion route to reach the United Kingdom and, from there, the rest of Europe, then Asia and Oceania. The results obtained here highlight the worldwide public health relevance of the ST213 genotype, which contains a great diversity of genetic elements associated with MAR.

Keywords: ST213 worldwide strains; genomic databases; antibiotic resistance plasmids; Typhimurium

1. Introduction

Salmonella enterica is one of the main pathogens associated with food contamination; it is considered responsible for around 94 million cases of gastrointestinal illnesses and 155,000 annual deaths worldwide [1–3]. The typing method used for the follow-up of outbreaks and epidemiological studies of *S. enterica* for nearly 90 years is serotyping [4]. Currently, more than 2600 serotypes are registered worldwide [5,6]. The Typhimurium and Enteritidis serotypes are considered of the greatest global public health relevance, because these have the widest geographical distribution and the highest incidence in clinical and food samples worldwide [7,8]. However, among the plethora of molecular genetic typing methods generated in the last few decades, comparative genomic analysis stands out for its greater discrimination power, making it possible to distinguish strains associated with an outbreak from those that are not [9–11]. This discrimination power is of epidemiological and public health relevance, since it allows the generation of strategies for the prevention and control of outbreaks [12–14].

Among the genotyping methods used as epidemiological tools for the study of *S. enterica*, multi locus sequence typing (MLST) for the assignment of sequence type (ST) through variations in seven loci efficiently identify clonal groups and founder genotypes [15,16]. Recently, MLST analysis was modified in accordance with the possibilities offered by whole-genome sequencing (WGS) to use a large number of genes from the core genome [17–19], enabling the differentiation between clonal groups to be more precise, allowing the description of emerging genotypes, and making it possible to distinguish between lineages within an ST [20–22]. This resolution power offered by MLST using WGS opens the possibility of comparing strains between very diverse space-time scales; in addition to its application in epidemiological studies, this allows evaluating the micro-evolutionary process of *S. enterica* and detecting the emergence of variants of relevance in the field of public health [21–23]. Beyond its excellent ability to discriminate between clonal groups and variants within an ST, comparative genomic analysis allows the study of dispersion and distribution of virulence and resistance to antibiotics-associated genes [24–28] and relevant genetic determinants of *S. enterica* emerging pathogenic variants.

MLST analysis has clearly established that the founding genotype of *S. enterica* Typhimurium is ST19, as the most prevalent genotype of this serotype around the world from which the vast majority of other STs have been derived [15,16]. However, it has been recently documented that some countries have experienced the ST19 being replaced with other STs that show a higher incidence in clinical and food samples. These replacement genotypes have genotypic and phenotypic characteristics that make them relevant in terms of epidemiology and public health. In this sense, the best documented cases of replacement are replacements by the ST313 genotype, which have been identified in Sub-Saharan Africa and are associated with systemic disease in HIV patients. Thus far, two sub-lineages have been identified, one of which has resistance to antibiotics, complicating the treatment of HIV [21,29]. The increased incidence of the ST34 genotype carrying antibiotic resistance genes/plasmids has been documented in some regions of China in both clinical and food samples [30,31].

In Mexico, a study on clinical and food samples carried out for four years in different states of the country revealed evidence of the replacement of ST19 by ST213 [32]. Subsequent analyses showed that this strain carries IncA/C plasmids, now changed to IncC [33], with genetic determinants for multiple resistance to antibiotics [34]. According to the genome metadata available in EnteroBase [35], the ST213 genotype has been isolated in recent years from different regions of the world. However, to date the epidemiological risk associated with the geographical dispersion of this genotype sequence is unknown, as is whether the patterns of resistance to antibiotics are shared between strains from different regions.

It has been documented that bioinformatic genomic analysis allows the robust prediction of phenotypic resistance to antibiotics in *S. enterica* [28,36]. Thanks to the considerable increase in the available genomes of ST213 from different regions of the world, it is feasible to perform an analysis that allows establishing solid hypotheses about the resistance of this genotype to antibiotics. Furthermore, it is possible to establish whether the strains from different geographical regions have the same set of genetic determinants of resistance to antibiotics and the presence of multi-resistance between them. Therefore, the objective of this work was to carry out a comparative genomic analysis of all ST213 strains whose genomes are available in public databases in order to analyze the presence and distribution of genes and mutations associated with antibiotic resistance. Differences in the presence/absence of genetic determinants associated with resistance to antibiotics in relation to the geographical origin, year of isolation, and type of samples of ST213 genotype strains are analyzed and discussed.

2. Results

2.1. Distribution of the Genomes of the Strains Analyzed by Country, Type of Sample and Year of Isolation

Of the 220 genomes retrieved from databases of S. Typhimurium strains belonging to the ST213 genotype and included in this study, 29% (n = 64) came from the United Kingdom, 25.9% (n = 57) came from Mexico, 24.5% (n = 54) came from the United States of America, and 9.5% (n = 21) came from Canada, while the remaining 11% (n = 24) were obtained from nine different countries, mainly from Europe (Figure 1).

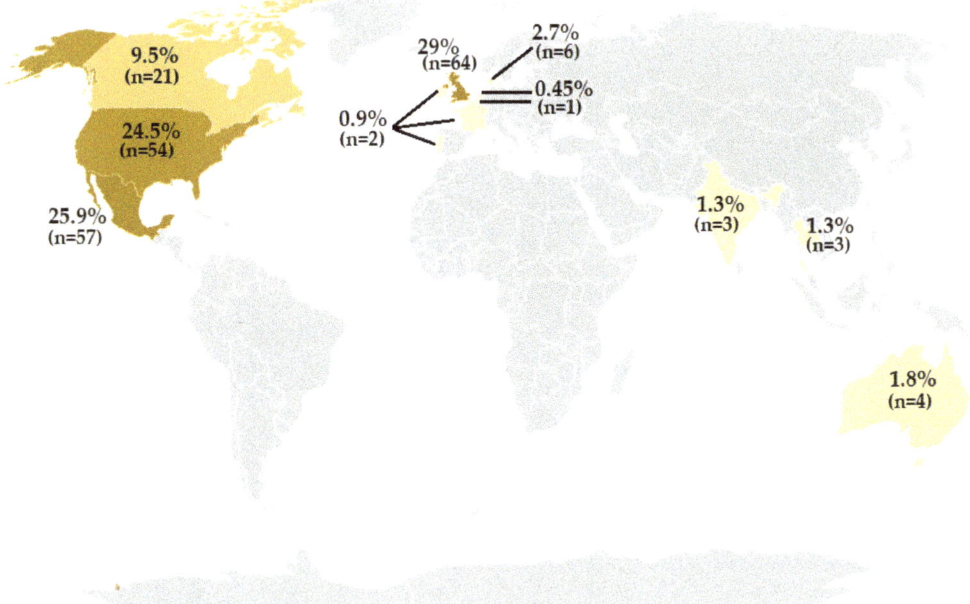

Figure 1. World distribution of the *Salmonella enterica* Typhimurium strain genotype ST213, the genomes of which were analyzed in this work.

A total of 70% (n = 154) of the strains were obtained from human clinical samples, 15% (n = 33) were obtained from samples of animal origin, and the remaining 15% (n = 33) were obtained from nine other sources in smaller percentages (Figure 2). In relation to the year of isolation, the genomes were derived from strains isolated between the years 1957 and 2021, with the years 2003, 2004, 2017, 2018, and 2019 being the ones in which the largest number of records were made. Between 2002 and 2005, most of the Mexican records were made; between 2014 and 2019, the largest number of records from the United Kingdom were obtained; and the strains from the United States were all registered in 19 years, apart from the years 1965, 1967, 2001, 2004, 2005, 2008, and 2009. It is interesting to note that, so far, during the year 2013, no isolates of ST213 have been recorded in any geographical area (Figure 3).

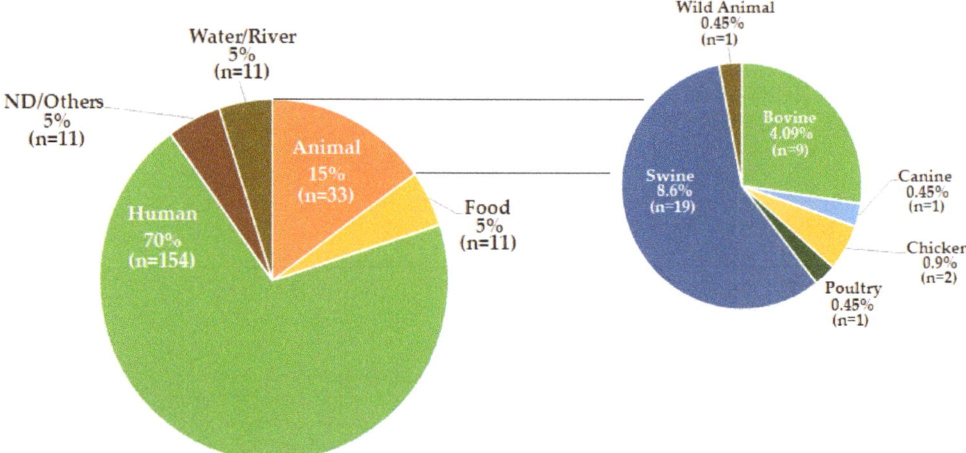

Figure 2. Sample type provenance of the ST213 strains from which genomes were analyzed in this work. The number of strains (*n*) for each sample type is given in parenthesis. The total number of genomes analyzed after filtering was 220 (see Figure 8).

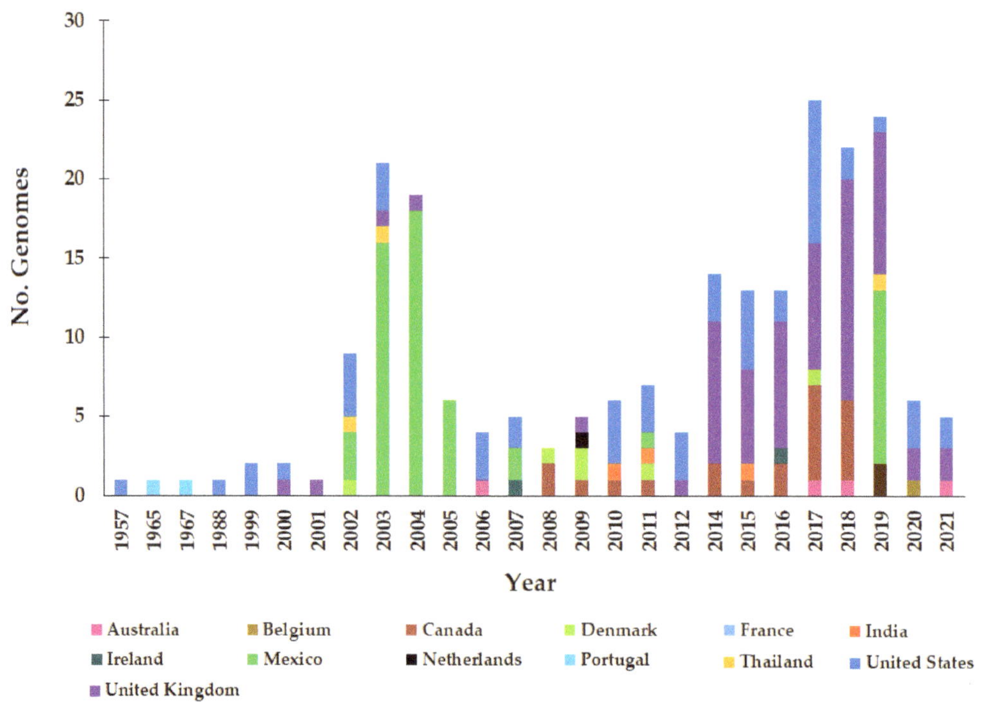

Figure 3. Year of isolation of the ST213 strains from which genomes were analyzed in this work. The total number of genomes analyzed after filtering was 220 (see Figure 8).

2.2. Presence of Antibiotic Resistance Genes

Forty-four antibiotic resistance genes were identified in the analyzed genomes (Table 1), of which *aac(6)-Iaa*, *golS*, *mdsA*, *mdsB*, *mdsC*, *mdtK*, and *sdiA* are present in the 220 (100%) genomes analyzed. The genes *aac(3)-IV*, *aph(3)-IIa*, *aph(4)-Ia*, *bla*$_{CARB-3}$, *catII*, *cmlA1*, *dfrA1*, *linG*, *mefB*, *qnrA1*, *qnrB19*, *qnrS1*, *tetM*, and *tetU* were found in <1% of the genomes studied. The *aadA5*, *ant(3)-IIa*, *aph(3)-Ia*, *bla*TEM-1, *dfrA17*, *qnrB5*, *sul1*, *tetB*, and *tetC* genes were found in between 1 and 15% of the genomes included in this study, while the *aac(3)-IId*, *aadA2*, *bla*CMY-59, *dfrA12*, *floR*, *oqxA*, *oqxB*, *qacH*, *sul2*, and *sul3* genes were identified in 20% to 40% of the study genomes. Finally, the *aph(3)-Ib*, *aph(6)-Id*, *tetA*, and *tetR* genes were found in 50–57% of all the analyzed genomes (Table 1). Regarding the type of sample from which the genomes were isolated, the *bla*CARB-3 gene was only found in strains from samples of animal origin, the *bla*CMY-59 gene was not found in strains from food, and *bla*TEM-1 was identified in strains taken from water samples. On the other hand, the *linG* gene was found in food-associated strains and *mefB* was identified in human clinical samples from the United Kingdom and the United States, respectively. Regarding the country from which the strains were obtained, ten of the genes that confer resistance to aminoglycosides were found in the genomes of strains from Australia, Canada, Denmark, Thailand, Mexico, the United Kingdom, and the United States. Genes related to resistance to β-lactams and fluoroquinolones were not found in the genomes of strains from Denmark, and genes related to resistance to chloramphenicol were not found in the genomes of strains from Thailand. Similarly, tetracycline and diaminopyrimidine resistance genes were not found in strains from Denmark and Thailand, while the *qacH* gene was not identified in the genomes of strains isolated in Australia.

Table 1. Resistance genes found in the analyzed genomes of ST213 strains.

Antibiotic Group/Resistance Gene(s)	Encoded [1]	% Frequency [2]
Aminoglycosides		
aph(4)-Ia, *aac(3)-IV*, *ant(3)-IIa*, *aac(3)-IId*	P (IncHI2)	0.4 ($n = 1$), 0.9 ($n = 2$), 10.9 ($n = 24$), 29 ($n = 64$)
aph(3)-IIa, *aph(3)-Ia*	T	0.9 ($n = 2$), 9.5 ($n = 21$)
aac(6)-Iaa	C	100 ($n = 220$)
aadA2	P/I (IncHI2/IncHI2A)	40 ($n = 89$)
aadA5	P/T/I	1.8 ($n = 4$)
aph(3)-Ib	P/T/C (IncC, IncFII/IA/IB,)	50 ($n = 111$)
aph(6)-Id	P/CG-I (IncC, IncFII/IA/IB,)	50 ($n = 110$)
Cephamicin		
*bla*CMY-59	P (IncC)	36.8 ($n = 81$)
Diaminopyrimidines		
dfrA1, *dfrA12*, *dfrA17*	I	0.4 ($n = 1$), 37.2 ($n = 82$), 1.8 ($n = 4$)
Penam		
*bla*CARB-3	P	0.4 ($n = 1$)
Penam, penem, cephalosporin, monobactam		
*bla*TEM-1	C/P (IncHI2/IncHI2A)	14 ($n = 31$)
Disinfecting agents and intercalating dyes		
qacH	P	26.3 ($n = 58$)
Fluoroquinolones		
mdtK	C	100 ($n = 220$)
qnrA1, *qnrB5*, *qnrB19*, *qnrS1*	P (IncHI2/IncHI2A)	0.4 ($n = 1$), 10 ($n = 22$), 0.4 ($n = 1$), 0.9 ($n = 2$)
Lincosamides		
linG	I-agc, with *aadA2*	0.4 ($n = 1$)
Macrolides		
mefB	P, located in the *sul3* vicinity	0.4 ($n = 1$)
Sulfonamides		
sul1	C-1 I	8.6 ($n = 19$)
sul2	SP (IncHI2, IncC, IncFII/IA/IB)	49.5 ($n = 109$)
sul3	P (IncHI2/IncHI2A)	27.2 ($n = 60$)
Tetracyclines		
tetA, *tetB*, *tetR*	C/P (IncHI2, IncC, IncFII/IA/IB)	56.8 ($n = 125$), 1.8 ($n = 4$), 55.9 ($n = 123$)
tetC	P (IncHI2/IncHI2A/IncI1_I_γ)	1.3 ($n = 3$)
tetU	P (pKQ10)	0.4 ($n = 1$)
tetM	T	0.4 ($n = 1$)
Phenicol		

Table 1. *Cont.*

Antibiotic Group/Resistance Gene(s)	Encoded [1]	% Frequency [2]
catII (E. coli K-12), cmlA1, floR	C/P (IncHI2/ IncHI2A/ IncI1_I_γ/IncQ1, IncC)	0.4 (n = 1), 0.9 (n = 2), 47.7 (n = 105)
Phenicol, β-lactams, diaminopyrimidines, fluoroquinolones, glycyl-cyclines, nitrofuran and tetracyclines, rifamycin, triclosan.		
golS, mdsA, mdsB, mdsC	C	100 (n = 220)
oqxA, oqxB	C/P (IncHI2)	24.5 (n = 54)
sdiA	C/P	100 (n = 220)

Note: [1] C, chromosome; CG-I, chromosome genomic islands; C-1 I, class 1 integron; I, integrons; Iagc, Integron-associated gene cassette; P, plasmids; SP, small plasmids; T, transposon. [2] Number of genomes in which each gen/plasmid was found is given in parentheses.

2.3. Antibiotic Resistance Mutations

A total of five point mutations associated with codon/amino acid changes that confer resistance to fluoroquinolones were found in the analyzed genomes, four of which were found in the *gyrA* gene and one of which was found in the *parC* gene (Table 2). The most frequent mutation was p.S83Y, which was carried by 23 genomes, whereas the least frequent were p.D87G and p.S80I, which were present in one genome each. Except for the genome of one Thailand strain, the rest of the genomes carrying such mutations belonged to North American strains, with one coming from Canada, nine from Mexico, and 19 from the USA (Table 2).

Table 2. Mutations associated with antibiotic resistance in the analyzed genomes.

Mutation	Codon Change	Amino Acid Change	Genome	% of Genomes [1]
gyrA p.D87G	GAC → GGC	D → G	MEX04C_2003	3.4 (n = 1)
parC p.S80I	AGC → ATC	S → I	US37H_2015	3.4 (n = 1)
gyrA p.S83F	TCC → TTC	S → F	MEX22F_2008	6.8 (n = 2)
			MEX51H_2011	
gyrA p.D87N	GAC → AAC	D → N	US17H_2011	10.3 (n = 3)
			US19H_2011	
			US43H_2011	
gyrA p.S83Y	TCC → TAC	S → Y	CAN20H_2017	79.3 (n = 23)
			MEX14F_2009	
			MEX15F_2009	
			MEX16F_2009	
			MEX18F_2008	
			MEX34H_2003	
			MEX56R_2019	
			THA01C_2019	
			US07B_2020	
			US12B_2021	
			US18H_2010	
			US20H_2015	
			US21H_2015	
			US22H_2015	
			US24H_2016	
			US28H_2012	
			US30H_2017	
			US37H_2015	
			US38H_2017	
			US44H_2010	
			US46H_2010	
			US47H_2017	
			US49B_2019	

Note: [1] Percentage of genomes that carry each plasmid. The number of genomes for each case is shown in parentheses. The total number of analyzed genomes was 220.

2.4. Plasmid Replicons Detection

Thirty-three different plasmid replicons were found in the analyzed genomes, of which 23 were identified in fewer than 5% of the study genomes, seven in more than 5% but fewer than 30%, two in 34%, and one in 47.7%. However, twelve of the analyzed genomes did not present plasmid replicons. Of the detected plasmids, it was recently proposed that the IncA/C2 denomination must be discarded and replaced by the IncC nomenclature [33]; thus, in the present paper, all bioinformatic detection of IncA/C2 in this work was designed as IncC. Despite the bioinformatics detection of plasmid replicons, it must be taken into account that, although this occurs infrequently, plasmids can be integrated in the bacterial chromosome or co-integrated in a plasmid with multiple replicons [37,38].

In relation to the type of sample used, the ColRNAI, IncC, and IncFIB(K)_1_Kpn3 plasmids were present in all the genomes of all the sample types analyzed here. Plasmids IncP1 and IncQ1 were only found in strains of samples of animal origin, while plasmids Col(BS512), IncFIC(FII), IncFII(pCTU2)_1_pCTU2, IncI2, IncI2_1_δ, IncX4, pESA2, pSL483, and rep14a_4_rep(AUS0004p3) were identified only in strains obtained from human clinical samples. In the case of the plasmids Col(MG828), Col440II, IncFIA, IncFIB(AP001918), IncFII, IncFII(pCoo)_1_pCoo, and IncX, these were found in genomes whose strains came from samples of both animal and human origin. Plasmids IncHI1A and IncHI1B(R27)_1_R27 were identified in strains from samples of animal and undefined origin, while Col156 and ColpVC were found in strains of animal and human origin as well as in food samples. Plasmid IncFIA(HI1)_1_HI1 was only found in strains from undefined samples; IncI_γ_1 was found in strains of animal, human, and undefined origin; and Col440I was not found in the genomes of food-associated strains. Plasmids IncFIB(S), IncFII(S), IncHI2A, IncHI2, and IncI1_1_α_1 were not identified in strains identified from water samples. On the other hand, the plasmids IncFIC(FII), IncI2_1_δ, pESA2 rep14a_4_rep(AUS0004p3), and IncFII(pCTU2)_1_pCTU2 were identified in the genomes of strains isolated in Mexico; pSL483 was identified only in strains from Canada; IncI2 was identified only in strains from the United Kingdom, and Col(BS512), IncFIA(HI1)_1_HI1, IncFIA, IncP1, and IncQ1 were identified only in strains from the United States. Plasmids IncX4 and IncI_γ_1 were found only in the genomes of strains from the United States and the United Kingdom; IncFII was found only in strains from Canada and the United States; and IncFIB(AP001918), IncFII(pCoo)_1_pCoo, IncHI1A, IncHI1B(R27)_1_R27, IncHI2, and IncHI2A were identified only in strains from Mexico and the United States. In the case of the Col(MG828) and IncX plasmids, they were identified in strains from Canada, Mexico, and the United States; Col156 and Col440II in strains from Canada, the United States, and the United Kingdom; ColpVC were identified in strains from Mexico, the United States, and the United Kingdom; Col440I and IncFIB(K)_1_Kpn3 were identified in strains from these three North American countries and the United Kingdom; and ColRNAI and IncC were identified in strains from North America, the UK, and Australia. Finally, the IncI1_1_α_1 plasmid was not found in strains from Australia, France, India, Ireland, the Netherlands, or Portugal, and the IncFIB(S), IncFII(S) plasmids were not present in the genomes of strains from Mexico and Portugal.

2.5. Grouping Patterns

When the presence/absence of resistance genes in the study genomes is used as a matrix for the generation of clustering patterns, interesting groups can be observed in relation to the country and the sample of origin of the strains. The presence of resistance genes brings together the strains from the three North American countries (Canada, Mexico, USA), as well as those from the United Kingdom and Australia. Mexico and the United States appear together in a subcluster while Canada, the United Kingdom, and Australia are grouped in another branch (Figure 4). In the other large cluster, the strains from European countries appear alongside strains from India and Thailand, with the strains from the latter and Denmark being distinguished by the presence of particular resistance genes that have previously been mentioned. Regarding the clustering pattern of the heat map generated when information on the presence/absence of resistance genes in the analyzed genomes is combined with the type of sample used, the genomes from chicken-, water-, swine-, and non-determined-origin strains cluster in the first group (Figure 5). Food and bovine samples are grouped together, and human samples are segregated in a single terminal ramification. Genomes derived from poultry and canine strains cluster in the last group.

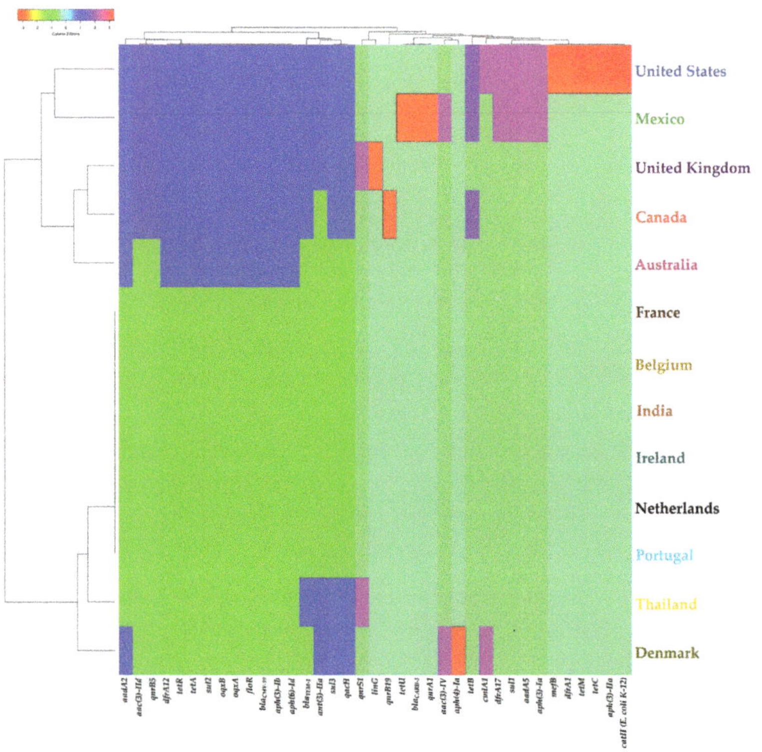

Figure 4. Grouping pattern by country of isolation and resistance genes identified in the genomes of the ST213 strains analyzed in this work. A binary matrix of presence (1)/absence (0) of each gene shown was used as an input for grouping using Manhattan and UPGMA for the calculation of distances and the generation of the grouping pattern, respectively. See the Section 4 for details.

Regarding the grouping pattern created by the presence/absence of resistance genes from each individual genome/strain, it can be seen that two larger clusters are clearly defined (Figure 6). The first includes mainly genomes from UK strains along with two Canadian genomes and all the European, Asian, and Australian samples (Figure 6, cluster A). This first cluster also includes, though in a different subgroup, several North American samples (Canada, Mexico, USA), two samples from the UK, and one sample from Thailand. The second main cluster contains genomes/strains exclusively from the three North American countries, and only three UK samples are dispersed in the three subgroups of this cluster (Figure 6, cluster B).

The plasmid replicons presence/absence grouping pattern shows a similar pattern to that observed for previously described resistance genes. Genomes of countries from North American (Canada, Mexico, USA) strains shape the first big cluster, with one strain from the UK and one from Australia (Figure 7, cluster A). The second big cluster is composed of European, Asian, and Australian genomes/strains, with five samples from Canada, one from Mexico, and one from the USA dispersed in the subgroups of the cluster (Figure 7, cluster B). A total of three small clusters are made up of genomes of strains from Mexico and the USA, as well as one from the UK (Figure 7, clusters C–E).

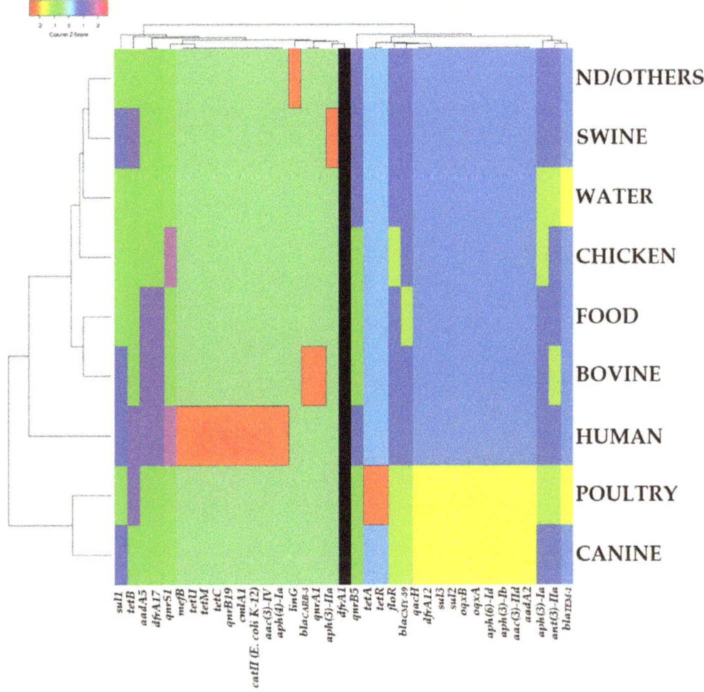

Figure 5. Grouping pattern by type of sample of isolation and resistance gene identified in the genomes of the ST213 strains analyzed in this work. A binary matrix of the presence (1)/absence (0) of each shown gene was used as an input for grouping using the Manhattan and UPGMA methods for the calculation of distances and the generation of the grouping pattern, respectively. See Section 4 for details.

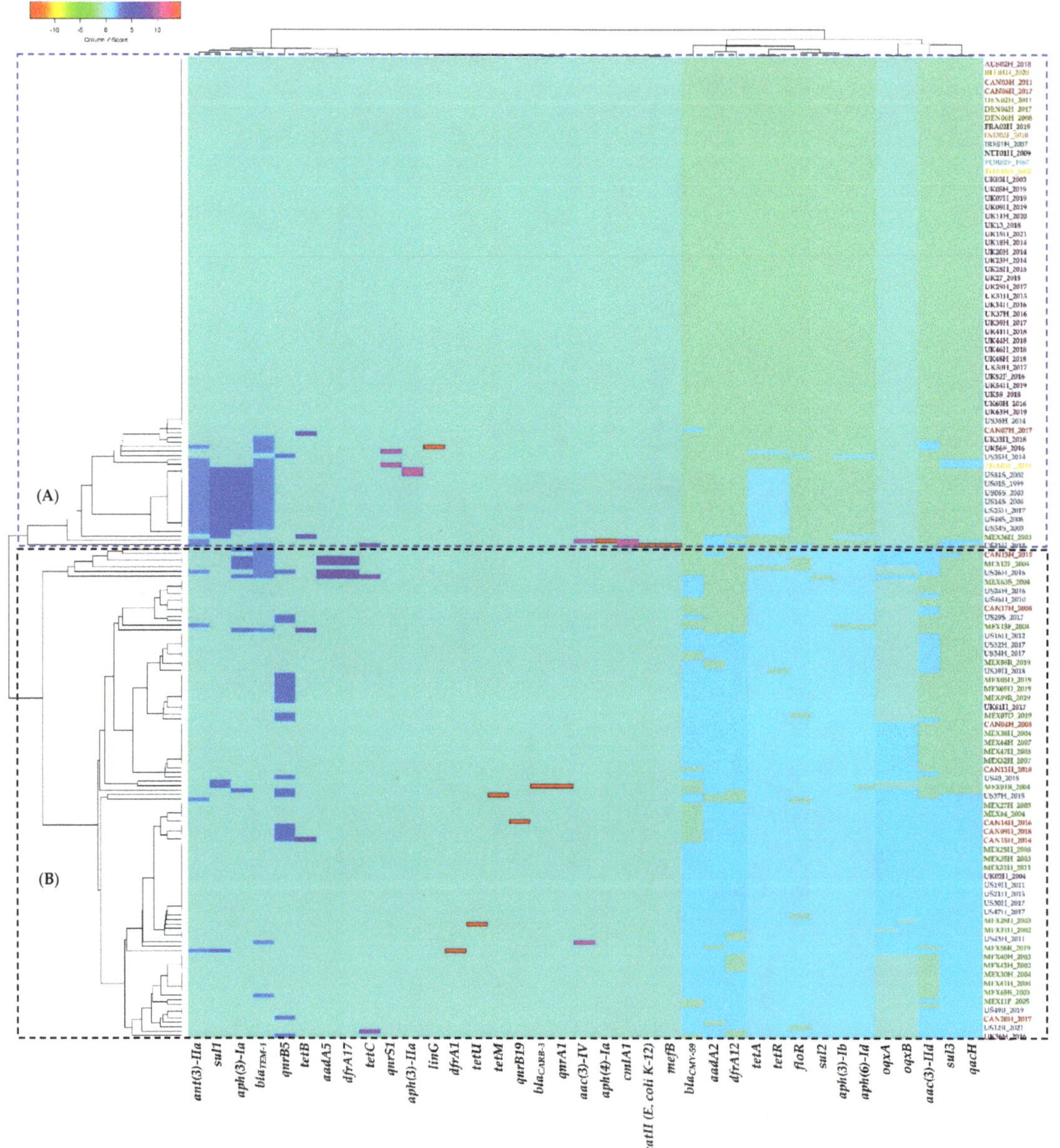

Figure 6. Grouping pattern by strain and resistance gene identified in the genomes of the ST213 strains analyzed in this work. A binary matrix of presence (1)/absence (0) of each shown gene was used as an input for grouping using the Manhattan and UPGMA methods for the calculation of distances and the generation of the grouping pattern, respectively. See Section 4 for details. (**A**) Grouping pattern of genome/strain from UK strains along with two Canadian ge-nomes and all the European, Asian, and Australian samples; (**B**) Grouping pattern of ge-nome/strain from the three North American countries, and only three UK. Country symbols: **AUS**, Australia; **BEL**, Belgium; **CAN**, Canada; **DEN**, Denmark; **FRA**, France; **IND**, India; **IRE**, Ireland; **MEX**, Mexico; **NET**, Netherlands; **POR**, Portugal; **THA**, Thailand; **UK**, United Kingdom; **US**, United States of America.

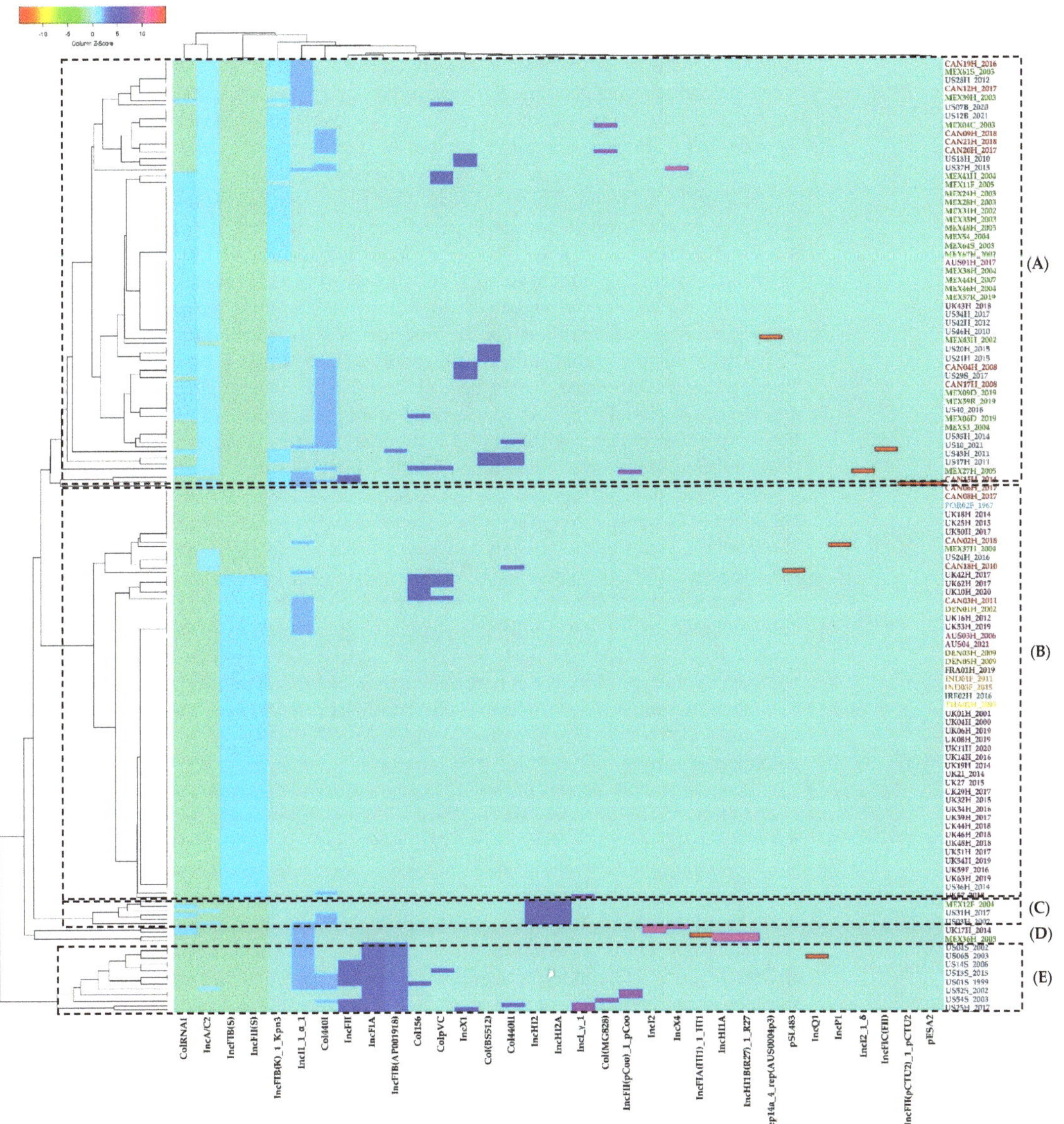

Figure 7. Grouping pattern by strain and plasmid presence/absence in the genomes of the ST213 strains analyzed in this work. A binary matrix of presence (1)/absence (0) of each shown plasmid was used as an input for grouping with the Manhattan and UPGMA methods for the calculation of distances and the generation of the grouping pattern, respectively. See Section 4 for details. (**A**) Grouping pattern of genome/strain from UK and one from Australia; (**B**) Grouping pattern of genome/strain European, Asian, and Australian, with five samples from Canada, one from Mexico, and one from the USA; (**C**) Grouping pattern of genome/strain from Mexico; (**D**) Grouping pattern of genome/strain from the USA; (**E**) Grouping pattern of genome/strain from the UK. Country symbols: **AUS**, Australia; **BEL**, Belgium; **CAN**, Canada; **DEN**, Denmark; **FRA**, France; **IND**, India; **IRE**, Ireland; **MEX**, Mexico; **NET**, Netherlands; **POR**, Portugal; **THA**, Thailand; **UK**, United Kingdom; **US**, United States of America.

3. Discussion

In the present work, the intragenotypic variability of *S*. Typhimurium ST213 strains in relation to the presence of genes associated with antibiotic resistance was evaluated by means of comparative genomic analysis. In the same way, this variation was documented in relation to the country and the type of sample of origin of the strains whose genomes were analyzed. Previous analyses related to the distribution of *S. enterica* strains in different countries and sample types, as well as the variability in the distribution of genetic determinants of antibiotic resistance, have focused mainly on evaluating the differences between serotypes or within the same serotype, as shown by several recent reviews on the subject [2,8,24,39,40]. Epidemiological studies documenting genomic variation associated with antibiotic resistance at a more subtle level, such as within the same *S. enterica* ST genotype, despite having been performed increasingly more frequently in recent years [29,41–43], are still relatively scarce. In this sense, the description of the variation in the resistance genes of *S*. Typhimurium strains of the ST313 genotype that causes systemic infection in Africa stands out, with differences having been observed in the phenotypic pattern of resistance to antibiotics associated with both the presence of the plasmid named pSLT-BT as well as with the composition of genes within it [21,29]. Similarly, strains of the emerging genotype ST34 present different patterns of resistance to antibiotics and show variation in the presence of genetic determinants of resistance [31,43]. As far as we have been able to document, these are the cases in which intragenotypic differences have been established in the epidemiology and patterns of antibiotic resistance in *S. enterica* in emergent/re-emergent STs replacing the ancestral ST19 genotype.

The emerging ST213 genotype was reported in Mexico to be associated with a process of displacement of the ST19 genotype [44], considered to be the founder within the Typhimurium serotype. The process of genome retrieval from public databases of strains of the ST213 genotype carried out here shows that this genotype is now frequently isolated in North America (Canada, Mexico, USA) and has recently spread to Europe, Asia, and Oceania, which is why it may represent a global public health problem as an emerging/re-emerging genotype.

It is noteworthy that six of the ST213 genomes retrieved from the databases belong to stored strains isolated before this century, with the oldest one in the USA dating back to 1957. This finding clearly shows that ST213 genotype was present long before its detection as an emerging public health concern in Mexico, replacing the founder ST19 genotype [44]. Additionally, the genomes retrieved indicate that ST213 strains have been widely dispersed throughout the North America region (Canada, Mexico, USA) since the beginning of this century. Furthermore, despite its presence in Europe (specifically Portugal) since 1965, strains of the ST213 genotype began to be frequently recovered from clinical and food samples at the beginning of this century, particularly in the United Kingdom, but also reached Thailand at the same time. The isolation year of ST213 strains whose genomes were analyzed here suggests its recent dispersion to India and Australia.

The fast spread of strains of the ST213 genotype in North America makes sense, given the geographical proximity and the constant flow of people and food between the three countries of this region [45–47]. Despite the need for a detailed epidemiological analysis, the years of isolation of the ST213 strains analyzed here suggest a dispersion route since its emergence and detection in Mexico. The available data strongly suggest that despite their global presence of at least 65 years, the ST213 strains have only recently become a public health issue because of its emergence in Mexico and/or the USA. The grouping presence/absence patterns of plasmids and resistance genes indicate that ST213 reached the UK from the USA and, from this country, spread to the rest of Europe, Asia, and Oceania. It is plausible that the ST213 genotype dispersed from the USA to the UK and, from there, to the rest of Europe, Asia, and Oceania through a traveler, rather than food or a vehicle. The traveler income/outcome of multi-resistant *S. enterica* strains of different serotypes has recently been documented in the regions and countries involved in the present analysis [48–52]. As previously stated, this epidemiological pattern is a hypothesis that emerged from the results obtained here and deserves further genomic epidemiological analysis.

The grouping pattern generated based on the sample of isolated strains whose genomes were analyzed here indicates that human strains possess a particular set of antibiotic resistance genetic determinants. However, the determinant genetics of antibiotic resistance from human samples share a bigger clade with animal and meat (food) samples. Thus, such a grouping pattern suggests that ST213 human strains are mainly acquired from farm animals and food and/or interchange some genetic subset(s) with strains from these sources. This hypothesis is consistent with reports of the frequent zoonotic transmission of non-invasive Typhimurium serotype strains in Mexico [32,53], Canada [54,55], and the United States [56,57], the three North American countries, which contributed the greatest number of strains to this study. Furthermore, in the zoonotic reports from these three countries, the transmission of strains of *S. enterica* from pigs, cattle, and poultry to humans is common, and these are the species samples whose genomes were analyzed in this work.

All antibiotic resistance genes detected in the ST213 strains genomes analyzed were predominantly relevant in the North American countries. In the same way, all the resistance genes/mutations were present predominantly in human samples. Both regional and sample type predominancies can be partially explained by the overrepresentation of genomes from strains at regional and sample levels. Furthermore, Mexico or North America are the probable origin of the ST213 as an emergent/reemergent pathogen of global health relevance. Thus, the discussion highlights results for North America and the human ST213 genomes. The ST213 genomes harbor a great diversity of genetic determinants for antibiotic resistance, and the gene/mutations/plasmid replicon combinations per genome are also diverse. The genetic determinants to resistance against quinolones and aminoglycosides are relevant because the World Health Organization classifies them as critically important antimicrobials which are the "Sole, or one of limited available therapies, to treat serious bacterial infections in people", along with other relevant criteria [58].

The *qnr* and *oqxAB* genes associated with quinolone resistance are mainly related to plasmids [59,60]. Despite being found in other serotypes in Mexico, the *qnr* genes were absent in Typhimurium strains of animal origin [61,62], but present in Mexican human strains of this serotype [63]. The *oqxAB* genes found here have also been previously reported for several serotypes of human isolates in Mexico [61,62]. Resistance to quinolones is considered low for *S. enterica* strains in the USA, although in this country it has been documented the presence of genes conferring resistance to such antimicrobial compounds in the serotype Typhimurium strains coming from clinical, pork meat, and livestock samples [64–67]. In contrast, in the case of Canada, the *qnrB* and *qnrS* genes were recently reported in ciprofloxacin-resistant human clinical strains of different serotypes of *S. enterica*, including Typhimurium, and the *oqxAB* gene has rarely been found [68]. Regarding the five point mutations that occur in the quinolone resistance-determining regions (QRDR) of DNA gyrase (*gyrA*) and topoisomerase IV (*parC*) [60] found in the genomes of ST213 strains, the double mutant S83Y in *gyrA* and S80I in *parC* is interesting, since *S.* Typhimurium mutations in *gyrA*, in addition to playing a dominant role in resistance to fluoroquinolones, have a synergistic effect with other resistance mechanisms, while the S80I mutation in *parC* appears to have no effect on quinolone resistance without *gyrA* mutations [69]. Whereas QRDR mutations are frequent in bovine *S. enterica* isolated strains of different serotypes in Mexico [62] and Canada [70], it appears to be absent or to occur with low frequency in the USA [36,71].

In relation to aminoglycoside resistance, the ST213 genomes carry the genes coding for the three types of modifying enzymes [72]. However, the CLSI has cautioned that aminoglycosides are not clinically effective against *S. enterica*, with this species being relevant as a carrier and potentially a means of transfer to other pathogenic bacteria [73]. The transference of genomic island one of *S. enterica* to *Escherichia coli*, a mobilizable element that carries multiple resistance antibiotics genes, has been experimentally demonstrated [74]. It has been documented that gentamicin resistance has increased significantly in the strains of eight serovars in human and retail meat in Canada in the last few decades [75]. Around 50% of Typhimurium strains from different sources from Mexico now show gentamicin resistance [32] Additionally, aminoglycoside resistance genes are highly prevalent in *S. enterica* strains isolated from farm animals in the USA, showing the geographical variations in its incidence [76].

In the case of tetracycline, sulfonamides, and phenicols, these antibiotics are considered by the WHO in the category of highly important antimicrobials, fulfilling the same first criterion of the previous antimicrobials but not adding more criteria [58]. The six tetracycline resistance genes detected in the ST213 strains are related to two resistance mechanisms, efflux pumps and ribosomal protection [77]. Regarding the sulfonamide resistance genes, the ST213 genome carries three of the four resistance genes associated with dihydropteroate synthetase modifications [78]. Interestingly, *sul2* and *sul3* predominate over the most frequently reported *sul1*. Additionally, the genomes carry three modifications of the dihydrofolate reductase enzyme for resistance to trimethoprim, of which *dfrA12* predominates. Although three genes that confer resistance to phenicols in the ST213 genomes are also present, the most widely distributed is the efflux bomb-coding gene *floR*. The strains of the Typhimurium serotype isolated from different sources show high percentages of resistance to these three types of antibiotics [32].

Besides all the previously commented resistance genes, all the analyzed genomes harbor *aac(6′)-Iaa*, *mdsA*, *mdsB*, *mdsC*, *golS*, *sdiA*, and the majority carried the *qacH* gene. The *aac(6′)-Iaa* gene for aminoglycosides resistance resides in the chromosome of Typhimurium strains and is present in the ancestral ST19 genotype [79]. The genes *mdsA*, *mdsB*, and *mdsC* code for the transporter efflux pump mdsABC *golS*, a promoter that is related to resistance to novobiocin [80], And it is also present in the ST19 genotype [81]. The *sdiA* gene codes for a quorum-sensing regulator that mediates the multi-drug resistance AcrAB efflux pump [82], the overproduction of which confers multidrug resistance; it is also present in the ST19 genotype. Finally, the *qacH* gene codes for an efflux bomb that confers resistance to the quaternary ammonium compounds, commonly used organic disinfectants [83]. This last gene is mainly found in different types of mobile genetic elements and confers resistance to aminoglycosides, chloramphenicol, erythromycin, and tetracyclines. As a whole, these antecedents suggest that such common ST213 genes are ancestral characters present in the ST19 genotype that can confer antibiotic multi-resistance to both genotypes.

All the antibiotics mentioned above are relevant for their use in veterinary and human medicine [58,84]. The samples of isolation of the ST213 strains for which genomes were analyzed here were predominantly from human and farm animals. This suggests that both farms and hospitals are relevant reservoirs for *S. enterica* ST213 strains carrying resistance genes that can be transferred to other pathogenic bacteria or directly to humans [85]. The diversity and distribution of the genes/mutations found in the analyzed genomes suggest that most of the North American ST213 strains are phenotypically multi-resistant to antibiotics.

In this work, 33 different plasmids were bioinformatically identified in the genomes of ST213 strains from around the world. Inc-type plasmids were found in 207 strains, while Col-type plasmids in 103; pESA2, pSL483 and rep14a_4_rep were found in a single strain each. The genomes of *S. enterica* North American ST123 strains from the US, Mexico, and Canada harbor 32 of the 33 plasmids, in contrast to the reduced variation in genomes of strains from Europe, but the low number of ST213 strains from Asia and Australia makes comparison difficult within these regions.

The Inc (C, F, H, I1) conjugative plasmids found in the analyzed genomes are frequently reported in different *S. enterica* serotypes, including Typhimurium, which together carry genes for all kind of antibiotics, some of which can be within an integron [37,38]. The simultaneous presence of IncHI2/IncHI2A and IncFIB(S)/IncFII(S) plasmids in the same ST213 genomes has previously been observed in *S. enterica* strains from Asia [86], Europe [87], and South America [88] for the first pair and in Africa for the second [89]. In addition, the IncHI2/IncHI2A pair was only found in the genomes of North American strains, while the presence of the IncFIB(S)/IncFII(S) pair was predominant in European strains. The plasmids IncI2 [90], ColpVC, IncHI2/IncHI2A, IncFIA, IncH1A, IncFIA(HI1)_1_HI1 [88,91], and ColRNAI, IncFIB(S)/IncFII(S) [41,92] have already been reported in strains of the ST19 founder genotype. The Col and IncX plasmids were present but less frequent in the ST213 genomes analyzed. The Col plasmids are mainly associated with the spread of *qnrS1* and *qnrB19* [37], and specifically the ColpVC detected here is associated with *aph(3″)-Ia* in bovine strains [93]. The IncX plasmids have been mainly isolated from human and animal samples and mainly encode genetic determinants against extended-spectrum β-lactams and quinolones [37,38].

The presence of the replicon rep14a_4_rep(AUS0004p3) is surprising because this plasmid belongs to a three-membered family of small mobilizable plasmids only described in the Gram-positive species *Enterococcus faecium*; thus, it must be considered to have a narrow host range [94]. Furthermore, the presence of these plasmids in *E. faecium* is associated with tetracycline resistance in strains isolated from human clinical samples, and it has been stated that such plasmids cannot be transferred by conjugation. This intriguing finding indicates the existence of a horizontal gene transfer (HGT) mechanism between *E. faecium* and *S. enterica*, which, to the best of our knowledge, has not previously been suggested. Further data on the plasmid sequence are needed to corroborate this result, and it will be relevant to analyze the HGT mechanism between the involved species.

The diversity of plasmid replicons found in the ST213 genomes indicates that such genotypes can be prone to the horizontal gene transfer (HGT) of antibiotic resistance genes. The plasmid diversity and distribution in the genomes analyzed here indicates that the plasmid interchange in ST213 strains features different dynamics between North America and Europe, apparently being more dynamic in the former region. Several plasmid replicons were found in the genomes of ST213 strains from all types of samples but showed different rates of incidence. However, most plasmids are present in samples related to the well-established zoonotic/food transmission chains of *S. enterica* to humans. Further work is needed to analyze a possible positive relationship between the type of sample and the presence/absence of a particular type of plasmid. Additionally, it is important that further studies define whether there is any significant relationship between the presence of these plasmids and the geographical region of origin of the ST213 strains, as well as its public health consequences.

It is important to consider that one of the weaknesses of the present study is that the analyzed genomes did not come from a systematic program based on the random sampling of strains. All ST213 genotype genomes available in public databases, of good quality, and with adequate metadata were analyzed. Thus, the trends and rates of prevalence reported here are the result of various factors, including the study of specific outbreaks by country as well as the capacity for genomic sequencing and its use as an epidemiological surveillance tool in different countries. The effect of these variables is reflected in the number of genomes of the ST213 genotype reported by year and by country (Figure 3), where the prevalence of genomes of strains from the USA, the UK, and other European countries, as well as from Australia, can be observed. Some of these countries routinely use genome sequencing in outbreak analyses and have adopted such strategies as the standard for the epidemiological surveillance of *S. enterica* or are in the process of doing so [95–99].

Another factor that may influence the results of the analysis carried out here is the quality of the sequences used, since low-level contamination can affect the detection of antibiotic resistance genes and point mutations associated with resistance [100]. However, the assembled genomes of EnteroBase undergo a quality control process that excludes the effects of such contamination [35], and the sets of reads used were filtered with bioinformatic tools to remove those that did not meet appropriate quality standards. Furthermore, it has been documented that the bioinformatic analysis of WGS to predict antibiotic resistance in *S. enterica* can generate results inconsistent with the observed phenotypic resistance [101]. However, in such works, only the ResFinder 3.2.0 tool is used for the location of antibiotic resistance genes, in contrast to the use of this software together with the ARG-ANNOT and CARD tools here to aid in locating the genetic determinants of antibiotic resistance, along with the use of databases recently curated to include genes and mutations previously not included in ResFinder [96–98]. In addition, the ResFinder update makes it possible to determine the number of copies of genes associated with resistance [102], another factor in antibiotic resistance that has previously not been considered by bioinformatics packages in resistance determination. Despite the predictive power of this bioinformatic analysis, it was desirable to contrast the results obtained here with the phenotypic antibiotic resistance/susceptibility data in order to obtain a clearer picture of the public health relevance of the ST213 genotype.

4. Materials and Methods

4.1. Data Retrieval

A total of 286 sets (572 reads pairs clusters) of Illumina sequencing reads available in EMBL's European Bioinformatics Institute (https://www.ebi.ac.uk/, accessed on 1 July 2022) and 33 EnteroBase assembled genomes (https://enterobase.warwick.ac.uk/, accessed on 1 July 2022) [35,103,104] assigned to ST213 genotype strains of *Salmonella enterica* were retrieved. All read sets and assembled genomes retrieved were filtered for a quality assessment to exclude duplicates, remove low-quality reads, and sampling coverage (Figure 8). Additionally, reads/genomes for which relevant metadata (country of isolation, year of isolation, sample kind) were not available were discarded. After filtering (see Section 4.2 below), 188 read sets and 32 assembled genomes were deemed to be adequate for analysis (Supplementary Table S1).

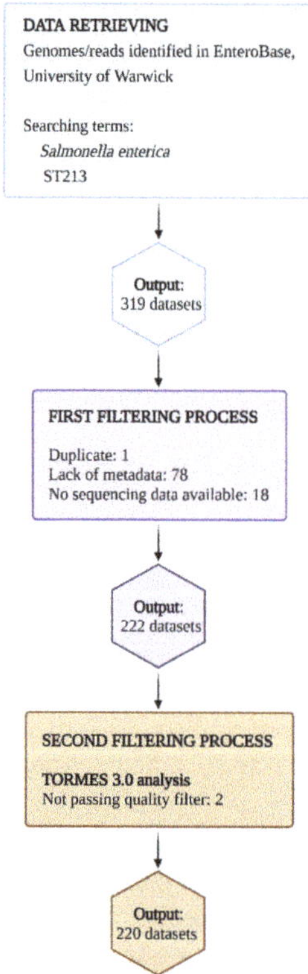

Figure 8. Data retrieval and exclusion criteria used for the analyzed genomes. The flowchart shows the number of genomes of read datasets initially found in EnteroBase and after the filtering criteria

were applied. Each filtering criterion shows the number of genomes/reads excluded. The metadata considered to include a genome/reads dataset were country of precedence, type of sample, and year of isolation. For details on the analyzed genomes, see Supplementary Tables S1–S3. For quality filter analysis, see the main text (figure created with Biorender.com, accessed on 1 July 2022).

4.2. Bioinformatic Analysis

The 188 sets of sequencing reads and 32 assembled genomes were processed with TORMES 3.0 [105] for sequence quality filtering, de novo genome assembly, and antibiotic resistance gene screening. All the genomes used in this analysis were at the level of *draft* genomes. The quality parameters used in the bioinformatic filtering to select both the genomes obtained from Enterobase (Supplementary Table S2) and for the genomes assembled in this work (Supplementary Table S3) were a high value of N50, an average length of contigs of greater than 5000 bases, and a low number of *contigs* [106], Additionally, the genome length obtained was in agreement with the genome length of different serotypes of *Salmonella enterica*, including Typhimurium [107]. Regarding the number of contigs, the mean of the assemblies obtained from the Enterobase was 73, while the mean for the assemblies carried out in this work was 89. This allowed us to ensure an adequate quality of the assemblies used for the gene search analysis and identification.

The identification of antibiotic resistance genes in the genomes of interest was carried out using ARG-ANNOT [108], CARD [109,110], and ResFinder 3.2.0 [111]. The PlasmidFinder tool [112] was used for the identification of plasmid replicons and PointFinder 3.1.0 for mutations in antibiotic resistance genes [113]. Gene nomenclature used here is provided by CARD.

4.3. Data Analysis

Binary (1,0) matrices were constructed to represent the presence (1)/absence (0) of resistance genes or plasmids, as appropriate. The genes/plasmids present in the genomes of all tested strains were excluded from this analysis. Both matrices were analyzed in the Heatmapper server [114] with the parameters of the Manhattan and UPGMA criteria for the calculation of distances and the generation of the grouping pattern, respectively. Manhattan distances were calculated based on the presence/absence of 37 identified antibiotic resistance genes, excluding seven genes that were present in 100% of the analyzed genomes (see Table 1). Additionally, those strains lacking all the genes/plasmids were excluded from this analysis.

5. Conclusions

The replacement of the ST19 genotype by other genotypes in different countries represents a serious public health concern worldwide because of the virulence and multiple antibiotic resistance of these emergent genotypes. Thus, the fast increase in the number of genomes deposited in public databases belonging to strains of the ST213 genotype throughout the world is a signal of the need to study this multiple-antibiotic-resistant and virulent *S*. Typhimurium variant. The great number of antibiotic resistance genes and plasmid replicons carried by ST213 strains and its simultaneous presence in the same genome indicates that most of them present a multiple antibiotic resistance phenotype and are prone to the HGT of antibiotic resistance determinants. Additionally, the genomes and grouping patterns obtained here suggest a route of dispersion of the ST213 emergent genotype beginning in North America (Canada, Mexico, USA) and moving to the United Kingdom, with further dispersion occurring from here to the rest of Europe and simultaneously to Asia and Oceania. Further detailed epidemiological analysis is necessary in order to clarify this dispersion hypothesis and to understand the mechanisms associated with the differences in the pattern of the presence/absence of antimicrobial resistance genetic determinants by the geographical origin of the strains and the type of sample of precedence.

Supplementary Materials: The following supporting information can be downloaded at: https://www.mdpi.com/article/10.3390/antibiotics11070925/s1. Supplementary Tables S1–S3: Data of the genomes analyzed in the present work.

Author Contributions: Conceptualization, G.V.-M. and M.S.V.-G.; methodology, E.A.H.-D. and A.M.N.-P.; formal analysis, E.A.H.-D. and A.M.N.-P.; investigation, G.V.-M. and M.S.V.-G.; resources, G.V.-M. and M.S.V.-G.; data curation, E.A.H.-D. and A.M.N.-P.; writing—original draft preparation, G.V.-M. and M.S.V.-G.; writing—review and editing, G.V.-M. and E.A.H.-D.; funding acquisition, G.V.-M. and M.S.V.-G. All authors have read and agreed to the published version of the manuscript.

Funding: This work was supported by the Coordinación de la Investigación Científica UMSNH, 2022 Research Program.

Institutional Review Board Statement: Not applicable.

Informed Consent Statement: Not applicable.

Data Availability Statement: Date is available in the supplementary materials.

Acknowledgments: Thanks is given to CONACYT for the graduate scholarship granted to E.A.H.-D. (No. 731999). We also acknowledge Elisa Vázquez-Vázquez for her support in translating this manuscript into English and editing it.

Conflicts of Interest: The authors declare no conflict of interest. The funders had no role in the design of the study; in the collection, analysis, or interpretation of data; in the writing of the manuscript; or in the decision to publish the results.

References

1. Alby, K.; Nachamkin, I. Gastrointestinal infections. *Microbiol. Spectr.* **2016**, *4*, 44. [CrossRef] [PubMed]
2. Eng, S.K.; Pusparajah, P.; Ab Mutalib, N.S.; Ser, H.L.; Chan, K.G.; Lee, L.H. *Salmonella*: A review on pathogenesis, epidemiology and antibiotic resistance. *Front. Life Sci.* **2015**, *8*, 284–293. [CrossRef]
3. Mafi, N.; Orenstein, R. Salmonella. In *Encyclopedia of Gastroenterology*, 2nd ed.; Johnson, L.R., Ed.; Elsevier: New York, NY, USA, 2020; pp. 384–391. [CrossRef]
4. St John-Brooks, R. The Genus *Salmonella* Lignieres, 1900: Issued by the *Salmonella* Subcommittee of the Nomenclature Committee of the International Society for Microbiology. *J. Hyg.* **1934**, *34*, 333–350.
5. Grimont, P.A.; Weill, F.X. *Antigenic Formulae of the Salmonella Serovars*, 9th ed.; WHO Collaborating Centre for Reference and Research on *Salmonella*; WHO: Paris, France, 2007; Volume 9, pp. 1–166.
6. Issenhuth-Jeanjean, S.; Roggentin, P.; Mikoleit, M.; Guibourdenche, M.; De Pinna, E.; Nair, S.; Fields, I.P.; Weill, F.X. Supplement 2008–2010 (no. 48) to the white–Kauffmann–Le minor scheme. *Res. Microbiol.* **2014**, *165*, 526–530. [CrossRef] [PubMed]
7. Hendriksen, R.S.; Vieira, A.R.; Karlsmose, S.; Lo Fo Wong, D.M.; Jensen, A.B.; Wegener, H.C.; Aarestrup, F.M. Global monitoring of *Salmonella* serovar distribution from the World Health Organization Global Foodborne Infections Network Country Data Bank: Results of quality assured laboratories from 2001 to 2007. *Foodborne Pathog. Dis.* **2011**, *8*, 887–900. [CrossRef] [PubMed]
8. Ferrari, R.G.; Rosario, D.K.; Cunha-Neto, A.; Mano, S.B.; Figueiredo, E.E.; Conte-Junior, C.A. Worldwide epidemiology of *Salmonella* serovars in animal-based foods: A meta-analysis. *Appl. Environ. Microbiol.* **2019**, *85*, e00591-19. [CrossRef]
9. Inns, T.; Ashton, P.M.; Herrera-Leon, S.; Lighthill, J.; Foulkes, S.; Jombart, T.; Rehman, Y.; Fox, A.; Dallman, T.; de Pinna, E.; et al. Prospective use of whole genome sequencing (WGS) detected a multi-country outbreak of *Salmonella enteritidis*. *Epidemiol. Infect.* **2017**, *145*, 289–298. [CrossRef] [PubMed]
10. Leekitcharoenphon, P.; Nielsen, E.M.; Kaas, R.S.; Lund, O.; Aarestrup, F.M. Evaluation of whole genome sequencing for outbreak detection of *Salmonella enterica*. *PLoS ONE* **2014**, *9*, e87991. [CrossRef] [PubMed]
11. Simon, S.; Trost, E.; Bender, J.; Fuchs, S.; Malorny, B.; Rabsch, W.; Prager, R.; Tietze, E.; Flieger, A. Evaluation of WGS based approaches for investigating a food-borne outbreak caused by *Salmonella enterica* serovar Derby in Germany. *Food Microbiol.* **2018**, *71*, 46–54. [CrossRef]
12. Deng, X.; den Bakker, H.C.; Hendriksen, R.S. Genomic epidemiology: Whole-genome-sequencing–powered surveillance and outbreak investigation of foodborne bacterial pathogens. *Annu. Rev. Food Sci. Technol.* **2016**, *7*, 353–374. [CrossRef]
13. Tang, P.; Croxen, M.A.; Hasan, M.R.; Hsiao, W.W.; Hoang, L.M. Infection control in the new age of genomic epidemiology. *Am. J. Infect. Control.* **2017**, *45*, 170–179. [CrossRef] [PubMed]
14. Rantsiou, K.; Kathariou, S.; Winkler, A.; Skandamis, P.; Saint-Cyr, M.J.; Rouzeau-Szynalski, K.; Amézquita, A. Next generation microbiological risk assessment: Opportunities of whole genome sequencing (WGS) for foodborne pathogen surveillance, source tracking and risk assessment. *Int. J. Food Microbiol.* **2018**, *287*, 3–9. [CrossRef] [PubMed]
15. Achtman, M.; Wain, J.; Weill, F.X.; Nair, S.; Zhou, Z.; Sangal, V.; Krauland, M.G.; Hale, J.L.; Harbottle, H.; Uesbeck, A.; et al. Multilocus sequence typing as a replacement for serotyping in *Salmonella enterica*. *PLoS Pathog.* **2012**, *8*, e1009040. [CrossRef]

16. Lan, R.; Reeves, P.R.; Octavia, S. Population structure, origins and evolution of major *Salmonella enterica* clones. *Infect. Genet. Evol.* **2009**, *9*, 996–1005. [CrossRef] [PubMed]
17. Feijao, P.; Yao, H.T.; Fornika, D.; Gardy, J.; Hsiao, W.; Chauve, C.; Chindelevitch, L. MentaLiST—A fast MLST caller for large MLST schemes. *Microb. Genom.* **2018**, *4*, e000146. [CrossRef] [PubMed]
18. Alba, P.; Leekitcharoenphon, P.; Carfora, V.; Amoruso, R.; Cordaro, G.; Di Matteo, P.; Ianzano, A.; Iurescia, M.; Diaconu, E.L.; ENGAGE-EURL-AR Network Study Group; et al. Molecular epidemiology of *Salmonella* Infantis in Europe: Insights into the success of the bacterial host and its parasitic pESI-like megaplasmid. *Microb Genomics.* **2020**, *6*, e000365. [CrossRef] [PubMed]
19. Bonifait, L.; Thépault, A.; Baugé, L.; Rouxel, S.; Le Gall, F.; Chemaly, M. Occurrence of *Salmonella* in the cattle production in France. *Microorganisms* **2021**, *9*, 872. [CrossRef]
20. Ashton, P.M.; Owen, S.V.; Kaindama, L.; Rowe, W.P.; Lane, C.R.; Larkin, L.; Nair, S.; Jenkins, C.; de Pinna, E.M.; Feasey, N.A.; et al. Public health surveillance in the UK revolutionises our understanding of the invasive *Salmonella* Typhimurium epidemic in Africa. *Genome Med.* **2017**, *9*, 92. [CrossRef]
21. Kingsley, R.A.; Msefula, C.L.; Thomson, N.R.; Kariuki, S.; Holt, K.E.; Gordon, M.A.; Harris, D.; Clarke, L.; Whitehead, S.; Sangal, V.; et al. Epidemic multiple drug resistant *Salmonella* Typhimurium causing invasive disease in sub-Saharan Africa have a distinct genotype. *Genome Res.* **2009**, *19*, 2279–2287. [CrossRef] [PubMed]
22. Okoro, C.K.; Barquist, L.; Connor, T.R.; Harris, S.R.; Clare, S.; Stevens, M.P.; Arends, M.J.; Hale, C.; Kane, L.; Pickard, D.J.; et al. Signatures of adaptation in human invasive *Salmonella* Typhimurium ST313 populations from sub-Saharan Africa. *PLOS Negl. Trop. Dis.* **2015**, *9*, e0003611. [CrossRef]
23. Hammarlöf, D.L.; Kröger, C.; Owen, S.V.; Canals, R.; Lacharme-Lora, L.; Wenner, N.; Schager, A.E.; Wells, T.J.; Henderson, I.R.; Wigley, P.; et al. Role of a single noncoding nucleotide in the evolution of an epidemic African clade of *Salmonella*. *Proc. Natl. Acad. Sci. USA* **2018**, *115*, E2614–E2623. [CrossRef] [PubMed]
24. Neuert, S.; Nair, S.; Day, M.R.; Doumith, M.; Ashton, P.M.; Mellor, K.C.; Jenkins, C.; Hopkins, K.L.; Woodford, N.; de Pinna, E.; et al. Prediction of phenotypic antimicrobial resistance profiles from whole genome sequences of non-typhoidal *Salmonella enterica*. *Front. Microbiol.* **2018**, *9*, 592. [CrossRef]
25. Tassinari, E.; Bawn, M.; Thilliez, G.; Charity, O.; Acton, L.; Kirkwood, M.; Petrovska, L.; Dallman, T.; Burgess, C.M.; Hall, N.; et al. Whole-genome epidemiology links phage-mediated acquisition of a virulence gene to the clonal expansion of a pandemic *Salmonella enterica* serovar Typhimurium clone. *Microb. Genom.* **2020**, *6*, mgen000456. [CrossRef]
26. Crouse, A.; Schramm, C.; Emond-Rheault, J.G.; Herod, A.; Kerhoas, M.; Rohde, J.; Gruenheid, S.; Kukavica-Ibrulj, I.; Boyle, B.; Greenwood, C.M.T.; et al. Combining Whole-Genome Sequencing and multimodel phenotyping to identify genetic predictors of *Salmonella* virulence. *mSphere* **2020**, *5*, e00293-20. [CrossRef] [PubMed]
27. Pires, J.; Huisman, J.S.; Bonhoeffer, S.; Van Boeckel, T.P. Multidrug resistance dynamics in *Salmonella* in food animals in the United States: An analysis of genomes from public databases. *Microbiol. Spectr.* **2021**, *9*, e00495-21. [CrossRef] [PubMed]
28. Yin, X.; Fu, Y.; Tate, H.; Pinto, C.; Dudley, E.G.; M'ikanatha, N.M. Genomic analysis of *Salmonella* Typhimurium from humans and food sources accurately predicts phenotypic multi-drug resistance. *Food Microbiol.* **2022**, *103*, 103957. [CrossRef] [PubMed]
29. Okoro, C.K.; Kingsley, R.A.; Connor, T.R.; Harris, S.R.; Parry, C.M.; Al-Mashhadani, M.N.; Kariuki, S.; Msefula, C.L.; Gordon, M.A.; de Pinna, E.; et al. Intracontinental spread of human invasive *Salmonella* Typhimurium pathovariants in sub-Saharan Africa. *Nat. Genet.* **2012**, *44*, 1215–1221. [CrossRef] [PubMed]
30. Wong, M.H.Y.; Yan, M.; Chan, E.W.C.; Liu, L.Z.; Kan, B.; Chen, S. Expansion of *Salmonella* enterica serovar Typhimurium ST34 clone carrying multiple resistance determinants in China. *Antimicrob. Agents Chemother.* **2013**, *57*, 4599–4601. [CrossRef] [PubMed]
31. Sun, J.; Ke, B.; Huang, Y.; He, D.; Li, X.; Liang, Z.; Ke, C. The molecular epidemiological characteristics and genetic diversity of *Salmonella* Typhimurium in Guangdong, China, 2007–2011. *PLoS ONE* **2014**, *9*, e113145. [CrossRef] [PubMed]
32. Zaidi, M.B.; Calva, J.J.; Estrada-Garcia, M.T.; Leon, V.; Vazquez, G.; Figueroa, G.; Lopez, E.; Contreras, J.; Abbott, J.; Zhao, S.; et al. Integrated food chain surveillance system for *Salmonella* spp. in Mexico. *Emerg. Infect. Dis.* **2008**, *14*, 429–435. [CrossRef] [PubMed]
33. Ambrose, S.J.; Harmer, C.J.; Hall, R.M. Compatibility and entry exclusion of IncA and IncC plasmids revisited: IncA and IncC plasmids are compatible. *Plasmid* **2018**, *96–97*, 7–12. [CrossRef] [PubMed]
34. Wiesner, M.; Calva, E.; Fernández-Mora, M.; Cevallos, M.A.; Campos, F.; Zaidi, M.B.; Silva, C. *Salmonella* Typhimurium ST213 is associated with two types of IncA/C plasmids carrying multiple resistance determinants. *BMC Microbiol.* **2011**, *11*, 9. [CrossRef] [PubMed]
35. Enterobase. Quality Assessment Evaluation—EnteroBase Documentation. 2018. Available online: https://enterobase.readthedocs. io/en/latest/pipelines/backend-pipeline-qaevaluation.html (accessed on 31 January 2022).
36. Lee, K.Y.; Atwill, E.R.; Pitesky, M.; Huang, A.; Lavelle, K.; Rickard, M.; Shafii, M.; Hung-Fan, M.; Li, X. Antimicrobial resistance profiles of non-typhoidal *Salmonella* from retail meat products in California, 2018. *Front. Microbiol.* **2022**, *13*, 835699. [CrossRef] [PubMed]
37. Rozwandowicz, M.; Brouwer MS, M.; Fischer, J.; Wagenaar, J.A.; Gonzalez-Zorn, B.; Guerra, B.; Mevius, D.J.; Hordijk, J. Plasmids carrying antimicrobial resistance genes in Enterobacteriaceae. *J. Antimicrob. Chemother.* **2018**, *73*, 1121–1137. [CrossRef] [PubMed]
38. McMillan, E.A.; Jackson, C.R.; Frye, J.G. Transferable plasmids of *Salmonella enterica* associated with antibiotic resistance genes. *Front. Microbiol.* **2020**, *11*, 562181. [CrossRef]

39. Lima, T.; Domingues, S.; Da Silva, G.J. Plasmid-mediated colistin resistance in *Salmonella enterica*: A review. *Microorganisms* **2019**, *7*, 55. [CrossRef] [PubMed]

40. Portes, A.B.; Rodrigues, G.; Leitão, M.P.; Ferrari, R.; Conte Junior, C.A.; Panzenhagen, P. Global distribution of plasmid-mediated colistin resistance *mcr* gene in *Salmonella*: A systematic review. *J. Appl. Microbiol.* **2022**, *132*, 872–889. [CrossRef]

41. Jain, P.; Sudhanthirakodi, S.; Chowdhury, G.; Joshi, S.; Anandan, S.; Ray, U.; Mukhopadhyay, A.; Dutta, S. Antimicrobial resistance, plasmid, virulence, multilocus sequence typing and pulsed-field gel electrophoresis profiles of *Salmonella enterica* serovar Typhimurium clinical and environmental isolates from India. *PLoS ONE* **2018**, *13*, e0207954. [CrossRef]

42. Elnekave, E.; Hong, S.L.; Lim, S.; Boxrud, D.; Rovira, A.; Mather, A.E.; Perez, A.; Alvarez, J. Transmission of multidrug-resistant *Salmonella enterica* subspecies *enterica* 4,[5], 12: I:- Sequence Type 34 between Europe and the United States. *Emerg. Infect. Dis.* **2020**, *26*, 3034–3038. [CrossRef] [PubMed]

43. Mather, A.E.; Phuong TL, T.; Gao, Y.; Clare, S.; Mukhopadhyay, S.; Goulding, D.A.; Hoang, N.T.D.; Tuyen, H.T.; Lan NP, H.; Thompson, C.N.; et al. New variant of multidrug-resistant *Salmonella enterica* serovar Typhimurium associated with invasive disease in immunocompromised patients in Vietnam. *mBio* **2018**, *9*, e01056-18. [CrossRef]

44. Wiesner, M.; Zaidi, M.B.; Calva, E.; Fernández-Mora, M.; Calva, J.J.; Silva, C. Association of virulence plasmid and antibiotic resistance determinants with chromosomal multilocus genotypes in Mexican *Salmonella enterica* serovar Typhimurium strains. *BMC Microbiol.* **2009**, *9*, 131. [CrossRef] [PubMed]

45. Holley, R.A. Food safety challenges within North American free trade agreement (NAFTA) partners. *Compr. Rev. Food Sci.* **2011**, *10*, 131–142. [CrossRef]

46. Johnson, L.R.; Gould, L.H.; Dunn, J.R.; Berkelman, R.; Mahon, B.E.; The FoodNet Travel Working Group. Salmonella infections associated with international travel: A Foodborne Diseases Active Surveillance Network (FoodNet) study. *Foodborne Pathog. Dis.* **2011**, *8*, 1031–1037. [CrossRef]

47. Tighe, M.K.; Savage, R.; Vrbova, L.; Toolan, M.; Whitfield, Y.; Varga, C.; Lee, B.; Allen, V.; Maki, A.; Walton, R.; et al. The epidemiology of travel-related *Salmonella* Enteritidis in Ontario, Canada, 2010–2011. *BMC Public Health* **2012**, *12*, 310. [CrossRef] [PubMed]

48. Coipan, C.E.; Westrell, T.; van Hoek, A.H.; Alm, E.; Kotila, S.; Berbers, B.; de Keersmaecker, S.C.J.; Ceyssens, P.J.; Borg, M.L.; Chattaway, M.; et al. Genomic epidemiology of emerging ESBL-producing *Salmonella* Kentucky *bla* CTX-M-14b in Europe. *Emerg. Microbes Infect.* **2020**, *9*, 2124–2135. [CrossRef] [PubMed]

49. Waldram, A.; Dolan, G.; Ashton, P.M.; Jenkins, C.; Dallman, T.J. Epidemiological analysis of *Salmonella* clusters identified by whole genome sequencing, England and Wales 2014. *Food Microbiol.* **2018**, *71*, 39–45. [CrossRef] [PubMed]

50. Williamson, D.A.; Lane, C.R.; Easton, M.; Valcanis, M.; Strachan, J.; Veitch, M.G.; Kirk, M.D.; Howden, B.P. Increasing antimicrobial resistance in nontyphoidal *Salmonella* isolates in Australia from 1979 to 2015. *Antimicrob. Agents Chemother.* **2018**, *62*, e02012–e02017. [CrossRef] [PubMed]

51. Medalla, F.; Gu, W.; Friedman, C.R.; Judd, M.; Folster, J.; Griffin, P.M.; Hoekstra, R.M. Increased incidence of antimicrobial-resistant nontyphoidal *Salmonella* infections, United States, 2004–2016. *Emerg. Infect. Dis.* **2021**, *27*, 1662. [CrossRef]

52. Varga, C.; John, P.; Cooke, M.; Majowicz, S.E. Spatial and space-time clustering and demographic characteristics of human nontyphoidal *Salmonella* infections with major serotypes in Toronto, Canada. *PLoS ONE* **2020**, *15*, e0235291. [CrossRef]

53. Zaidi, M.B.; Campos, F.D.; Estrada-García, T.; Gutierrez, F.; León, M.; Chim, R.; Calva, J.J. Burden and transmission of zoonotic foodborne disease in a rural community in Mexico. *Clin. Infect. Dis.* **2012**, *55*, 51–60. [CrossRef]

54. Parmley, E.J.; Pintar, K.; Majowicz, S.; Avery, B.; Cook, A.; Jokinen, C.; Gannon, V.; Lapen, D.R.; Topp, E.; Edge, T.A.; et al. A Canadian application of One Health: Integration of *Salmonella* data from various Canadian surveillance programs (2005–2010). *Foodborne Pathog. Dis.* **2013**, *10*, 747–756. [CrossRef] [PubMed]

55. Flockhart, L.; Pintar, K.; Cook, A.; McEwen, S.; Friendship, R.; Kelton, D.; Pollari, F. Distribution of *Salmonella* in humans, production animal operations and a watershed in a FoodNet Canada sentinel site. *Zoonoses Public Health* **2017**, *64*, 41–52. [CrossRef]

56. Marus, J.R.; Magee, M.J.; Manikonda, K.; Nichols, M.C. Outbreaks of *Salmonella enterica* infections linked to animal contact: Demographic and outbreak characteristics and comparison to foodborne outbreaks—United States, 2009–2014. *Zoonoses Public Health* **2019**, *66*, 370–376. [CrossRef] [PubMed]

57. Zhang, S.; Li, S.; Gu, W.; den Bakker, H.; Boxrud, D.; Taylor, A.; Roe, C.; Driebe, E.; Engelthaler, D.M.; Allard, M.; et al. Zoonotic source attribution of *Salmonella enterica* serotype Typhimurium using genomic surveillance data, United States. *Emerg. Infect. Dis.* **2019**, *25*, 82–91. [CrossRef] [PubMed]

58. World Health Organization (WHO). *Critically Important Antimicrobials for Human Medicine*, 6th ed.; World Health Organization (WHO): Geneva, Switzerland, 2018.

59. Ruiz, J. Transferable mechanisms of quinolone resistance from 1998 onward. *Clin. Microbiol. Rev.* **2019**, *32*, e00007–e00019. [CrossRef] [PubMed]

60. Correia, S.; Poeta, P.; Hébraud, M.; Capelo, J.L.; Igrejas, G. Mechanisms of quinolone action and resistance: Where do we stand? *J. Med. Microbiol.* **2017**, *66*, 551–559. [CrossRef] [PubMed]

61. Delgado-Suárez, E.J.; Ortíz-López, R.; Gebreyes, W.A.; Allard, M.W.; Barona-Gómez, F.; Rubio-Lozano, M.S. Genomic surveillance links livestock production with the emergence and spread of multi-drug resistant non-typhoidal *Salmonella* in Mexico. *J. Microbiol.* **2019**, *57*, 271–280. [CrossRef] [PubMed]

62. Delgado-Suárez, E.J.; Palós-Guitérrez, T.; Ruíz-López, F.A.; Hernández Pérez, C.F.; Ballesteros-Nova, N.E.; Soberanis-Ramos, O.; Méndez-Medina, R.D.; Allard, M.W.; Rubio-Lozano, M.S. Genomic surveillance of antimicrobial resistance shows cattle and poultry are a moderate source of multi-drug resistant non-typhoidal *Salmonella* in Mexico. *PLoS ONE* **2021**, *16*, e0243681. [CrossRef]

63. Zaidi, M.B.; Estrada-García, T.; Campos, F.D.; Chim, R.; Arjona, F.; Leon, M.; Michell, A.; Chaussabel, D. Incidence, clinical presentation, and antimicrobial resistance trends in *Salmonella* and *Shigella* infections from children in Yucatan, Mexico. *Front. Microbiol.* **2013**, *4*, 288. [CrossRef]

64. McDermott, P.F.; Zhao, S.; Tate, H. Antimicrobial resistance in nontyphoidal *Salmonella*. *Microbiol. Spectr.* **2018**, *6*, 780–790. [CrossRef]

65. Sjölund-Karlsson, M.; Folster, J.P.; Pecic, G.; Joyce, K.; Medalla, F.; Rickert, R.; Whichard, J.M. Emergence of plasmid-mediated quinolone resistance among non-Typhi *Salmonella enterica* isolates from humans in the United States. *Antimicrob. Agents Chemother.* **2009**, *53*, 2142–2144. [CrossRef] [PubMed]

66. Sjölund-Karlsson, M.; Howie, R.; Rickert, R.; Krueger, A.; Tran, T.T.; Zhao, S.; Ball, T.; Haro, J.; Pecic, G.; Joyce, K.; et al. Plasmid-mediated quinolone resistance among non-typhi *Salmonella enterica* isolates, USA. *Emerg. Infect. Dis.* **2010**, *16*, 1789–1791. [CrossRef] [PubMed]

67. Cuypers, W.L.; Jacobs, J.; Wong, V.; Klemm, E.J.; Deborggraeve, S.; Van Puyvelde, S. Fluoroquinolone resistance in *Salmonella*: Insights by whole-genome sequencing. *Microbial. Genomics.* **2018**, *4*, e000195. [CrossRef]

68. Kim, J.; Han, X.; Bae, J.; Chui, L.; Louie, M.; Finley, R.; Mulvey, M.R.; Ferrato, C.J.; Jeon, B. Prevalence of plasmid-mediated quinolone resistance (PMQR) genes in non-typhoidal *Salmonella* strains with resistance and reduced susceptibility to fluoroquinolones from human clinical cases in Alberta, Canada, 2009–2013. *J. Antimicrob. Chemother.* **2016**, *71*, 2988–2990. [CrossRef] [PubMed]

69. Luo, Y.; Li, J.; Meng, Y.; Ma, Y.; Hu, C.; Jin, S.; Zhang, Q.; Ding, H.; Cui, S. Joint effects of topoisomerase alterations and plasmid-mediated quinolone-resistant determinants in *Salmonella enterica* Typhimurium. *Microb. Drug. Resist.* **2011**, *17*, 1–5. [CrossRef] [PubMed]

70. Bharat, A.; Petkau, A.; Avery, B.P.; Chen, J.C.; Folster, J.P.; Carson, C.A.; Kearney, A.; Nadon, C.; Mabon, P.; Thiessen, J.; et al. Correlation between phenotypic and in silico detection of antimicrobial resistance in *Salmonella enterica* in Canada using Staramr. *Microorganisms* **2022**, *10*, 292. [CrossRef] [PubMed]

71. Hawkey, J.; Le Hello, S.; Doublet, B.; Granier, S.A.; Hendriksen, R.S.; Fricke, W.F.; Ceyssens, P.J.; Gomart, C.; Billman-Jacobe, H.; Holt, K.E.; et al. Global phylogenomics of multidrug-resistant *Salmonella enterica* serotype Kentucky ST198. *Microbial. Genomics.* **2019**, *5*, e000269. [CrossRef] [PubMed]

72. Van Hoek, A.H.; Mevius, D.; Guerra, B.; Mullany, P.; Roberts, A.P.; Aarts, H.J. Acquired antibiotic resistance genes: An overview. *Front. Microbiol.* **2011**, *2*, 203. [CrossRef]

73. CLSI. *Performance Standards for Antimicrobial Susceptibility Testing*, 32nd ed.; CLSI Supplement M100; Clinical and Laboratory Standards Institute: Wayne, PA, USA; Available online: http://em100.edaptivedocs.net/dashboard.aspx (accessed on 1 May 2022).

74. Van Duijkeren, E.; Schwarz, C.; Bouchard, D.; Catry, B.; Pomba, C.; Baptiste, K.E.; Moreno, M.A.; Rantala, M.; Ružauskas, M.; Sanders, P.; et al. The use of aminoglycosides in animals within the EU: Development of resistance in animals and possible impact on human and animal health: A review. *J. Antimicrob. Chemother.* **2019**, *74*, 2480–2496. [CrossRef]

75. Cox, G.W.; Parmley, E.J.; Avery, B.P.; Irwin, R.J.; Reid-Smith, R.J.; Deckert, A.E.; Finley, R.L.; Daignault, D.; Alexander, D.C.; Allen, V.; et al. A One-Health genomic investigation of gentamicin resistance in *Salmonella* from human and chicken sources in Canada, 2014 to 2017. *Antimicrob. Agents Chemother.* **2021**, *65*, e00966-21. [CrossRef]

76. Nyirabahizi, E.; Tyson, G.H.; Tate, H.; Strain, E. The Western United States has greater antibiotic resistance among *Salmonella* recovered from intestinal cecal samples of food animals. *J. Food Protect.* **2020**. [CrossRef] [PubMed]

77. Markley, J.L.; Wencewicz, T.A. Tetracycline-inactivating enzymes. *Front. Microbiol.* **2018**, *9*, 1058. [CrossRef] [PubMed]

78. Nunes, O.C.; Manaia, C.M.; Kolvenbach, B.A.; Corvini PF, X. Living with sulfonamides: A diverse range of mechanisms observed in bacteria. *Appl. Microbiol. Biotechnol.* **2020**, *104*, 10389–10408. [CrossRef] [PubMed]

79. Monte, D.F.; Sellera, F.P.; Lopes, R.; Keelara, S.; Landgraf, M.; Greene, S.; Fedorka-Cray, P.J.; Thakur, S. Class 1 integron-borne cassettes harboring *bla CARB-2* gene in multidrug-resistant and virulent *Salmonella* Typhimurium ST19 strains recovered from clinical human stool samples, United States. *PLoS ONE* **2020**, *15*, e0240978. [CrossRef] [PubMed]

80. Nishino, K.; Latifi, T.; Groisman, E.A. Virulence and drug resistance roles of multidrug efflux systems of *Salmonella enterica* serovar typhimurium. *Mol. Microbiol.* **2006**, *59*, 126–141. [CrossRef] [PubMed]

81. Park, C.J.; Li, J.; Zhang, X.; Gao, F.; Benton, C.S.; Andam, C.P. Diverse lineages of multidrug resistant clinical *Salmonella enterica* and a cryptic outbreak in New Hampshire, USA revealed from a year-long genomic surveillance. *Infect. Genet. Evol.* **2021**, *87*, 104645. [CrossRef]

82. Rahmati, S.; Yang, S.; Davidson, A.L.; Zechiedrich, E.L. Control of the AcrAB multidrug efflux pump by quorum-sensing regulator SdiA. *Mol. Microbiol.* **2002**, *43*, 677–685. [CrossRef]

83. Tezel, U.; Pavlostathis, S.G. Quaternary ammonium disinfectants: Microbial adaptation, degradation and ecology. *Curr. Opin. Biotechnol.* **2015**, *33*, 296–304. [CrossRef]

84. OIE. OIE Strategy on Antimicrobial Resistance and the Prudent Use of Antimicrobials. 2016. Available online: http://www.oie.int/fileadmin/Home/eng/Media_Center/docs/pdf/PortailAMR/EN_OIE-AMRstrategy.pdf (accessed on 20 May 2022).

85. Monte, D.F.; Lincopan, N.; Fedorka-Cray, P.J.; Landgraf, M. Current insights on high priority antibiotic-*resistant Salmonella enterica* in food and foodstuffs: A review. *Curr. Opin. Food Sci.* **2019**, *26*, 35–46. [CrossRef]

86. Elbediwi, M.; Tang, Y.; Shi, D.; Ramadan, H.; Xu, Y.; Xu, S.; Li, Y.; Yue, M. Genomic investigation of antimicrobial-resistant *Salmonella enterica* isolates from dead chick embryos in China. *Front. Microbiol.* **2021**, *12*, 684400. [CrossRef]

87. Diaconu, E.L.; Alba, P.; Feltrin, F.; Di Matteo, P.; Iurescia, M.; Chelli, E.; Donati, V.; Marani, I.; Giacomi, A.; Franco, A.; et al. Emergence of IncHI2 plasmids with mobilized colistin resistance (mcr)-9 gene in esbl-producing, multidrug-resistant *Salmonella* Typhimurium and its monophasic variant ST34 from food-producing animals in Italy. *Front. Microbiol.* **2021**, *12*, 705230. [CrossRef] [PubMed]

88. Monte, D.F.; Lincopan, N.; Berman, H.; Cerdeira, L.; Keelara, S.; Siddhartha Thakur, S.; Fedorka-Cray, P.; Landgraf, M. Genomic features of high-priority *Salmonella enterica* serovars circulating in the food production chain, Brazil, 2000–2016. *Sci. Rep.* **2019**, *9*, 11058. [CrossRef] [PubMed]

89. Ball, T.; Monte, D.; Aidara-Kane, A.; Matheu, J.; Ru, H.; Thakur, S.; Ejobi, F.; Fedorka-Cray, P. International lineages of Salmonella enterica serovars isolated from chicken farms, Wakiso District, Uganda. *PLoS ONE* **2020**, *15*, e0220484. [CrossRef] [PubMed]

90. Moon, D.C.; Kim, S.-J.; Mechesso, A.F.; Kang, H.Y.; Song, H.-J.; Choi, J.-H.; Yoon, S.-S.; Lim, S.-K. Mobile colistin resistance gene mcr-1 detected on an IncI2 plasmid in *Salmonella* Typhimurium Sequence Type 19 from a healthy pig in South Korea. *Microorganisms* **2021**, *9*, 398. [CrossRef]

91. Wu, B.; Ed-Dra, A.; Pan, H.; Dong, C.; Jia, C.; Yue, M. Genomic investigation of *Salmonella* isolates recovered from a pig slaughtering process in Hangzhou, China. *Front. Microbiol.* **2021**, *12*, 704636. [CrossRef]

92. Viana, C.; Grossi, J.L.; Sereno, M.J.; Yamatogi, R.S.; Bersot, L.D.S.; Call, D.R.; Nero, L.A. Phenotypic and genotypic characterization of non-typhoidal *Salmonella* isolated from a Brazilian pork production chain. *Food Res. Int.* **2020**, *137*, 109406. [CrossRef]

93. Carroll, L.M.; Wiedmann, M.; den Bakker, H.; Siler, J.; Warchocki, S.; Kent, D.; Lyalina, S.; Davis, M.; Sischo, W.; Besser, T.; et al. Whole-genome sequencing of drug-resistant *Salmonella enterica* isolates from dairy cattle and humans in New York and Washington states reveals source and geographic associations. *Appl. Environ. Microbiol.* **2017**, *83*, e00140-17. [CrossRef]

94. Jensen, L.B.; Garcia-Migura, L.; Valenzuela, A.J.; Løhr, M.; Hasman, H.; Aarestrup, F.M. A classification system for plasmids from enterococci and other Gram-positive bacteria. *J. Microbiol. Methods.* **2010**, *80*, 25–43. [CrossRef]

95. Ashton, P.M.; Nair, S.; Peters, T.M.; Bale, J.A.; Powell, D.G.; Painset, A.; Tewolde, R.; Schaefer, U.; Jenkins, C.; Dallman, T.J.; et al. *Salmonella* Whole Genome Sequencing Implementation Group. Identification of *Salmonella* for public health surveillance using whole genome sequencing. *PeerJ* **2016**, *4*, e1752. [CrossRef]

96. Chattaway, M.A.; Dallman, T.J.; Larkin, L.; Nair, S.; McCormick, J.; Mikhail, A.; Hartman, H.; Godbole, G.; Powell, D.; Day, M.; et al. The transformation of reference microbiology methods and surveillance for *Salmonella* with the use of whole genome sequencing in England and Wales. *Public. Health Front.* **2019**, *7*, 317. [CrossRef]

97. Van Goethem, N.; Van Den Bossche, A.; Ceyssens, P.J.; Lajot, A.; Coucke, W.; Vernelen, K.; Roosens, N.C.H.; De Keersmaecker, S.C.J.; Van Cauteren, D.; Mattheus, W. Coverage of the national surveillance system for human *Salmonella* infections, Belgium, 2016–2020. *PLoS ONE* **2021**, *16*, e0256820. [CrossRef] [PubMed]

98. Brown, E.; Dessai, U.; McGarry, S.; Gerner-Smidt, P. Use of whole-genome sequencing for food safety and public health in the United States. *Foodborne Pathog. Dis.* **2019**, *16*, 441–450. [CrossRef] [PubMed]

99. Hoang, T.; da Silva, A.G.; Jennison, A.V.; Williamson, D.A.; Howden, B.P.; Seemann, T. AusTrakka: Fast-tracking nationalized genomics surveillance in response to the COVID-19 pandemic. *Nat. Commun.* **2022**, *13*, 865. [CrossRef] [PubMed]

100. Goig, G.A.; Blanco, S.; Garcia-Basteiro, A.L.; Comas, I. Contaminant DNA in bacterial sequencing experiments is a major source of false genetic variability. *BMC Biol.* **2020**, *18*, 1–15. [CrossRef]

101. Zwe, Y.H.; Chin, S.F.; Kohli, G.S.; Aung, K.T.; Yang, L.; Yuk, H.G. Whole genome sequencing (WGS) fails to detect antimicrobial resistance (AMR) from heteroresistant subpopulation of *Salmonella enterica*. *Food Microbiol.* **2020**, *91*, 103530. [CrossRef] [PubMed]

102. Zankari, E. Comparison of the web tools ARG-ANNOT and ResFinder for detection of resistance genes in bacteria. *Antimicrob. Agents Chemother.* **2014**, *58*, 4986. [CrossRef]

103. Achtman, M.; Zhou, Z.; Alikhan, N.F.; Tyne, W.; Parkhill, J.; Cormican, M.; Chiou, C.S.; Torpdahl, M.; Litrup, E.; Prendergast, D.M.; et al. Genomic diversity of *Salmonella enterica*—The UoWUCC 10 K genomes project. *Wellcome Open Res.* **2021**, *5*, 223. [CrossRef]

104. Zhou, Z.; Alikhan, N.F.; Mohamed, K.; Fan, Y.; Achtman, M.; Brown, D.; Agama Study Group; Achtman, M. The EnteroBase user's guide, with case studies on *Salmonella* transmissions, *Yersinia pestis* phylogeny, and *Escherichia* core genomic diversity. *Genome Res.* **2020**, *30*, 138–152. [CrossRef]

105. Quijada, N.M.; Rodríguez-Lázaro, D.; Eiros, J.M.; Hernández, M. TORMES: An automated pipeline for whole bacterial genome analysis. *Bioinformatics* **2019**, *35*, 4207–4212. [CrossRef]

106. Wajid, B.; Serpedin, E. Do it yourself guide to genome assembly. *Brief. Funct. Genomics.* **2016**, *15*, 1–9. [CrossRef] [PubMed]

107. Jacobsen, A.; Hendriksen, R.S.; Aaresturp, F.M.; Ussery, D.W.; Friis, C. The *Salmonella enterica* pan-genome. *Microb. Ecol.* **2011**, *62*, 487–504. [CrossRef] [PubMed]

108. Gupta, S.K.; Padmanabhan, B.R.; Diene, S.M.; Lopez-Rojas, R.; Kempf, M.; Landraud, L.; Rolain, J.M. ARG-ANNOT, a new bioinformatic tool to discover antibiotic resistance genes in bacterial genomes. *Antimicrob. Agents Chemother.* **2014**, *58*, 212–220. [CrossRef] [PubMed]

109. Alcock, B.P.; Raphenya, A.R.; Lau, T.T.; Tsang, K.K.; Bouchard, M.; Edalatmand, A.; Huynh, W.; Nguyen, A.V.; Cheng, A.A.; Liu, S.; et al. CARD 2020: Antibiotic resistome surveillance with the comprehensive antibiotic resistance database. *Nucleic Acids Res.* **2020**, *48*, D517–D525. [CrossRef] [PubMed]
110. McArthur, A.G.; Waglechner, N.; Nizam, F.; Yan, A.; Azad, M.A.; Baylay, A.J.; Bhullar, K.; Canova, M.J.; De Pascale, G.; Ejim, L.; et al. The comprehensive antibiotic resistance database. *Antimicrob. Agents Chemother.* **2013**, *57*, 3348–3357. [CrossRef] [PubMed]
111. Zankari, E.; Hasman, H.; Cosentino, S.; Vestergaard, M.; Rasmussen, S.; Lund, O.; Aarestrup, F.M.; Larsen, M.V. Identification of acquired antimicrobial resistance genes. *J. Antimicrob. Chemother.* **2012**, *67*, 2640–2644. [CrossRef] [PubMed]
112. Carattoli, A.; Zankari, E.; García-Fernández, A.; Voldby Larsen, M.; Lund, O.; Villa, L.; Aarestrup, F.M.; Hasman, H. In silico detection and typing of plasmids using PlasmidFinder and plasmid multilocus sequence typing. *Antimicrob. Agents Chemother.* **2014**, *58*, 3895–3903. [CrossRef] [PubMed]
113. Zankari, E.; Allesøe, R.; Joensen, K.G.; Cavaco, L.M.; Lund, O.; Aarestrup, F.M. PointFinder: A novel web tool for WGS-based detection of antimicrobial resistance associated with chromosomal point mutations in bacterial pathogens. *J. Antimicrob. Chemother.* **2017**, *72*, 2764–2768. [CrossRef] [PubMed]
114. Babicki, S.; Arndt, D.; Marcu, A.; Liang, Y.; Grant, J.R.; Maciejewski, A.; Wishart, D.S. Heatmapper: Web-enabled heat mapping for all. *Nucleic Acids Res.* **2016**, *44*, W147–W153. [CrossRef]

Article

Genomic Analysis of Two MDR Isolates of *Salmonella enterica* Serovar Infantis from a Spanish Hospital Bearing the $bla_{CTX-M-65}$ Gene with or without *fosA3* in pESI-like Plasmids

Xenia Vázquez [1,2], Javier Fernández [2,3,4,5], Jesús Rodríguez-Lozano [6], Jorge Calvo [6,7], Rosaura Rodicio [2,8] and M. Rosario Rodicio [1,2,*]

[1] Área de Microbiología, Departamento de Biología Funcional, Universidad de Oviedo (UO), 33006 Oviedo, Spain; xenia_grao@hotmail.com
[2] Instituto de Investigación Sanitaria del Principado de Asturias (ISPA), 33011 Oviedo, Spain; javifdom@gmail.com (J.F.); mrosaura@uniovi.es (R.R.)
[3] Servicio de Microbiología, Hospital Universitario Central de Asturias (HUCA), 33011 Oviedo, Spain
[4] Research & Innovation, Artificial Intelligence and Statistical Department, Pragmatech AI Solutions, 33003 Oviedo, Spain
[5] Centro de Investigación Biomédica en Red-Enfermedades Respiratorias, 20029 Madrid, Spain
[6] Servicio de Microbiología, Hospital Universitario Marqués de Valdecilla (IDIVAL), 39008 Santander, Spain; jesus.rodriguez@scsalud.es (J.R.-L.); jorge.calvo@scsalud.es (J.C.)
[7] CIBER de Enfermedades Infecciosas, Instituto de Salud Carlos III, 28029 Madrid, Spain
[8] Departamento de Bioquímica y Biología Molecular, Universidad de Oviedo (UO), 33006 Oviedo, Spain
* Correspondence: rrodicio@uniovi.es; Tel.: +34-985103562

Citation: Vázquez, X.; Fernández, J.; Rodríguez-Lozano, J.; Calvo, J.; Rodicio, R.; Rodicio, M.R. Genomic Analysis of Two MDR Isolates of *Salmonella enterica* Serovar Infantis from a Spanish Hospital Bearing the $bla_{CTX-M-65}$ Gene with or without *fosA3* in pESI-like Plasmids. *Antibiotics* **2022**, *11*, 786. https://doi.org/10.3390/antibiotics11060786

Academic Editor: Teresa V. Nogueira

Received: 12 May 2022
Accepted: 5 June 2022
Published: 9 June 2022

Publisher's Note: MDPI stays neutral with regard to jurisdictional claims in published maps and institutional affiliations.

Abstract: *Salmonella enterica* serovar Infantis (*S.* Infantis) is a broiler-associated pathogen which ranks in the fourth position as a cause of human salmonellosis in the European Union. Here, we report a comparative genomic analysis of two clinical *S.* Infantis isolates recovered in Spain from children who just returned from Peru. The isolates were selected on the basis of resistance to cefotaxime, one of the antibiotics of choice for treatment of *S. enterica* infections. Antimicrobial susceptibility testing demonstrated that they were resistant to eight classes of antimicrobial agents: penicillins, cephalosporins, phenicols, aminoglycosides, tetracyclines, inhibitors of folate synthesis, (fluoro)quinolones and nitrofurans, and one of them was also resistant to fosfomycin. As shown by whole-genome sequence analysis, each isolate carried a pESI-like megaplasmid of ca. 300 kb harboring multiple resistance genes [$bla_{CTX-M-65}$, *aph(4)-Ia*, *aac(3)-IVa*, *aph(3′)-Ia*, *floR*, *dfrA14*, *sul1*, *tet*(A), *aadA1* ± *fosA3*], as well as genes for resistance to heavy metals and disinfectants (*mer*, *ars* and *qacEΔ1*). These genes were distributed in two complex regions, separated by DNA belonging to the plasmid backbone, and associated with a wealth of transposable elements. The two isolates had a D87Y amino acid substitution in the GyrA protein, and truncated variants of the nitroreductase genes *nfsA* and *nsfB*, accounting for chromosomally encoded resistances to nalidixic acid and nitrofurantoin, respectively. The two *S.* Infantis isolates were assigned to sequence type ST32 by in silico multilocus sequence typing (MLST). Phylogenetic analysis revealed that they were closely related, differing only by 12 SNPs, although they were recovered from different children two years apart. They were also genetically similar to $bla_{CTX-M-65}$-positive ± *fosA3* isolates obtained from humans and along the poultry production chain in the USA, South America, as well as from humans in several European countries, usually associated with a travel history to America. However, this is the first time that the *S.* Infantis $bla_{CTX-M-65}$ ± *fosA3* MDR clone has been reported in Spain.

Keywords: *S.* Infantis; multidrug-resistance; $bla_{CTX-M-65}$; *fosA3*; heavy-metal resistance; pESI-like megaplasmid; whole-genome sequencing

1. Introduction

Salmonella enterica serovar Infantis (*S.* Infantis) is one of the main non-typhoid serovars of *S. enterica*. In 2020, it ranked in the fourth position as a cause of human salmonellosis in the European Union, only preceded by *S.* Enteritidis, *S.* Typhimurium and the monophasic 1,4,[5],12:i:- variant of the latter serovar. In addition, it was the most common serovar in broiler flocks and broiler meat [1]. Apart from the European Union, *S.* Infantis has increasingly been found in clinical, animal and food samples from many other geographical regions [2–6]. The epidemiological success of *S.* Infantis is strongly associated with the spread of isolates harboring a large conjugative plasmid which confers resistance, virulence and fitness properties. This plasmid, termed pESI (plasmid of Emerging *S.* Infantis), was first reported in a multidrug-resistant (MDR) clone of *S.* Infantis detected since 2006 in Israel [7,8]. pESI was identified as a chimeric megaplasmid of ca. 280 kb, found to contain the origin of replication (*oriV*) and incompatibility region of IncP-1α plasmids, and the IncI leading and transfer region. Later on, pESI-like plasmids were also detected in *S.* Infantis from many other countries, including Italy, Switzerland, Denmark, England and Wales, the USA and South America, and they were shown to contain the IncFIB replicon [2,9–15]. All pSE1-like plasmids confer resistance against multiple antimicrobial agents, including clinically relevant antibiotics such as third-generation cephalosporins. So far, two distinct types of pESI-like plasmids bearing genes encoding extended spectrum β-lactamases (ESBL) have been reported: $bla_{\text{CTX-M-1}}$-positive and $bla_{\text{CTX-M-65}}$-positive plasmids, found in European and American isolates, respectively [16].

The molecular epidemiology of *S.* Infantis and its pESI-like plasmids in Europe has been investigated by whole-genome sequencing and bioinformatics analysis, using isolates from America as an outgroup [16]. This study revealed the heterogeneity of the European *S.* Infantis population, which is composed by multiple clusters defined at the core-genome level, confirming previous proposals of the polyphyletic nature of this serovar [11,17]. However, pESI-like variants, which were present in 64.1% of the isolates analyzed (a total of 362; [16]), were genetically more homogeneous and were distributed among different clonal lineages in most countries. It has been proposed that once acquired by an *S.* Infantis isolate, the conjugative megaplasmid would rapidly become established in the local population because of the selective advantages it confers to the host bacteria [16]. Such advantages are not only provided by the presence of antibiotics resistance genes, but also by genes coding for resistances to quaternary ammonium compounds and heavy metals, which are likewise widespread in other successful clones of non-typhoid serovars of *S. enterica* [13,18–21]. Moreover, pESI-like megaplasmids also provide genes involved in colonization, virulence and fitness, including those encoding two types of fimbriae (Fim and K88-like) and the siderophore yersiniabactin, as well as toxin–antitoxin systems (like CcdBA, PemKI) [7,16].

In a screen for *S. enterica* resistant to third-generation cephalosporins in Spanish hospitals, two isolates of *S.* Infantis resistant to cefotaxime were identified. The present study aimed to combine epidemiological information with experimental research and whole-genome sequencing (WGS) analysis for in-depth characterization of these isolates. This provided evidence for the migration of the *S.* Infantis clone carrying pESI-like plasmids of the "American type" from Peru to Spain.

2. Results

2.1. Origin of the Isolates, Genome Sequencing and Resistance Phenotypes

The two isolates, HUMV 13-6278 and HUMV 15-5476, were recovered in 2013 and 2015 at the "Hospital Universitario Marqués de Valdecilla" (HUMV) from fecal samples of small children with gastroenteritis (Table 1). The children did not require hospitalization and were attended to at Primary Care Centers associated with the hospital. In both cases, the onset of the disease coincided with their return to Spain after a trip to Peru.

As shown in Tables 1 and 2, the latter compiling MIC values for relevant compounds, the two isolates were resistant to 14 out of the 21 antimicrobials tested, including cefotaxime (MIC of 32 µg/mL; used as the bases for selection of the isolates), nalidixic acid (MIC of 128 and >256 µg/mL for HUMV 13-6278 and HUMV 15-5476, respectively), pefloxacin (selected by EUCAST as a surrogate of ciprofloxacin to disclose clinical resistance to fluoroquinolones; https://eucast.org/clinical_breakpoints; accessed on 8 April 2022), ciprofloxacin (MIC of 0.125 μg/mL, resistant according to EUCAST criteria), and nitrofurantoin (MIC of 512 μg/mL). HUMV 15-5476 but not HUMV 13-6278 was also resistant to fosfomycin, displaying MIC values of 512 and 0.19 μg/mL, respectively. With regard to heavy metals, the two *S.* Infantis isolates were resistant to mercury as well as to inorganic and organic arsenic compounds (Table 2).

Table 1. Origin and properties of clinical cefotaxime-resistant isolates of *Salmonella enterica* serovar Infantis ST32.

HUMV Isolate [a]	Patient Sex [b]/Age	Travel History	Clinical Sample	Resistance Pattern [c] Antibiotic Resistance Genes [d]	Chromosome (Size bp) Plasmid [e] (Size bp)
13-6278	M/23 mth	Peru	Feces	[NAL, CIP, PEF], NIT [*gyrA* D87Y, *parC* T57S], [Δ*nfsA*, Δ*nfsB*] [AMP, CTX], CHL, [GEN, TOB, KAN, STR], TET, [SUL, TMP]	Chromosome (4,686,236)
				*bla*CTX-M-65, *floR*, [*aac(3)-IVa*, *aph(3′)-Ia*, *aph(4)-Ia*, *aadA1*], *tet*(A), [*sul1*, *dfrA14*]	IncFIB (313,645)
15-5476	M/12 mth	Peru	Feces	[NAL, CIP, PEF], NIT [*gyrA* D87Y, *parC* T57S], [Δ*nfsA*, Δ*nfsB*] [AMP, CTX], CHL [GEN, TOB, KAN, STR], TET, [SUL, TMP], FOS	Chromosome (4,682,901)
				*bla*CTX-M-65, *floR*, [*aac(3)-IVa*, *aph(3′)-Ia*, *aph(4)-Ia*, *aadA1*], *tet*(A), [*sul1*, *dfrA14*], *fosA3*	IncFIB (317,684)

[a], HUMV, "Hospital Universitario Marqués de Valdecilla", Santander, Cantabria, Spain. [b], M, male; mth, months. [c], NAL, nalidixic acid; CIP, ciprofloxacin; PEF, pefloxacin; NIT, nitrofurantoin; AMP, ampicillin; CTX, cefotaxime; CHL, chloramphenicol; GEN, gentamycin; TOB, tobramycin; KAN, kanamycin; STR, streptomycin; TET, tetracycline; SUL, sulfonamides; TMP, trimethoprim; FOS: fosfomycin (note that resistance to hygromycin, which could be conferred by the *aph(4)-1a* gene, was not experimentally tested). Antimicrobials belonging to the same class are combined in brackets. [d], Genes conferring resistance to antimicrobials of the same class are combined in brackets. [e], Inc, plasmid incompatibility group. The IncFIB megaplasmids of the isolates under study were named pHUMV 13-6278 and pHUMV 15-5476.

Table 2. Minimum inhibitory concentration (MIC, given in μg/mL) of relevant antibiotics and heavy metals determined for *Samonella enterica* serovar Infantis strains from a Spanish hospital.

Strain [a]	CTX [b]	NAL [b]	CIP [b]	NIT [b]	FOS [b]	HgCl$_2$	NaAsO$_2$	Na$_2$HAsO$_4$	Phenylarsine Oxide
HUMV 13-6278	32	128	0.125	512	0.19	64	128	2048	8
HUMV 15-5476	32	>256	0.125	512	512	32	128	512	8
LT2	4	3	0.016	32	0.019	4	64	128	0.25
LSP 146/02	nd	nd	nd	256	nd	32	32	256	0.5
LSP 389/97	nd	nd	nd	nd	nd	32	64	128	4

[a], HUMV, "Hospital Universitario Marqués de Valdecilla", Santander, Cantabria, Spain; LSP, "Laboratorio de Salud Pública", Asturias, Spain. *S. enterica* serovar Typhimurium strain LT2 was included as negative control. *S. enterica* serovar Typhimurium strain LSP 146/02 and *S. enterica* serovar 4,5,12:i:- strain LSP 389/97 were included as positive controls of resistance to some of the antimicrobials tested [21–23]. [b], CTX, cefotaxime; NAL, nalidixic acid; CIP, ciprofloxacin; NIT, nitrofurantoin; FOS: fosfomycin.

The draft genomes of HUMV 13-6278 and HUMV 15-5476 contained a total of 88 and 86 contigs, respectively, with an assembly size of about 4700 kb (Table 1). They shared the antigenic formula 7:r:1,5 and the sequence type ST32, both associated with *S.* Infantis. Each isolate harbored a single pESI-like plasmid of more than 310 kb (Table 1), termed pHUMV 13-6278 and pHUMV 15-5476, respectively. The two plasmids carried the IncFIB replicon, the IncP origin of replication (*oriV*) and a gene for a MOBP-family relaxase. Additionally, plasmid typing identified the *ardA_2*, *pilL_3*, *sogS_9*, *trbA_21* alleles of the pMLST scheme designed for the IncI1 incompatibility group, but lacked the *repI1* gene. Based on the detected alleles, the closest related sequence type within the IncI pMLST scheme was ST71.

2.2. Genetic Bases of Antimicrobial Drug Resistance

The resistance genes detected in the genomes of the two isolates by bioinformatics analysis were in agreement with the observed phenotypes (Table 1). Both isolates contained *bla*~CTX-M-65~ (ampicillin and cefotaxime), *floR* (chloramphenicol), *aac(3)-IVa*, *aph(3′)-Ia*, *aph(4)-Ia*, *aadA1* (aminoglycosides), *tet*(A) (tetracycline), *sul1* and *dfrA14* (sulfonamides and trimethoprim, respectively), together with genes conferring resistance to heavy metals: locus *merRTPCADE* (mercury), *arsA* and *arsD* (inorganic arsenic compounds), *arsH* (organic arsenicals), *arsR2* (metal-responsive transcriptional regulator), and *qacEΔ1* (although resistance to disinfectants was not experimentally tested). In the case of HUMV 15-5476, fosfomycin resistance was mediated by *fosA3*. All these genes were located on the pESI-like plasmids, as part of two complex resistance regions: R1 and R2, comprising approximately 36 or 32 kb (depending on the presence or absence of *fosA3*) and 21 kb, respectively. R1 and R2 were separated by DNA of the pESI-like backbone, and contained a wealth of transposable elements belonging to multiple families, including several copies of IS26 (Figure 1).

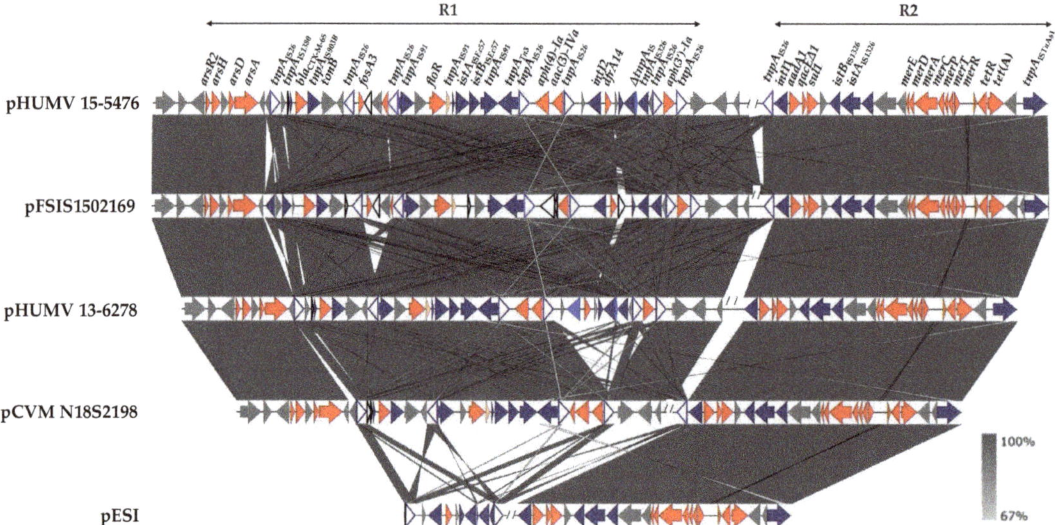

Figure 1. Comparison between the assembled resistance regions (R1 and R2) of the pESI-like plasmids pHUMV 13-6278 and pHUMV 15-5476 found in the *Salmonella enterica* serovar Infantis isolates under study with related regions from other pESI-like plasmids (accession No. CP016407, CP082522 and GCA_000506925.1 for plasmids of the FSIS1502169, CVM N18S2198 isolates and plasmid pES1 of the 119944 isolate, respectively). Coding regions are represented by arrows indicating the direction of transcription and colored according to their function: red, resistance; blue, insertion sequences, with the multiple copies of IS26 highlighted by arrowheads with white background and blue border lines; grey, other roles; white with black border lines, hypothetical proteins. The alignments were created with EasyFig BLASTn. The gray shading between regions reflects nucleotide sequence identities according to the scale shown at the right lower corner of the figure.

The resistance regions of the two plasmids differed only by the presence of the IS26-*fosA3*-IS26 pseudo-compound transposon in R1 of pHUMV 15-5476, which was lacking in the corresponding region of pHUMV 13-6278, and by a copy of IS26 located at the 5′-end of R2 in the former but not in the latter. Consistent with the high conservation of the pESI-like plasmids [16], regions identical to those of the plasmids under study were found in other plasmids of this group. However, there was also considerable variation, as shown by their comparison with some key examples (Figure 1), which include the original pESI plasmid, detected in a human isolate from Israel (119944), which lacks most of the R1 region [7,8]. A higher degree of similarity was found with two $bla_{\text{CTX-M-65}}$-positive plasmids, with or without *fosA3* (FSIS1502169 and CVM N18S2198, respectively; [14,24]), identified in isolates from chicken samples in the USA.

Apart from the plasmid-encoded genes, the two isolates displayed a point mutation in the chromosomal *gyrA* gene (G259T), resulting in D87Y exchange in the protein, which could account for nalidixic acid resistance. In addition, both strains also carried a single-nucleotide polymorphism in their chromosomal *parC* gene (C170G), leading to T57S substitution in the protein. Finally, nitrofurantoin resistance of the two isolates was associated with premature stop codons in the *nsfA* and *nsfB* genes, producing truncated forms of the NsfA and NsfB nitroreductases, with only 158 and 136 amino acids, as opposed to 240 and 217 amino acids in the wild type, respectively.

2.3. Phylogenetic Analysis

Figure 2 shows the phylogenetic position of the two HUMV isolates in the context of other *S*. Infantis isolates. Since *S*. Infantis is a broiler-associated pathogen, we chose a selective comparison of sequences of isolates from humans, chickens or the chicken farm environment in different countries of Europe and America, and in Israel (Supplementary Materials Table S2). All isolates harbored a pESI-like plasmid, which carried the $bla_{\text{CTX-M-1}}$, the $bla_{\text{CTX-M-65}}$ gene, or neither. The Spanish isolates, which differed by only 12 SNPs, belonged to a clonal lineage that groups all $bla_{\text{CTX-M-65}}$-positive isolates together with isolates lacking a $bla_{\text{CTX-M}}$ gene. The minimum and maximum number of SNPs between isolates of this cluster ranged from 3 to 60, and the distance with respect to the 119944 strain from Israel, which contains the original pESI plasmid, ranged from 93 to 127 SNPs.

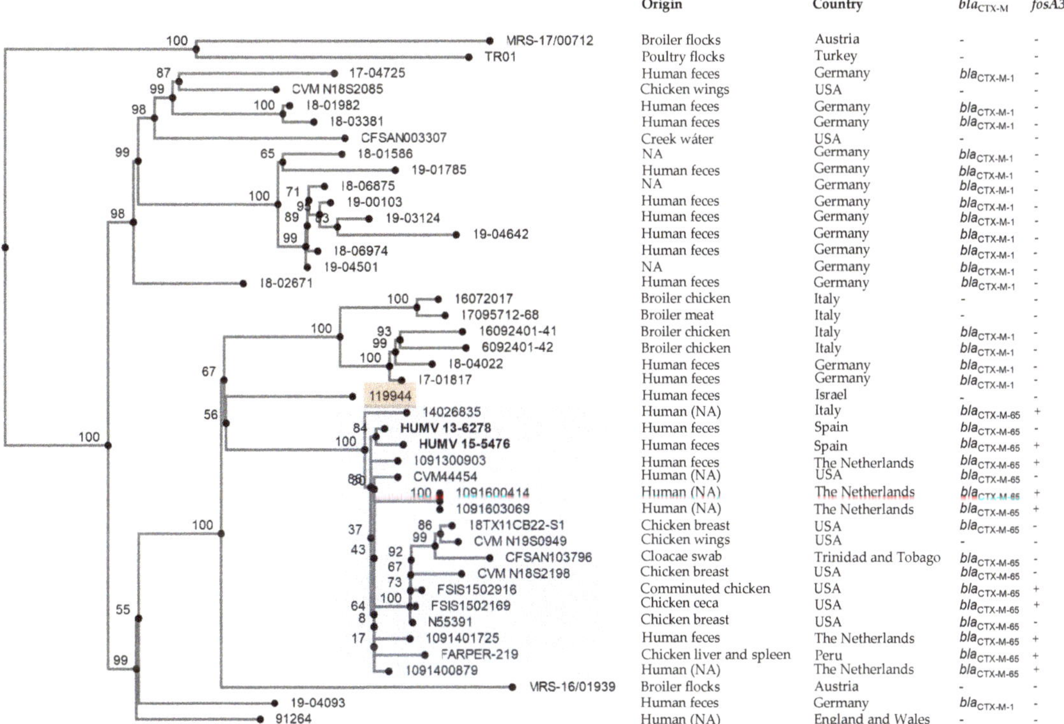

Isolate	Origin	Country	$bla_{\text{CTX-M}}$	*fosA3*
MRS-17/00712	Broiler flocks	Austria	-	-
TR01	Poultry flocks	Turkey	-	-
17-04725	Human feces	Germany	$bla_{\text{CTX-M-1}}$	-
CVM N18S2085	Chicken wings	USA	-	-
18-01982	Human feces	Germany	$bla_{\text{CTX-M-1}}$	-
18-03381	Human feces	Germany	$bla_{\text{CTX-M-1}}$	-
CFSAN003307	Creek water	USA	-	-
18-01586	NA	Germany	$bla_{\text{CTX-M-1}}$	-
19-01785	Human feces	Germany	$bla_{\text{CTX-M-1}}$	-
18-06875	NA	Germany	$bla_{\text{CTX-M-1}}$	-
19-00103	Human feces	Germany	$bla_{\text{CTX-M-1}}$	-
19-03124	Human feces	Germany	$bla_{\text{CTX-M-1}}$	-
19-04642	Human feces	Germany	$bla_{\text{CTX-M-1}}$	-
18-06974	Human feces	Germany	$bla_{\text{CTX-M-1}}$	-
19-04501	NA	Germany	$bla_{\text{CTX-M-1}}$	-
18-02671	Human feces	Germany	$bla_{\text{CTX-M-1}}$	-
16072017	Broiler chicken	Italy	-	-
17095712-68	Broiler meat	Italy	-	-
16092401-41	Broiler chicken	Italy	$bla_{\text{CTX-M-1}}$	-
6092401-42	Broiler chicken	Italy	$bla_{\text{CTX-M-1}}$	-
18-04022	Human feces	Germany	$bla_{\text{CTX-M-1}}$	-
17-01817	Human feces	Germany	$bla_{\text{CTX-M-1}}$	-
119944	Human feces	Israel	-	-
14026835	Human (NA)	Italy	$bla_{\text{CTX-M-65}}$	+
HUMV 13-6278	Human feces	Spain	$bla_{\text{CTX-M-65}}$	-
HUMV 15-5476	Human feces	Spain	$bla_{\text{CTX-M-65}}$	+
1091300903	Human feces	The Netherlands	$bla_{\text{CTX-M-65}}$	+
CVM44454	Human (NA)	USA	$bla_{\text{CTX-M-65}}$	-
1091600414	Human (NA)	The Netherlands	$bla_{\text{CTX-M-65}}$	+
1091603069	Human (NA)	The Netherlands	$bla_{\text{CTX-M-65}}$	+
18TX11CB22-S1	Chicken breast	USA	$bla_{\text{CTX-M-65}}$	-
CVM N19S0949	Chicken wings	USA	-	-
CFSAN103796	Cloacae swab	Trinidad and Tobago	$bla_{\text{CTX-M-65}}$	-
CVM N18S2198	Chicken breast	USA	$bla_{\text{CTX-M-65}}$	-
FSIS1502916	Comminuted chicken	USA	$bla_{\text{CTX-M-65}}$	+
FSIS1502169	Chicken ceca	USA	$bla_{\text{CTX-M-65}}$	+
N55391	Chicken breast	USA	$bla_{\text{CTX-M-65}}$	-
1091401725	Human feces	The Netherlands	$bla_{\text{CTX-M-65}}$	+
FARPER-219	Chicken liver and spleen	Peru	$bla_{\text{CTX-M-65}}$	+
1091400879	Human (NA)	The Netherlands	$bla_{\text{CTX-M-65}}$	+
MRS-16/01939	Broiler flocks	Austria	-	-
19-04093	Human feces	Germany	$bla_{\text{CTX-M-1}}$	-
91264	Human (NA)	England and Wales	-	-

Figure 2. Phylogenetic position of *Salmonella enterica* serovar Infantis HUMV 13-6278 and HUMV

15-5476 from a Spanish hospital, in the context of *S*. Infantis isolates from different sources and countries. The tree was constructed with the CSI Phylogeny 1.4 (https://cge.cbs.dtu.dk/services/CSIPhylogeny/, accessed on 6 May 2022) using the genome of *S*. Infantis 119944 as the reference for SNP calling. Values at each node represent percent bootstrap support based on 1000 replicates. The cluster containing the Spanish isolates (shown in bold) is highlighted in blue, and the 119944 isolate is emphasized in red. Relevant information related to the isolates is shown at the right of the figure. Accession numbers of the genomes and the pairwise distance matrix used to construct the phylogenetic tree are shown in Supplementary Materials Tables S2 and S3, respectively.

3. Discussion

The present study examined two clinical isolates of *S*. Infantis harboring the $bla_{\text{CTX-M-65}}$ gene on pESI-like megaplasmids. As other $bla_{\text{CTX-M}}$ genes, it confers resistance to most β-lactam antibiotics, including penicillins, broad spectrum cephalosporins and monobactams, but not to amoxicillin-clavulanic acid, cephamycins and carbapenems [25]. As such, it conveys resistance to third-generation cephalosporins, which currently are first-line drugs for the management of severe *S. enterica* infections. Thus, the presence of $bla_{\text{CTX-M-65}}$ is a cause of concern, particularly when associated with resistance to other classes of medically relevant antimicrobials. In fact, the two isolates were resistant to antimicrobials belonging to eight classes: penicillins, cephalosporins, phenicols, aminoglycosides, tetracyclines, inhibitors of folate synthesis, (fluoro)quinolones, and nitrofurans, as confirmed by disk diffusion assays complemented with MIC determinations for selected compounds. HUMV 15-5476, but not HUMV 13-6278, was additionally resistant to fosfomycin.

Most of the resistance genes found were located on pESI-like megaplasmids, but resistances to (fluoro)quinolones and nitrofurantoin were mediated by chromosomal loci. Thus, the two isolates had single-point mutations both in *gyrA* (G259T→D87Y) and *parC* (C170G→T57S). The former mutation, which is now widespread among *S*. Infantis, was originally identified as responsible for nalidixic acid resistance in isolates from Israel. This trait was already present in a limited number of pre-emergent isolates, which were obtained before 2006 and lacked pESI, while becoming fixed in the *gyrA* gene of the emergent pESI-positive clone (recorded after 2006), adding to its MDR phenotype [3,7]. In addition to D87Y, three other point mutations in *gyrA* (D87G, S83Y and S83F) were later associated with (fluoro)quinolone resistance in an extensive collection of *S*. Infantis recovered in Europe from animals, meat, feed and humans [16]. In that work, a single-(fluoro)quinolone-resistant isolate tested negative for known mutations in *gyrA*, but presented the T57S substitution in the ParC protein. The latter change has been previously found both in resistant and in susceptible strains of different serovars of *S. enterica*, with or without concomitant mutations in *gyrA* [16,26–29]. The contribution of the T57S substitution to (fluoro)quinolone resistance in the isolates thus remains unclear. PMQR (Plasmid-Mediated Quinolone Resistance) mechanisms, which confer low level, albeit clinically relevant, resistance to fluoroquinolones, were not present in our isolates but have already been reported in *S*. Infantis from different countries [16].

In *S. enterica*, as well as in *Escherichia coli*, resistance to nitrofurans has been associated with alterations affecting the *nfsA* and *nfsB* genes that encode type I (i.e., oxygen insensitive) nitroreductases. Amongst other substrates, these enzymes appear to reduce nitrofurans, generating active intermediates with antimicrobial activity. Nitrofurans are used for treating uncomplicated urinary tract infections caused by a wide spectrum of Gram-positive and Gram-negative bacteria [30,31]. In both *E. coli* and *S*. Typhimurium, a stepwise increase in resistance to nitrofurantoin has been correlated with sequential inactivation first of *nfsA* and then of *nfsB*, with alterations in the two genes being required to achieve full resistance [22,32,33]. The *S*. Infantis isolates in the present study were highly resistant to nitrofurantoin (Table 2). In agreement with this, both carried premature stop codons in the two *nfs* genes, leading to truncated forms of the proteins. The same mutation in *nfsA* was previously detected in *S*. Infantis from Israel, without reference to alterations in the *nfsB* gene [7]. Yet, sequence comparisons of the *nsfB* gene from our isolates and strain 119944, obtained in 2008 from a clinical sample in Israel [3], revealed the presence of the mutation (data not shown). Accordingly, the isolate from Israel was likely to display high-level resistance to nitrofurantoin, although the MIC of nitrofurantoin was not reported.

As previously shown for other *S.* Infantis isolates, the pESI-like megaplasmids played a major role in the MDR phenotype of the isolates under study. They not only harbor numerous antibiotic resistance genes, but also genes for resistance to mercury (*merRTP-CADE* locus) and arsenic (*arsA, arsD, arsH* and *arsR2*) compounds. In the case of arsenic, toxicity depends on the chemical form (inorganic or organic) and oxidation state (trivalent or pentavalent species) of the compound, with As(III) arsenite being more toxic than As(V) arsenate. The product of *arsA* is an ATPase that interacts with ArsB to form an arsenite efflux pump energized by ATP hydrolysis. It is of note that *arsA*, but not *arsB*, is present in the pESI-like plasmids of the isolates under study, which were nevertheless resistant to NaAsO$_2$ [As(III)]. However, besides canonical *ars* genes, additional genes have been discovered that broaden the spectrum of arsenic tolerance [34]. Apart from interacting with ArsB, the ArsA ATPase has been proposed to form primary arsenite transporters in association with different membrane proteins [35]. The most likely candidate would be Acr3, but the *acr3* gene was not found in the genome of our isolates. The product of *arsD*, a gene also present in pESI-like plasmids, acts as a metallochaperone that binds arsenite and transfers it to the ArsA ATPase for export [36]. Finally, the ArsH protein, which is under the control of the ArsR2 transcriptional regulator, encodes an organoarsenical oxidase that transforms trivalent methylated and aromatic arsenicals into their less toxic pentavalent species [37]. Accordingly, the two isolates described herein were resistant to the aromatic As(III) phenylarsine oxide. Of note, although *mer* and *ars* genes are frequently found in pESI-like plasmids, their association with phenotypic resistance of *S.* Infantis has not been previously tested [38]. The selective pressure caused by the use of heavy metals, and also of nalidixic acid, tetracyclines, sulfonamides and nitrofurans in livestock production, could have contributed to the selection of pESI-positive isolates of *S.* Infantis, as it was probably the case for other successful clones of different non-typhoidal serovars of *S. enterica* [13,18–21].

Of the multiple resistance genes harbored by the pESI-like plasmids under study, only *fosA3* was differentially present in HUMV 15-5476, which was highly resistant to fosfomycin (Table 2). This broad-spectrum bacteriolytic antibiotic was originally approved for treating uncomplicated urinary tract infections in the early 1970s. However, the major threat posed by antimicrobial resistance to human health, and the shortage of new antimicrobial agents, has led to re-evaluating older antibiotics, including fosfomycin, for the treatment of infections caused by multidrug-resistant bacteria [39]. Therefore, the increasing rate of fosfomycin resistance, which is mainly mediated by inactivating enzymes, is a cause of concern. Like other *fosA* genes, the *fosA3* gene of the pESI-like plasmid of HUMV 15-5476 encodes a glutathione-S-transferase, which catalyzes the opening of the epoxide ring of fosfomycin by adding glutathione [40]. In *S.* Infantis, *fosA3* has been found in the subgroup of pESI-like plasmids positive for *bla*$_{CTX-M-65}$, but not in those carrying *bla*$_{CTX-M-1}$ [16].

It is interesting to note that, as previously reported for other pESI-like plasmids, those of the HUMV isolates not only contained antimicrobial-resistant genes, but also genes associated with virulence, colonization and fitness, including genes for fimbriae, for the siderophore yersiniabactin, and for toxin-antitoxin systems which are involved in the post segregational killing of bacteria that lose the plasmid. Taken together, these determinants confer selective advantages which explain the epidemiological success of the isolates that have managed to acquire this self-transmissible plasmid [7,41].

It is finally of note that isolates harboring pESI-like plasmids with *bla*$_{CTX-M-65}$ $\pm$ *fosA3*, like those reported herein, form a well-defined cluster within the phylogenetic tree of *S.* Infantis ([13,16]; Figure 2). They have been reported in humans and along the poultry production chain in the USA and South America, with only a limited number of isolates being detected in other livestock animals [14,24,42–46]. Isolates carrying pESI-like plasmids positive for *bla*$_{CTX-M-65}$ $\pm$ *fosA3* have also been associated with human clinical cases in Europe, specifically in patients with a travel history to America [10,12,16]. However, to the best of our knowledge, this is the first report of such isolates in Spain, with a clearly established link to Peru. In Spain, information is also lacking on the possible presence of *S.* Infantis carrying pESI-like plasmids of the "European type", positive for $bla_{CTX-M-1}$ instead of $bla_{CTX-M-65}$, which are circulating in food-animal production systems of many other European countries [16]. Within the One Health Concept, promoted by the World Health Organization, further studies are required to investigate the occurrence of such isolates in Spanish livestock, and particularly along the poultry food chain. WGS analyses could certainly assist in the detection and detailed characterization of isolates belonging to the different MDR CTX-M-producing lineages of *S.* Infantis, and to establish their phylogenetic relationships with those from other countries.

4. Materials and Methods

4.1. Bacterial Isolates and Antimicrobial Susceptibility Testing

The two cefotaxime-resistant isolates of *S.* Infantis, HUMV 13-6278 and HUMV 15-5476, selected for the present study, were detected at the "Hospital Universitario Marqués de Valdecilla" (HUMV), Santander, Cantabria, Spain (Table 1). They were obtained from fecal samples of small children with gastroenteritis, using selective culture media (Selenite broth and Hecktoen agar; bioMerieux, Marcy l'Etoile, France). Identification was performed with MALDI-TOF MS, following the manufacturer's instructions (Bruker Daltonics, Billerica, MA, USA).

Susceptibility to antimicrobial agents was determined by automated MicroScan NC 53 (Beckman Coulter, CA, USA), complemented with disk diffusion assays. For the latter, the following compounds (Oxoid, Madrid, Spain, except for pefloxacin purchased from Bio-Rad, Alcobendas, Madrid, Spain) were used, with the amount per disk in µg shown in parenthesis: ampicillin (10), amoxicillin-clavulanic acid (30), cefepime (30), cefotaxime (30), cefoxitin (30), erthapenem (10), chloramphenicol (30), amikacin (30), gentamicin (10), kanamycin (30), streptomycin (10), tobramycin (10), azithromycin (15), nalidixic acid (30), ciprofloxacin (5), pefloxacin (5), sulfonamides (300), tetracycline (30), trimethoprim (5), fosfomycin (300) and nitrofurantoin (300). The minimum inhibitory concentrations (MIC) of cefotaxime, ciprofloxacin and fosfomycin were determined by E-test (bioMérieux, Marcy l'Étoile, France), and the MIC of nitrofurantoin by a broth microdilution assay with concentrations ranging from 4 to 512 µg/mL. Results were interpreted according to EUCAST (European Committee on Antimicrobial Susceptibility Testing) guidelines (https://eucast.org/clinical_breakpoints/; accessed on 8 April 2022) or to CLSI [47], in the case of nalidixic acid which is not contemplated by EUCAST. To obtain additional information on the resistance properties of the *S.* Infantis isolates, the MIC of heavy metals were also determined by broth microdilution assays, using $HgCl_2$ (0–256 µg/mL) and the following inorganic and organic arsenic compounds (Sigma-Aldrich, Merck Life Science, Madrid, Spain): $NaAsO_2$ (0–256 µg/mL), $Na_2HAsO_4 \cdot 7H_2O$ (0–256 µg/mL) and phenylarsine oxide (0–16 µg/mL). *S.* Typhimurium LT2, *S.* Typhimurium LSP 146/02 and *S.* Typhimurium 4,5,12:i:- (monophasic) LSP 389/97 were included as negative or positive controls [21–23].

4.2. Whole-Genome Sequencing and Bioinformatics Analysis

For WGS, the genomic DNA of the isolates was purified from overnight cultures grown in Luria-Bertani (LB) broth, using the GenEluteTM Bacterial Genomic DNA Kit, according to the manufacturer's protocol (Sigma-Aldrich). WGS was performed with Illumina technology, at the "Centro de Investigación Biomédica", La Rioja (CIBIR), Spain. Paired-end reads of 100 nt, obtained from PCR-free fragment libraries of ca. 500 bp, were assembled de novo using the VelvetOptimiser.pl script implemented in the on line version of PLACNETw (https://omictools.com/placnet-tool; accessed on 6 June 2019). This tool also allows distinguishing contigs of chromosomal or plasmid origin [48]. The assembled genomes were deposited in GenBank under the accession numbers included in the "Data Availability Statement" (see below), and annotated by the NCBI Prokaryotic Genome Annotation Pipeline (PGAP; https://www.ncbi.nlm.nih.gov/genome/annotation_prok/;

accessed on 4 August 2021). Relevant parameters regarding the quality of the assemblies are shown in Table S1. Bioinformatics analysis was performed with PLACNETw, MOBscan (a web application for the identification of relaxase MOB families [49]), and with different tools available at the Center for Genomic Epidemiology (CGE) of the Technical University of Denmark (DTU) (https://cge.cbs.dtu.dk/services/; accessed on 7 June 2020). These included SeqSero 1.2 [50], MLST 2.0 [51], ResFinder 4.1 [52,53], PlasmidFinder 2.1 and pMLST 2.0 [54].

Once identified by PLACNETw, plasmid contigs carrying resistance genes were joined by PCR amplification (using the primers shown in Table S4), followed by Sanger sequencing of the obtained amplicons, when required. The annotation of the assembled resistance regions were manually curated with the aid of blastn, blastp (https://blast.ncbi.nlm.nih.gov/Blast.cgi; accessed on 28 April 2022) and CLONE Manager Professional (CmSuit9). The genetic organization of these regions was represented with EasyFig BLASTn (https://mjsull.github.io/Easyfig/; accessed on 7 May 2022).

4.3. Phylogenetic Analysis

A phylogenetic tree based on single-nucleotide polymorphisms (SNPs) was built with the CSI phylogeny 1.4 tool (https://cge.cbs.dtu.dk/services/CSIPhylogeny/; accessed on 6 May 2022 [55]). The pipeline was run with default parameters, setting a bootstrap replication value of 1,000 to generate the consensus tree [56]. Apart from the genomes of the HUMV isolates, 42 other *S.* Infantis genomes retrieved from databases were included in the analysis (see Table S3 for accession numbers and additional details). The genome of isolate 119944 from Israel (accession No. GCA_000506925.1) was used as the reference for SNP calling, and the resulting SNP matrix is shown in Table S4.

5. Conclusions

This study documents the import of the *S.* Infantis MDR clone which carries the *bla*CTX-M-65 within pESI-like megaplasmids from Peru into Spain. To the best of our knowledge, this is the first report on the presence of this clone in Spain, which has probably emerged in South America [46]. We also first experimentally demonstrated the resistance of the clone to mercury and to inorganic and organic arsenicals. Further studies will be required to establish the actual epidemiology of these and other CTX-M-producing *S.* Infantis lineages in Spain.

Supplementary Materials: The following supporting information can be downloaded at: https://www.mdpi.com/article/10.3390/antibiotics11060786/s1, Table S1: Accession numbers of the genomes cefotaxime resistance plasmid-mediated of *Salmonella enterica* serovar Infantis isolates from clinical samples and parameters related to the quality of the assemblies; Table S2: Origin and accession numbers of the genomes of *Salmonella enterica* serovar Infantis isolates used for phylogenetic analysis in the present study. Table S3: Pairwise SNP distance matrix calculated from the genomes of two *Salmonella enterica* serovar Infantis isolates from a Spanish hospital and 42 additional isolates obtained from different sources and countries. Table S4: Oligonucleotides designed for this work and used for the as-57 sembly of the resistance regions (R1 and R2) of pESI-like plasmids found 58 in *Salmonella enterica* serovar Infantis isolates from a Spanish hospital. References [57–60] are cited in the supplementary materials.

Author Contributions: Conceptualization, J.F., R.R. and M.R.R.; methodology, X.V., J.F., J.R.-L., J.C. and R.R.; investigation, X.V., J.F., J.R.-L., J.C. and R.R.; visualization, X.V., R.R. and M.R.R.; validation, all authors; formal analysis, all authors.; resources, J.F., J.R.-L., J.C. and M.R.R.; data curation, X.V., R.R. and M.R.R.; writing—original draft preparation, X.V., R.R. and M.R.R.; writing—review and editing, all authors.; supervision, R.R. and M.R.R.; project administration, M.R.R.; funding acquisition, M.R.R. All authors have read and agreed to the published version of the manuscript.

Funding: This research was funded by project FIS PI17/00474 of the "Fondo de Investigación Sanitaria, Instituto de Salud Carlos III, Ministerio de Economía y Competitividad", Spain, cofunded by the European Regional Development Fund of the European Union: a way to making Europe. X.V. was the recipient of grant BP17-018 from the Program "Severo Ochoa" for support of Research and Teaching in the Principality of Asturias, Spain.

Institutional Review Board Statement: Not applicable.

Informed Consent Statement: Not applicable.

Data Availability Statement: The genome sequences generated in the present study were deposited in GenBank under accession numbers JAICDN000000000 and JAICDM000000000 for *Salmonella enterica* serovar Infantis HUMV 13-6278 and HUMV 15-5476, respectively.

Acknowledgments: The authors are grateful to the Regional Ministry of Science of Asturias for supporting GRUPIN IDI/2022/000033.

Conflicts of Interest: The funders had no role in the design of the study; in the collection, analyses, or interpretation of data; in the writing of the manuscript, or in the decision to publish the results.

References

1. EFSA and ECDC (European Food Safety Authority and European Centre for Disease Prevention and Control). The European Union One Health 2020 Zoonoses Report. *EFSA J.* **2021**, *19*, e06971. [CrossRef]
2. Burnett, E.; Ishida, M.; de Janon, S.; Naushad, S.; Duceppe, M.O.; Gao, R.; Jardim, A.; Chen, J.C.; Tagg, K.A.; Ogunremi, D.; et al. Whole-Genome Sequencing Reveals the Presence of the $bla_{CTX-M-65}$ Gene in Extended-Spectrum β-Lactamase-Producing and Multi-Drug-Resistant Clones of *Salmonella* Serovar Infantis Isolated from Broiler Chicken Environments in the Galapagos Islands. *Antibiotics* **2021**, *10*, 267. [CrossRef] [PubMed]
3. Gal-Mor, O.; Valinsky, L.; Weinberger, M.; Guy, S.; Jaffe, J.; Schorr, Y.I.; Raisfeld, A.; Agmon, V.; Nissan, I. Multidrug-resistant *Salmonella enterica* serovar Infantis, Israel. *Emerg. Infect. Dis.* **2010**, *16*, 1754–1757. [CrossRef] [PubMed]
4. Marder Mph, E.P.; Griffin, P.M.; Cieslak, P.R.; Dunn, J.; Hurd, S.; Jervis, R.; Lathrop, S.; Muse, A.; Ryan, P.; Smith, K.; et al. Preliminary Incidence and Trends of Infections with Pathogens Transmitted Commonly Through Food—Foodborne Diseases Active Surveillance Network, 10 U.S. Sites, 2006–2017. *MMWR. Morb. Mortal. Wkly. Rep.* **2018**, *67*, 324–328. [CrossRef]
5. Ross, I.L.; Heuzenroeder, M.W. A comparison of three molecular typing methods for the discrimination of *Salmonella enterica* serovar Infantis. *FEMS Immunol. Med. Microbiol.* **2008**, *53*, 375–384. [CrossRef]
6. Shahada, F.; Sugiyama, H.; Chuma, T.; Sueyoshi, M.; Okamoto, K. Genetic analysis of multi-drug resistance and the clonal dissemination of β-lactam resistance in *Salmonella* Infantis isolated from broilers. *Vet. Microbiol.* **2010**, *140*, 136–141. [CrossRef]
7. Aviv, G.; Tsyba, K.; Steck, N.; Salmon-Divon, M.; Cornelius, A.; Rahav, G.; Grassl, G.A.; Gal-Mor, O. A unique megaplasmid contributes to stress tolerance and pathogenicity of an emergent *Salmonella enterica* serovar Infantis strain. *Environ. Microbiol.* **2014**, *16*, 977–994. [CrossRef]
8. Cohen, E.; Rahav, G.; Gal-Mor, O. Genome Sequence of an Emerging *Salmonella enterica* Serovar Infantis and Genomic Comparison with Other *S.* Infantis Strains. *Genome Biol. Evol.* **2020**, *12*, 151–159. [CrossRef]
9. Brown, A.C.; Chen, J.C.; Watkins, L.K.F.; Campbell, D.; Folster, J.P.; Tate, H.; Wasilenko, J.; Van Tubbergen, C.; Friedman, C.R. CTX-M-65 Extended-Spectrum beta-Lactamase-Producing *Salmonella enterica* Serotype Infantis, United States(1). *Emerg. Infect. Dis.* **2018**, *24*, 2284–2291. [CrossRef]
10. Franco, A.; Leekitcharoenphon, P.; Feltrin, F.; Alba, P.; Cordaro, G.; Iurescia, M.; Tolli, R.; D'Incau, M.; Staffolani, M.; Di Giannatale, E.; et al. Emergence of a Clonal Lineage of Multidrug-Resistant ESBL-Producing *Salmonella* Infantis Transmitted from Broilers and Broiler Meat to Humans in Italy between 2011 and 2014. *PLoS ONE* **2015**, *10*, e0144802. [CrossRef]
11. Gymoese, P.; Kiil, K.; Torpdahl, M.; Osterlund, M.T.; Sorensen, G.; Olsen, J.E.; Nielsen, E.M.; Litrup, E. WGS based study of the population structure of *Salmonella enterica* serovar Infantis. *BMC Genom.* **2019**, *20*, 870. [CrossRef] [PubMed]
12. Hindermann, D.; Gopinath, G.; Chase, H.; Negrete, F.; Althaus, D.; Zurfluh, K.; Tall, B.D.; Stephan, R.; Nuesch-Inderbinen, M. *Salmonella enterica* serovar Infantis from Food and Human Infections, Switzerland, 2010-2015: Poultry-Related Multidrug Resistant Clones and an Emerging ESBL Producing Clonal Lineage. *Front. Microbiol.* **2017**, *8*, 1322. [CrossRef] [PubMed]
13. Lee, W.W.Y.; Mattock, J.; Greig, D.R.; Langridge, G.C.; Baker, D.; Bloomfield, S.; Mather, A.E.; Wain, J.R.; Edwards, A.M.; Hartman, H.; et al. Characterization of a pESI-like plasmid and analysis of multidrug-resistant *Salmonella enterica* Infantis isolates in England and Wales. *Microb. Genom.* **2021**, *7*, 000658. [CrossRef] [PubMed]
14. Tate, H.; Folster, J.P.; Hsu, C.H.; Chen, J.; Hoffmann, M.; Li, C.; Morales, C.; Tyson, G.H.; Mukherjee, S.; Brown, A.C.; et al. Comparative Analysis of Extended-Spectrum-beta-Lactamase CTX-M-65-Producing *Salmonella enterica* Serovar Infantis Isolates from Humans, Food Animals, and Retail Chickens in the United States. *Antimicrob. Agents Chemother.* **2017**, *61*, e00488-17. [CrossRef]

15. Vallejos-Sanchez, K.; Tataje-Lavanda, L.; Villanueva-Perez, D.; Bendezu, J.; Montalvan, A.; Zimic-Peralta, M.; Fernandez-Sanchez, M.; Fernandez-Diaz, M. Whole-Genome Sequencing of a *Salmonella enterica* subsp. *enterica* Serovar Infantis Strain Isolated from Broiler Chicken in Peru. *Microbiol. Resour. Announc.* **2019**, *8*, e00826-19. [CrossRef]
16. Alba, P.; Leekitcharoenphon, P.; Carfora, V.; Amoruso, R.; Cordaro, G.; Di Matteo, P.; Ianzano, A.; Iurescia, M.; Diaconu, E.L.; Study Group, E.N.; et al. Molecular epidemiology of *Salmonella* Infantis in Europe: Insights into the success of the bacterial host and its parasitic pESI-like megaplasmid. *Microb. Genom.* **2020**, *6*, e000365. [CrossRef]
17. Yokoyama, E.; Murakami, K.; Shiwa, Y.; Ishige, T.; Ando, N.; Kikuchi, T.; Murakami, S. Phylogenetic and population genetic analysis of *Salmonella enterica* subsp. *enterica* serovar Infantis strains isolated in Japan using whole genome sequence data. *Infect. Genet. Evol.* **2014**, *27*, 62–68. [CrossRef]
18. Antunes, P.; Mourao, J.; Campos, J.; Peixe, L. Salmonellosis: The role of poultry meat. *Clin. Microbiol. Infect.* **2016**, *22*, 110–121. [CrossRef]
19. Clark, C.G.; Landgraff, C.; Robertson, J.; Pollari, F.; Parker, S.; Nadon, C.; Gannon, V.P.J.; Johnson, R.; Nash, J. Distribution of heavy metal resistance elements in Canadian *Salmonella* 4,[5],12:i:- populations and association with the monophasic genotypes and phenotype. *PLoS ONE* **2020**, *15*, e0236436. [CrossRef]
20. Mastrorilli, E.; Pietrucci, D.; Barco, L.; Ammendola, S.; Petrin, S.; Longo, A.; Mantovani, C.; Battistoni, A.; Ricci, A.; Desideri, A.; et al. A comparative genomic analysis provides novel insights into the ecological success of the monophasic *Salmonella* serovar 4,[5],12:i. *Front. Microbiol.* **2018**, *9*, 715. [CrossRef]
21. Vazquez, X.; Garcia, P.; Garcia, V.; de Toro, M.; Ladero, V.; Heinisch, J.J.; Fernandez, J.; Rodicio, R.; Rodicio, M.R. Genomic analysis and phylogenetic position of the complex IncC plasmid found in the Spanish monophasic clone of *Salmonella enterica* serovar Typhimurium. *Sci. Rep.* **2021**, *11*, 11482. [CrossRef] [PubMed]
22. Garcia, V.; Montero, I.; Bances, M.; Rodicio, R.; Rodicio, M.R. Incidence and Genetic Bases of Nitrofurantoin Resistance in Clinical Isolates of Two Successful Multidrug-Resistant Clones of *Salmonella enterica* Serovar Typhimurium: Pandemic "DT 104" and pUO-StVR2. *Microb. Drug Resist.* **2017**, *23*, 405–412. [CrossRef] [PubMed]
23. McClelland, M.; Sanderson, K.E.; Spieth, J.; Clifton, S.W.; Latreille, P.; Courtney, L.; Porwollik, S.; Ali, J.; Dante, M.; Du, F.; et al. Complete genome sequence of *Salmonella enterica* serovar Typhimurium LT2. *Nature* **2001**, *413*, 852–856. [CrossRef] [PubMed]
24. Li, C.; Tyson, G.H.; Hsu, C.H.; Harrison, L.; Strain, E.; Tran, T.T.; Tillman, G.E.; Dessai, U.; McDermott, P.F.; Zhao, S. Long-Read Sequencing Reveals Evolution and Acquisition of Antimicrobial Resistance and Virulence Genes in *Salmonella enterica*. *Front. Microbiol.* **2021**, *12*, 777817. [CrossRef]
25. Canton, R.; Gonzalez-Alba, J.M.; Galan, J.C. CTX-M Enzymes: Origin and Diffusion. *Front. Microbiol.* **2012**, *3*, 110. [CrossRef]
26. Baucheron, S.; Chaslus-Dancla, E.; Cloeckaert, A.; Chiu, C.H.; Butaye, P. High-level resistance to fluoroquinolones linked to mutations in *gyrA*, *parC*, and *parE* in *Salmonella enterica* serovar Schwarzengrund isolates from humans in Taiwan. *Agents Chemother.* **2005**, *49*, 862–863. [CrossRef]
27. Kim, K.Y.; Park, J.H.; Kwak, H.S.; Woo, G.J. Characterization of the quinolone resistance mechanism in foodborne *Salmonella* isolates with high nalidixic acid resistance. *Food Microbiol.* **2011**, *146*, 52–56. [CrossRef]
28. Vazquez, X.; Fernandez, J.; Hernaez, S.; Rodicio, R.; Rodicio, M.R. Plasmid-Mediated Quinolone Resistance (PMQR) in Two Clinical Strains of *Salmonella enterica* Serovar Corvallis. *Microorganisms* **2022**, *10*, 579. [CrossRef]
29. Weill, F.X.; Bertrand, S.; Guesnier, F.; Baucheron, S.; Cloeckaert, A.; Grimont, P.A. Ciprofloxacin-resistant *Salmonella* Kentucky in travelers. *Emerg. Infect. Dis.* **2006**, *12*, 1611–1612. [CrossRef]
30. Guay, D.R. An update on the role of nitrofurans in the management of urinary tract infections. *Drugs* **2001**, *61*, 353–364. [CrossRef]
31. McCalla, D.R.; Olive, P.; Tu, Y.; Fan, M.L. Nitrofurazone-reducing enzymes in *E. coli* and their role in drug activation in vivo. *Can. J. Microbiol.* **1975**, *21*, 1484–1491. [CrossRef] [PubMed]
32. Sandegren, L.; Lindqvist, A.; Kahlmeter, G.; Andersson, D.I. Nitrofurantoin resistance mechanism and fitness cost in *Escherichia coli*. *J. Antimicrob. Chemother.* **2008**, *62*, 495–503. [CrossRef] [PubMed]
33. Whiteway, J.; Koziarz, P.; Veall, J.; Sandhu, N.; Kumar, P.; Hoecher, B.; Lambert, I.B. Oxygen-insensitive nitroreductases: Analysis of the roles of *nfsA* and *nfsB* in development of resistance to 5-nitrofuran derivatives in *Escherichia coli*. *J. Bacteriol.* **1998**, *180*, 5529–5539. [CrossRef] [PubMed]
34. Ben Fekih, I.; Zhang, C.; Li, Y.P.; Zhao, Y.; Alwathnani, H.A.; Saquib, Q.; Rensing, C.; Cervantes, C. Distribution of Arsenic Resistance Genes in Prokaryotes. *Front. Microbiol.* **2018**, *9*, 2473. [CrossRef]
35. Castillo, R.; Saier, M.H. Functional Promiscuity of Homologues of the Bacterial ArsA ATPases. *Int. J. Microbiol.* **2010**, *2010*, 187373. [CrossRef]
36. Yang, H.C.; Fu, H.L.; Lin, Y.F.; Rosen, B.P. Pathways of arsenic uptake and efflux. *Curr. Top. Membr.* **2012**, *69*, 325–358. [CrossRef]
37. Yang, H.C.; Rosen, B.P. New mechanisms of bacterial arsenic resistance. *Biomed. J.* **2016**, *39*, 5–13. [CrossRef]
38. McMillan, E.A.; Wasilenko, J.L.; Tagg, K.A.; Chen, J.C.; Simmons, M.; Gupta, S.K.; Tillman, G.E.; Folster, J.; Jackson, C.R.; Frye, J.G. Carriage and Gene Content Variability of the pESI-Like Plasmid Associated with *Salmonella* Infantis Recently Established in United States Poultry Production. *Genes* **2020**, *11*, 1516. [CrossRef]
39. Zurfluh, K.; Treier, A.; Schmitt, K.; Stephan, R. Mobile fosfomycin resistance genes in Enterobacteriaceae-An increasing threat. *Microbiologyopen* **2020**, *9*, e1135. [CrossRef]
40. Yang, T.Y.; Lu, P.L.; Tseng, S.P. Update on fosfomycin-modified genes in Enterobacteriaceae. *J. Microbiol. Immunol. Infect.* **2019**, *52*, 9–21. [CrossRef]

41. Aviv, G.; Rahav, G.; Gal-Mor, O. Horizontal Transfer of the *Salmonella enterica* Serovar Infantis Resistance and Virulence Plasmid pESI to the Gut Microbiota of Warm-Blooded Hosts. *mBio* **2016**, *7*, e01395-16. [CrossRef] [PubMed]
42. Granda, A.; Riveros, M.; Martínez-Puchol, S.; Ocampo, K.; Laureano-Adame, L.; Corujo, A.; Reyes, I.; Ruiz, J.; Ochoa, T.J. Presence of Extended-Spectrum β-lactamase, CTX-M-65 in *Salmonella enterica* serovar Infantis Isolated from Children with Diarrhea in Lima, Peru. *J. Pediatr. Infect. Dis.* **2019**, *14*, 194–200. [CrossRef]
43. Lapierre, L.; Cornejo, J.; Zavala, S.; Galarce, N.; Sanchez, F.; Benavides, M.B.; Guzman, M.; Saenz, L. Phenotypic and Genotypic Characterization of Virulence Factors and Susceptibility to Antibiotics in *Salmonella* Infantis Strains Isolated from Chicken Meat: First Findings in Chile. *Animals* **2020**, *10*, 1049. [CrossRef] [PubMed]
44. Maguire, M.; Khan, A.S.; Adesiyun, A.A.; Georges, K.; Gonzalez-Escalona, N. Genomic comparison of eight closed genomes of multidrug resistant *Salmonella enterica* strains isolated from broiler farms and processing plants in Trinidad and Tobago. *Front. Microbiol.* **2022**, *13*, 863104. [CrossRef] [PubMed]
45. Martinez-Puchol, S.; Riveros, M.; Ruidias, K.; Granda, A.; Ruiz-Roldan, L.; Zapata-Cachay, C.; Ochoa, T.J.; Pons, M.J.; Ruiz, J. Dissemination of a multidrug resistant CTX-M-65 producer *Salmonella enterica* serovar Infantis clone between marketed chicken meat and children. *Int. J. Food Microbiol.* **2021**, *344*, 109109. [CrossRef] [PubMed]
46. Tyson, G.H.; Li, C.; Harrison, L.B.; Martin, G.; Hsu, C.H.; Tate, H.; Tran, T.T.; Strain, E.; Zhao, S. A Multidrug-Resistant *Salmonella* Infantis Clone is Spreading and Recombining in the United States. *Drug Resist.* **2021**, *27*, 792–799. [CrossRef]
47. CLSI. *Performance Standards for Antimicrobial Susceptibility Testing*; CLSI supplement M100; Clinical and Laboratory Standards Institute: Wayne, PA, USA, 2019.
48. Vielva, L.; de Toro, M.; Lanza, V.F.; de la Cruz, F. PLACNETw: A web-based tool for plasmid reconstruction from bacterial genomes. *Bioinformatics* **2017**, *33*, 3796–3798. [CrossRef]
49. Garcillan-Barcia, M.P.; Redondo-Salvo, S.; Vielva, L.; de la Cruz, F. MOBscan: Automated Annotation of MOB Relaxases. *Methods Mol. Biol.* **2020**, *2075*, 295–308. [CrossRef]
50. Zhang, S.; Yin, Y.; Jones, M.B.; Zhang, Z.; Deatherage Kaiser, B.L.; Dinsmore, B.A.; Fitzgerald, C.; Fields, P.I.; Deng, X. *Salmonella* serotype determination utilizing high-throughput genome sequencing data. *J. Clin. Microbiol.* **2015**, *53*, 1685–1692. [CrossRef]
51. Larsen, M.V.; Cosentino, S.; Rasmussen, S.; Friis, C.; Hasman, H.; Marvig, R.L.; Jelsbak, L.; Sicheritz-Ponten, T.; Ussery, D.W.; Aarestrup, F.M.; et al. Multilocus sequence typing of total-genome-sequenced bacteria. *J. Clin. Microbiol.* **2012**, *50*, 1355–1361. [CrossRef]
52. Bortolaia, V.; Kaas, R.S.; Ruppe, E.; Roberts, M.C.; Schwarz, S.; Cattoir, V.; Philippon, A.; Allesoe, R.L.; Rebelo, A.R.; Florensa, A.F.; et al. ResFinder 4.0 for predictions of phenotypes from genotypes. *J. Antimicrob. Chemother.* **2020**, *75*, 3491–3500. [CrossRef] [PubMed]
53. Zankari, E.; Allesoe, R.; Joensen, K.G.; Cavaco, L.M.; Lund, O.; Aarestrup, F.M. PointFinder: A novel web tool for WGS-based detection of antimicrobial resistance associated with chromosomal point mutations in bacterial pathogens. *J. Antimicrob. Chemother.* **2017**, *72*, 2764–2768. [CrossRef] [PubMed]
54. Carattoli, A.; Zankari, E.; Garcia-Fernandez, A.; Voldby Larsen, M.; Lund, O.; Villa, L.; Moller Aarestrup, F.; Hasman, H. In silico detection and typing of plasmids using PlasmidFinder and plasmid multilocus sequence typing. *Agents Chemother.* **2014**, *58*, 3895–3903. [CrossRef] [PubMed]
55. Kaas, R.S.; Leekitcharoenphon, P.; Aarestrup, F.M.; Lund, O. Solving the problem of comparing whole bacterial genomes across different sequencing platforms. *PLoS ONE* **2014**, *9*, e104984. [CrossRef] [PubMed]
56. Felsenstein, J. Confidence limits on phylogenies: An approach using the bootstrap. *Evolution* **1985**, *39*, 783–791. [CrossRef] [PubMed]
57. Drauch, V.; Kornschober, C.; Palmieri, N.; Hess, M.; Hess, C. Infection dynamics of *Salmonella* Infantis strains displaying different genetic backgrounds—with or without pESI-like plasmid—Vary considerably. *Emerg. Microbes. Infect.* **2021**, *10*, 1471–1480. [CrossRef] [PubMed]
58. Pietsch, M.; Simon, S.; Meinen, A.; Trost, E.; Banerji, S.; Pfeifer, Y.; Flieger, A. Third generation cephalosporin resistance in clinical non-typhoidal *Salmonella enterica* in Germany and emergence of bla_{CTX-M}-harbouring pESI plasmids. *Microb. Genom.* **2021**, *7*. [CrossRef]
59. Carfora, V.; Alba, P.; Leekitcharoenphon, P.; Ballaro, D.; Cordaro, G.; Di Matteo, P.; Donati, V.; Ianzano, A.; Iurescia, M.; Stravino, F.; et al. Colistin Resistance Mediated by *mcr-1* in ESBL-Producing, Multidrug Resistant *Salmonella* Infantis in Broiler Chicken Industry, Italy (2016–2017). *Front. Microbiol.* **2018**, *9*, 1880. [CrossRef]
60. Müstak, I.B.; Yardimci, H. Construction and in vitro characterisation of *aroA* defective (*aroA*Δ) mutant *Salmonella* Infantis. *Arch. Microbiol.* **2019**, *201*, 1277–1284. [CrossRef]

 *antibiotics*

Article

Are Virulence and Antibiotic Resistance Genes Linked? A Comprehensive Analysis of Bacterial Chromosomes and Plasmids

Helena Darmancier [1], Célia P. F. Domingues [1,2], João S. Rebelo [2], Ana Amaro [1], Francisco Dionísio [2,3], Joël Pothier [4], Octávio Serra [5] and Teresa Nogueira [1,2,*]

[1] Bacteriology and Mycology Laboratory, INIAV—National Institute for Agrarian and Veterinary Research, 2780-157 Oeiras, Portugal; helenadarmancier@gmail.com (H.D.); celiapfd@hotmail.com (C.P.F.D.); ana.amaro@iniav.pt (A.A.)

[2] cE3c—Center for Ecology, Evolution and Environmental Change & CHANGE—Global Change and Sustainability Institute, Faculdade de Ciências, Universidade de Lisboa, 1749-016 Lisboa, Portugal; joaorebelo_4@hotmail.com (J.S.R.); dionisio@fc.ul.pt (F.D.)

[3] Departamento de Biologia Vegetal, Faculdade de Ciências, Universidade de Lisboa, 1749-016 Lisboa, Portugal

[4] Atelier de Bioinformatique, ISYEB, UMR 7205 CNRS MNHN UPMC EPHE, Muséum National d'Histoire Naturelle, CP 50, 45 Rue Buffon, F-75005 Paris, France; joel.pothier@gmail.com

[5] INIAV—National Institute for Agrarian and Veterinary Research, Portuguese Plant Germoplasm Bank, 4700-859 Braga, Portugal; octavio.serra@iniav.pt

* Correspondence: teresa.nogueira@iniav.pt

Abstract: Although pathogenic bacteria are the targets of antibiotics, these drugs also affect hundreds of commensal or mutualistic species. Moreover, the use of antibiotics is not only restricted to the treatment of infections but is also largely applied in agriculture and in prophylaxis. During this work, we tested the hypothesis that there is a correlation between the number and the genomic location of antibiotic resistance (AR) genes and virulence factor (VF) genes. We performed a comprehensive study of 16,632 reference bacterial genomes in which we identified and counted all orthologues of AR and VF genes in each of the locations: chromosomes, plasmids, or in both locations of the same genome. We found that, on a global scale, no correlation emerges. However, some categories of AR and VF genes co-occur preferentially, and in the mobilome, which supports the hypothesis that some bacterial pathogens are under selective pressure to be resistant to specific antibiotics, a fact that can jeopardize antimicrobial therapy for some human-threatening diseases.

Keywords: antibiotic resistance; virulence; plasmid; Integrative and Conjugative Element; co-selection; genomics; evolution

Citation: Darmancier, H.; Domingues, C.P.F.; Rebelo, J.S.; Amaro, A.; Dionísio, F.; Pothier, J.; Serra, O.; Nogueira, T. Are Virulence and Antibiotic Resistance Genes Linked? A Comprehensive Analysis of Bacterial Chromosomes and Plasmids. *Antibiotics* **2022**, *11*, 706. https://doi.org/10.3390/antibiotics11060706

Academic Editor: Anna Psaroulaki

Received: 30 April 2022
Accepted: 21 May 2022
Published: 24 May 2022

Publisher's Note: MDPI stays neutral with regard to jurisdictional claims in published maps and institutional affiliations.

1. Introduction

Antimicrobial resistance is a natural phenomenon, yet its emergence has been driven by antimicrobial exposure in healthcare, agriculture, veterinary settings, and the environment [1]. In the case of antibiotic use to cure bacterial infections, the onset of the disease dictates the prescription [2] and so antibiotics are specifically targeted to the bacterial pathogen. Recently, a study referred to a large increase in the prevalence of carbapenem-resistant *Klebsiella pneumoniae* and hypervirulent *K. pneumoniae*, revealing a worrying increase in this association between antibiotic resistance and virulence in human pathogens [3–5]. Another recent study also showed that 31% of 56 clinical isolates of *Pseudomonas aeruginosa* (an important agent of nosocomial infections worldwide), collected from different medical centers in Kenya between 2015 and 2020, harbor both virulence genes and a multidrug resistance phenotype. The authors report novel strains with extensive antimicrobial resistance genes (hereafter named AR) and the highest number of virulence factor genes (hereafter named VF) [6–8]. Therefore, natural selection might have been favoring the co-localization of virulence and resistance genes in the same cell genome or even in the same replicon (plasmid or chromosome). However, even when targeting pathogens, antibiotics also strike hundreds of commensal or mutualistic bacterial species comprising the trillions of cells present in the human or animal bodies.

Besides their therapeutic use, antibiotics have been used in medicine and veterinary care for decades, for prophylactic and metaphylactic uses [9]. In agriculture and animal production systems, for example, antibiotics are used on a very large scale, driven by economic interests [10]. In addition, there has been the reported release of antibiotics and active antibiotic residues into the environment, issuing from agriculture and the pharmaceutical industry [11]. This widespread and large-scale use has made antibiotics an emerging environmental contaminant [12]. Sewage waters issuing from urban areas, hospitals, and agricultural fields reach rivers [13] and contaminate agricultural fields. In this scenario, antibiotic resistance can be easily transferred between the environment, humans, and animals, as envisaged in the one health approach [1]. Thus, in the environment, all bacteria could potentially be equally exposed to these antibiotics, regardless of whether they are a human or an animal pathogen. However, as is the case for antibiotic exposure in infection therapy, environmental contamination from antibiotic use in agriculture can also be biased towards some antibiotics due to usage guidelines and regulations in place.

Previous research has shown that the diversity of virulence and antibiotic resistance genes correlate positively across human gut microbial communities and their genomes (metagenomes) [14]. In other words, metagenomes with a higher diversity of drug resistance genes are also the ones with a higher diversity of virulence genes, and vice versa—metagenomes with a lower diversity of drug resistance genes are also the ones with a lower diversity of virulence genes [15]. A previous work has shown that this positive correlation arises whenever the transmission probability of bacteria, along with their genes, between people is higher than the probability of losing resistance genes. This is a likely relationship because fitness costs imposed by resistance determinants often disappear in a few tens or hundreds of generations after a new gene has arisen [16–18]. Therefore, the positive correlation did not arise due to co-selection by antibiotics on both gene types in the metagenome [15]. Human beings interact and exchange bacteria and resistance and virulence genes through physical contact—perhaps unexpectedly, individuals with a higher diversity of both gene types are the ones that have not taken antibiotics for a long time. Those individuals who have recently used antibiotics show a decreased diversity in these genes [15]. However, these works do not elucidate what to expect at the level of individual bacterial cells or replicon localization.

Recent studies on the co-localization of ARs and VFs have focused on the dynamics of antibiotic resistance in bacterial pathogens and its comparison with non-pathogenic bacteria [19,20]. It is particularly important, however, to understand the co-evolutionary paths of antibiotic resistance and bacterial virulence at a bacterial genome level, though, to evaluate potentially critical combinations that could hinder the treatment of human and animal bacterial infections. During this study, we tested the co-selection hypothesis at different genomic locations in an exceptionally large and comprehensive database and identified the most favored and the most disadvantaged combinations of AR and VF. It is an innovative work because it generates knowledge that allows the study of evolutionary forces and those driving the co-mobilization of every AR and VF type or class, in bacterial genomes, on a very large scale.

We consider the bacterial genome as two types of molecular replicon: the chromosome and the plasmids. Most of the housekeeping genes are chromosomally encoded. Plasmids, however, are especially important mobile genetic elements, part of the accessory genome responsible for local adaptation. They can encode for supplementary biochemical pathways, which allow the expression of the different phenotypes and lifestyles, adaptative traits enabling cells to respond to local competitive or environmental pressures, such as exposure to antibiotics, bacteriocins, heavy metals, or other xenobiotics. Plasmids are enriched with genes encoding extracellular traits, and for proteins targeted to the cell envelope, such as those in the bacterial secretion systems [21].

Plasmids can be mobilized from one bacterium cell into another and can carry Integrative Conjugative Elements (ICE), which are mobile genetic elements able to both integrate bacterial chromosomes by site-specific recombination or exist as autonomous plasmid-like conjugative elements [22]. ICEs are also known as conjugative transposons as they encode for the type IV secretion system (which are also bacterial virulence factors) that is necessary for horizontal gene transfer between cells. Moreover, ICEs can carry insertion sequences and/or transposons, Integrases, and the Relaxase enzyme, which is critical for conjugation [23].

However, as they also stay in the chromosome, we are considering here all the genes that can be on either type of replicon, such as ICE, for example, in a different category named 'both'. It is believed that they mostly remain integrated into the chromosome, where they are transferred vertically. Nevertheless, under a lethal challenge, they can be horizontally transferred when the host cells are subject to DNA damage and SOS stress or under a starvation/stationary phase [23], in a small fraction of the bacterial population. ICEs can encode for adaptative niche traits as they harbor cargo genes conferring the following phenotypes: antibiotic resistance, the ability to metabolize a new carbon source, pathogenicity, or symbiosis islands [23].

The aim of this work is to understand whether evolution may have shaped bacterial genomes favoring a genetic link between virulence factors and antibiotic resistance-encoding genes, and to understand the overly complex relation between VF and AR at the genomic level.

2. Results

To try to capture evolutionary patterns on an overly broad scale, we retrieved all 16,632 complete closed bacterial genomes from the NCBI RefSeq database and classified all their replicons as chromosomal or plasmid. An additional genomic category was considered that included genes that have copies in the chromosome and in a plasmid of the same genome, and which could potentially be ICE. Each gene can only belong to one of these three categories—chromosome, plasmid, or 'both'—as every gene that appears on both chromosome and plasmid is only considered in the category 'both'. Note that a single cell may harbor more than one plasmid; therefore, even if two genes co-localize in plasmids, they may localize in different plasmids.

2.1. Genomic Organization of Antibiotic Resistance-Encoding Genes in Bacterial Genomes

To assess the identification and localization of antibiotic resistance-encoding genes within the bacterial genomes, we identified and classified all proteins from the dataset. To do so, we aligned all these protein sequences against antibiotic resistance databases: ResFinder, which consists of 3160 genes of acquired antibiotic resistance organized into 17 categories, including disinfectant resistance [24], and MUSTARD, which is a catalogue of 3.9 million proteins from the human gut microbiome organized into 41 categories [25]. For each genome, we identified all orthologs of each antibiotic resistance category, in each replicon type: chromosome or plasmid. We then counted the number of AR orthologues of each category that are located only in the chromosome, only in their plasmids, or simultaneously in both replicons. One may expect that larger plasmids or chromosomes would contain more AR orthologues. To remove this effect, we proceeded by dividing the counts by the total number of coding sequences (CDS) that are in each of the genomic locations: on the chromosome only, on the plasmids only, or on both locations in the same genome (57020371, 1531377, and 29622, respectively).

Figure 1 represents the distribution of the normalized number of antibiotic resistance gene orthologues per antibiotic class. According to ResFinder (Figure 1a,b), the largest amount of ARs belong to the category of beta-lactamase, sulphonamide, quinolone, macrolide, phenicol, trimethoprim, and tetracycline, which are present in the chromosome, plasmids, or in both chromosomes and plasmids belonging to the same bacterial genome. We also found large amounts of disinfectant-resistant genes. Of note, no fosfomycin, nitroimidazole, pseudomonicacid, or glycopeptide orthologues were found in the 'both' location. Interestingly, most orthologues for disinfectant resistance also appear to be preferentially encoded in the bacterial mobilome (here represented by the plasmids and ICE). Figure 1b represents the relative number of AR orthologues per genomic location.

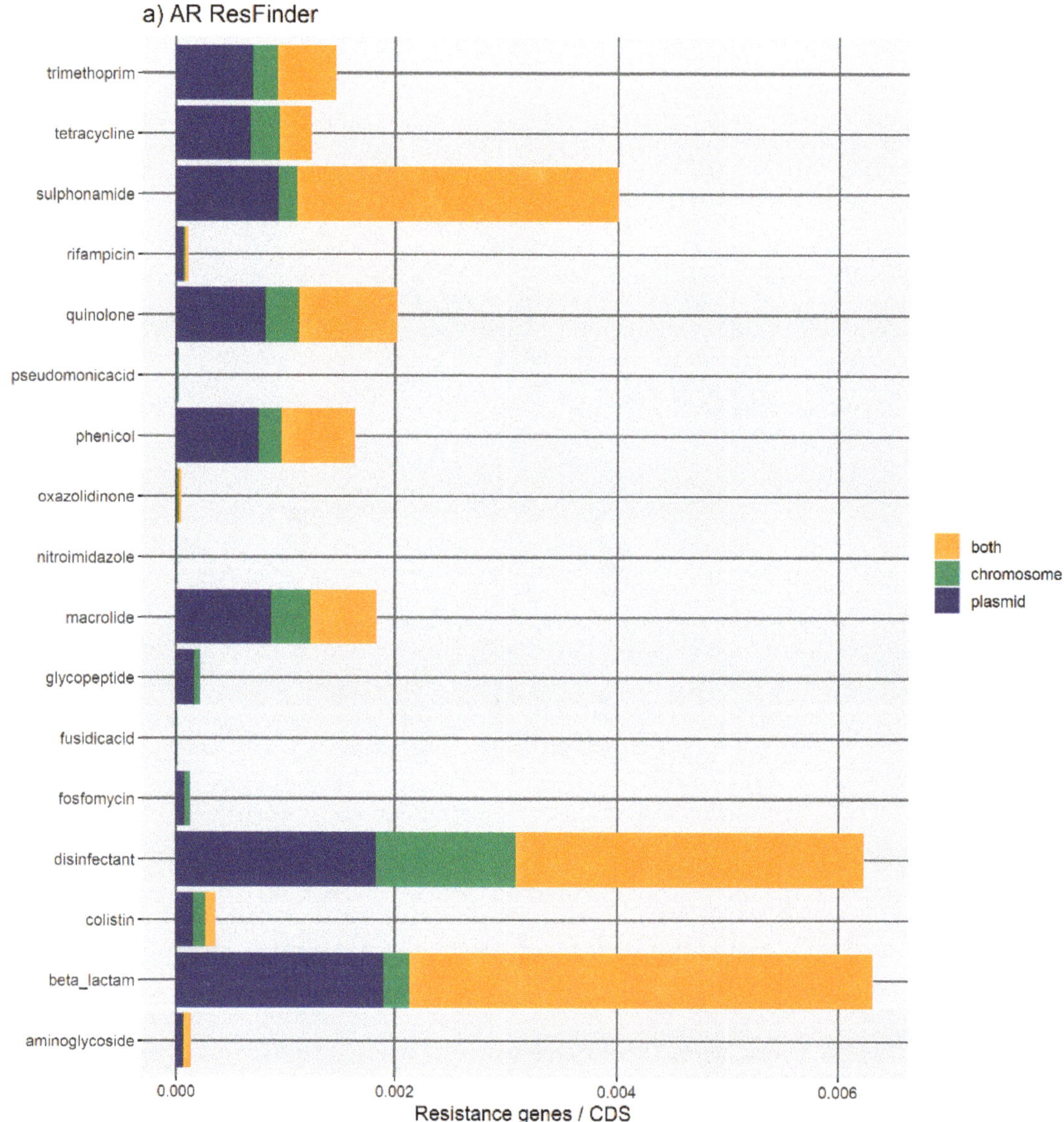

Figure 1. *Cont.*

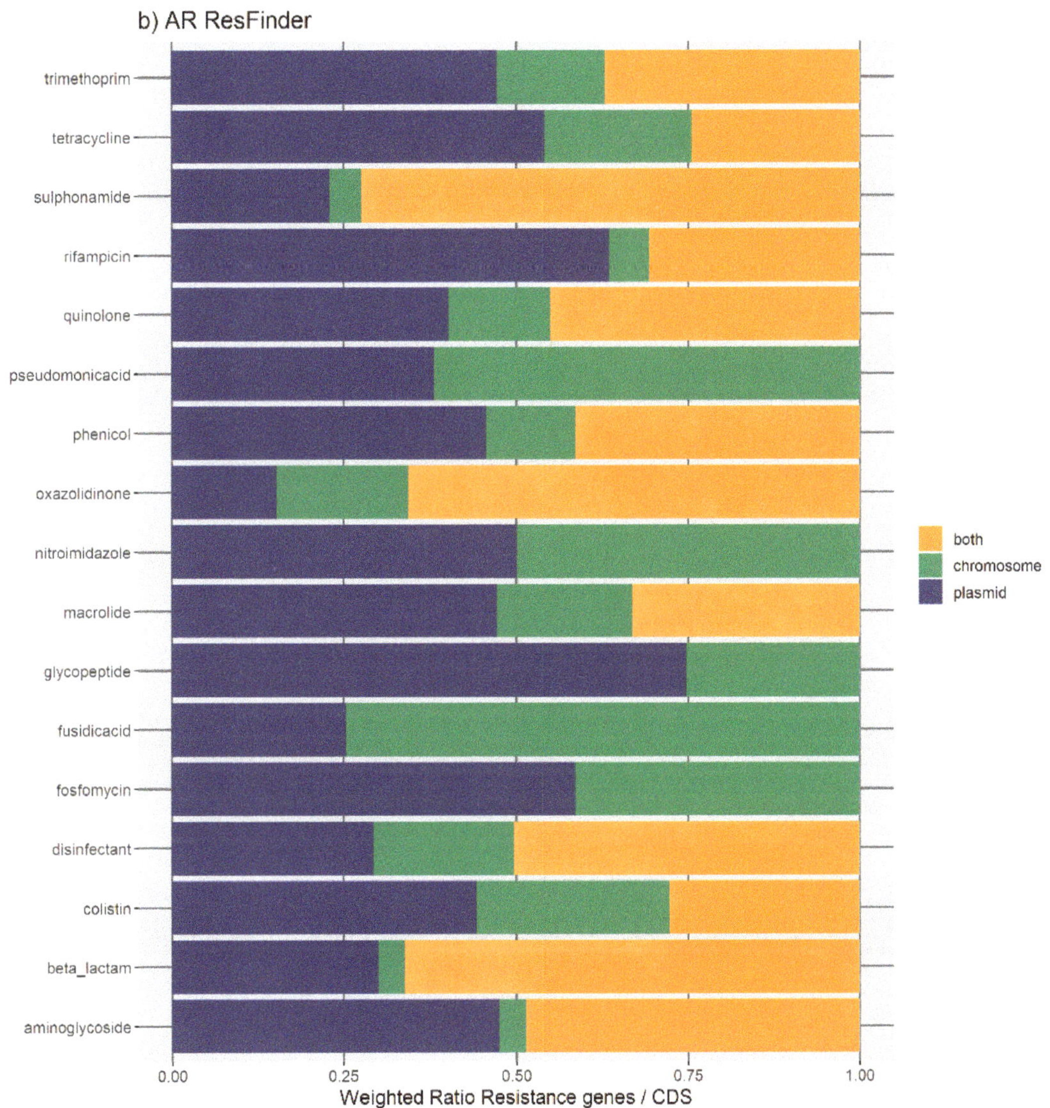

Figure 1. *Cont.*

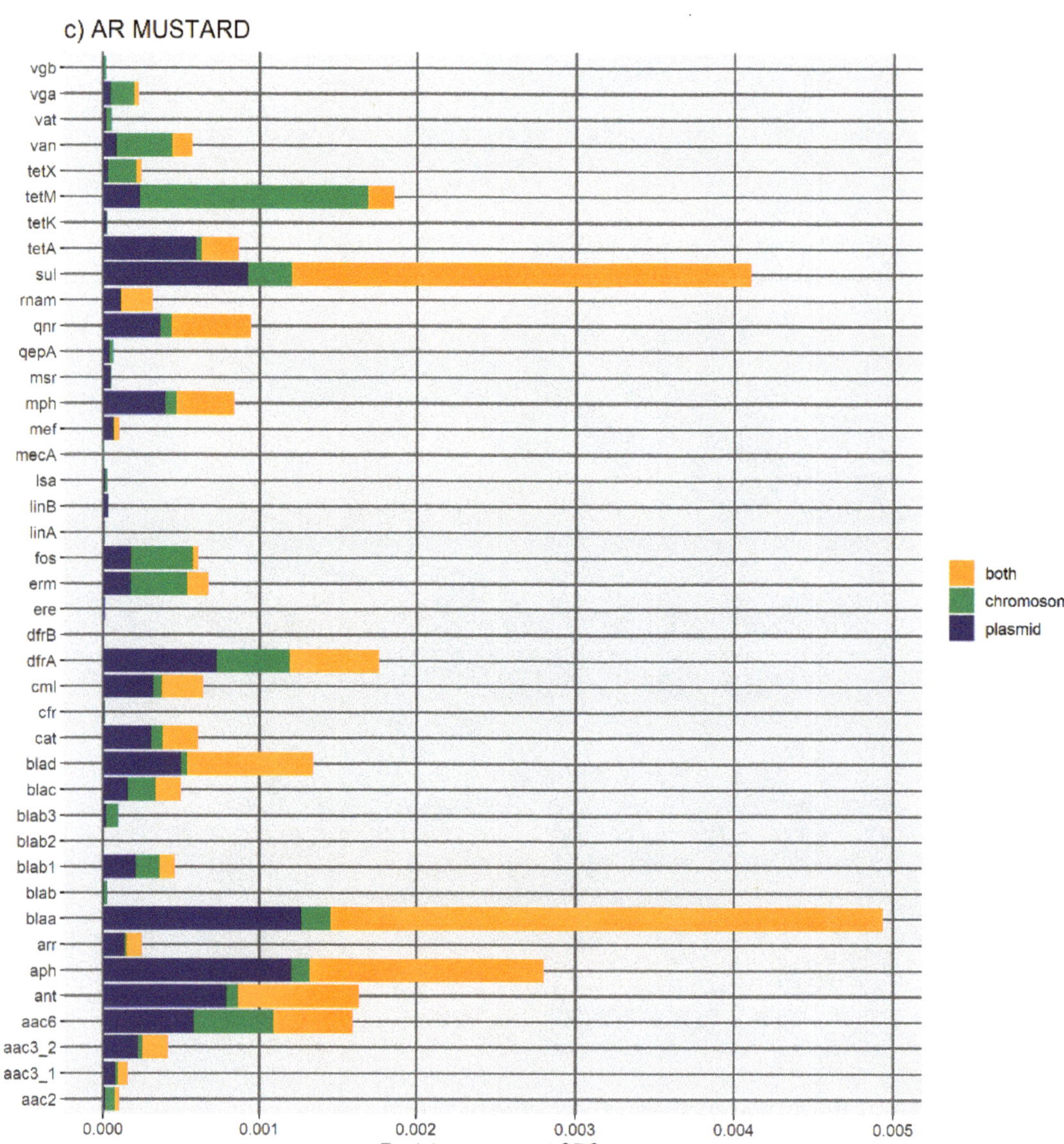

Figure 1. *Cont.*

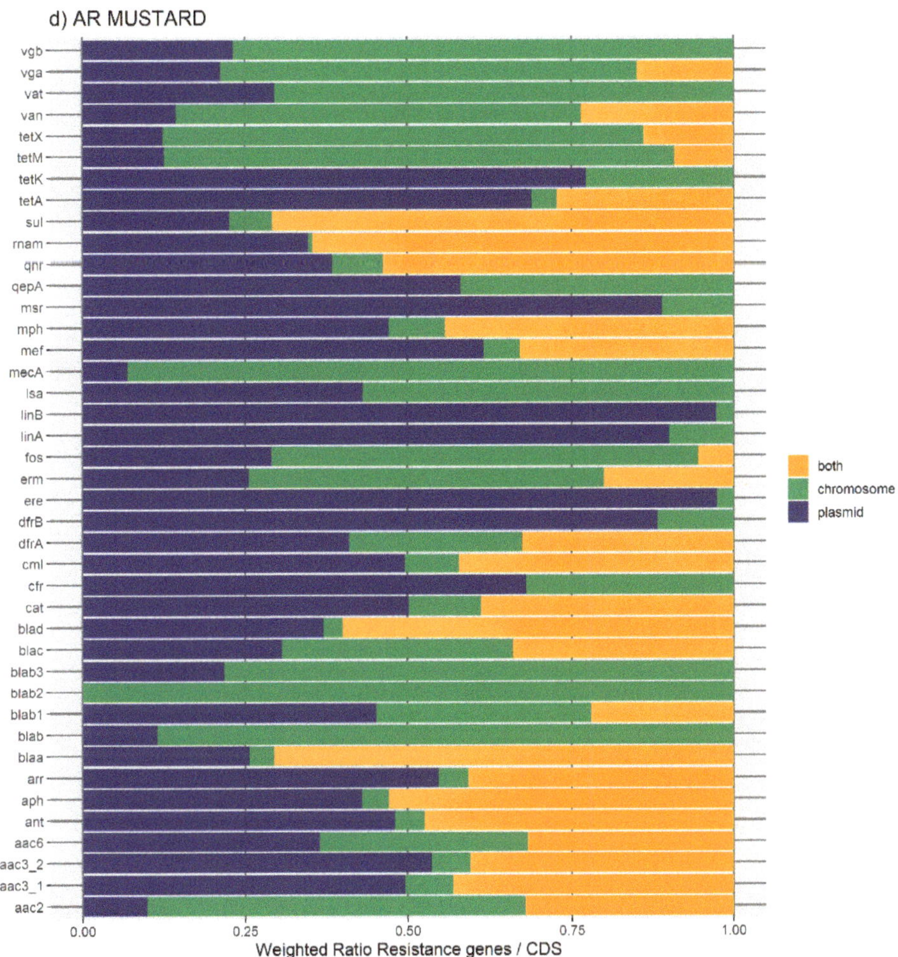

Figure 1. Bar plots and weighted bar plots of the number of antibiotic resistance orthologues. Because replicons vary in size, the number of resistance orthologues in each genome was divided by the amount of coding DNA sequences (CDS) in that genome (in that plasmid, in that chromosome, or in both). (**a,b**) Data from the ResFinder database; (**c,d**) data from the MUSTARD database. (**a**) For the ResFinder orthologues, the highest amount of antibiotic resistance orthologues is in the 'both' location (yellow). Beta-lactamase, sulphonamides, and disinfectant orthologues are more often (more than 50%) located in both replicons than in only a plasmid or the chromosome. (**b**) For the ResFinder orthologues, weighted bar plot shows that there are no fosfomicin, nitroimidazole, pseudomonicacid, fusidic acid, or glycopeptide orthologues in the 'both' location (yellow). Orthologues located in chromosomes (green) have a lower proportion compared to the other two genomic locations. (**c**) For the MUSTARD orthologues, there is a higher number of antibiotic resistance orthologues located simultaneously in both a plasmid and the chromosome ('both' location in yellow). There is a higher amount of *blaa*, *rnam*, and *sul* and orthologues. (**d**) For the MUSTARD database, there is a higher proportion of orthologues located in plasmids (blue). There is a lower proportion of orthologues located in 'both' a plasmid and the chromosome (yellow) than in a plasmid only or in the chromosome only. Moreover, 15 out of 45 orthologues (e.g., the orthologues of *vgb*, *vat*, or *tetk*) are present only in a plasmid or in the chromosome (i.e., without a 'both' location).

Overall, the results obtained with the two databases showed a high number of antibiotic resistance orthologues located in both genomic locations of the same genome, suggesting that ARs are carried in the bacterial mobilome. AR genes encoding beta-lactamases showed the highest numbers of orthologues in both the ResFinder and MUSTARD databases. This latter showed that the *blaa* category encodes Class A beta-lactamases, which are the predominant beta-lactamases found in the bacterial genome. All beta-lactamases can be found in plasmids and both plasmids and chromosome, except subclass B2 beta-lactamases, which are found only in the chromosome.

We have also generated boxplots of the antibiotic resistance orthologues per CDS for each genomic location (Figure 2). The boxplots do not show significant differences in the relative frequencies of ARs between the different genomic locations. The Kruskal–Wallis test was conducted to examine the differences between the relative number of antibiotic resistance orthologues according to the three types of genomic location. No significant differences were found (ResFinder: Chi square = 50.00, p-value = 0.2815, df = 16; and MUSTARD: Chi square = 116.42, p-value = 0.05838, df = 40).

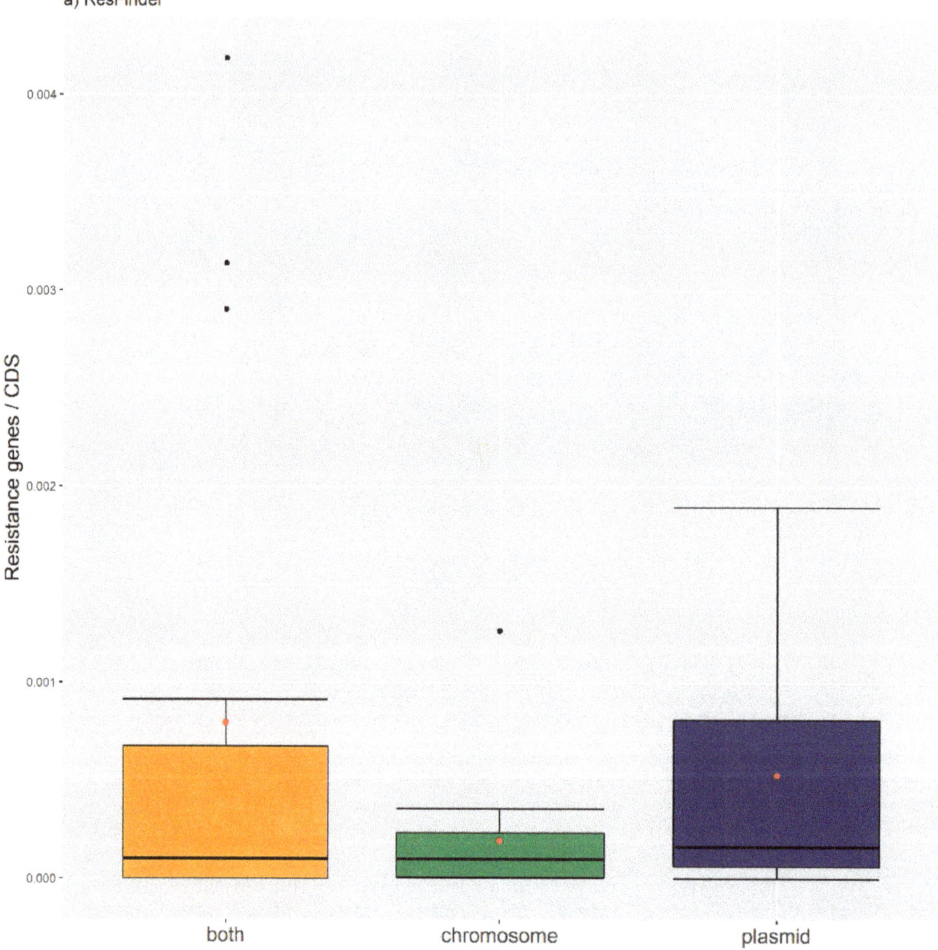

Figure 2. *Cont.*

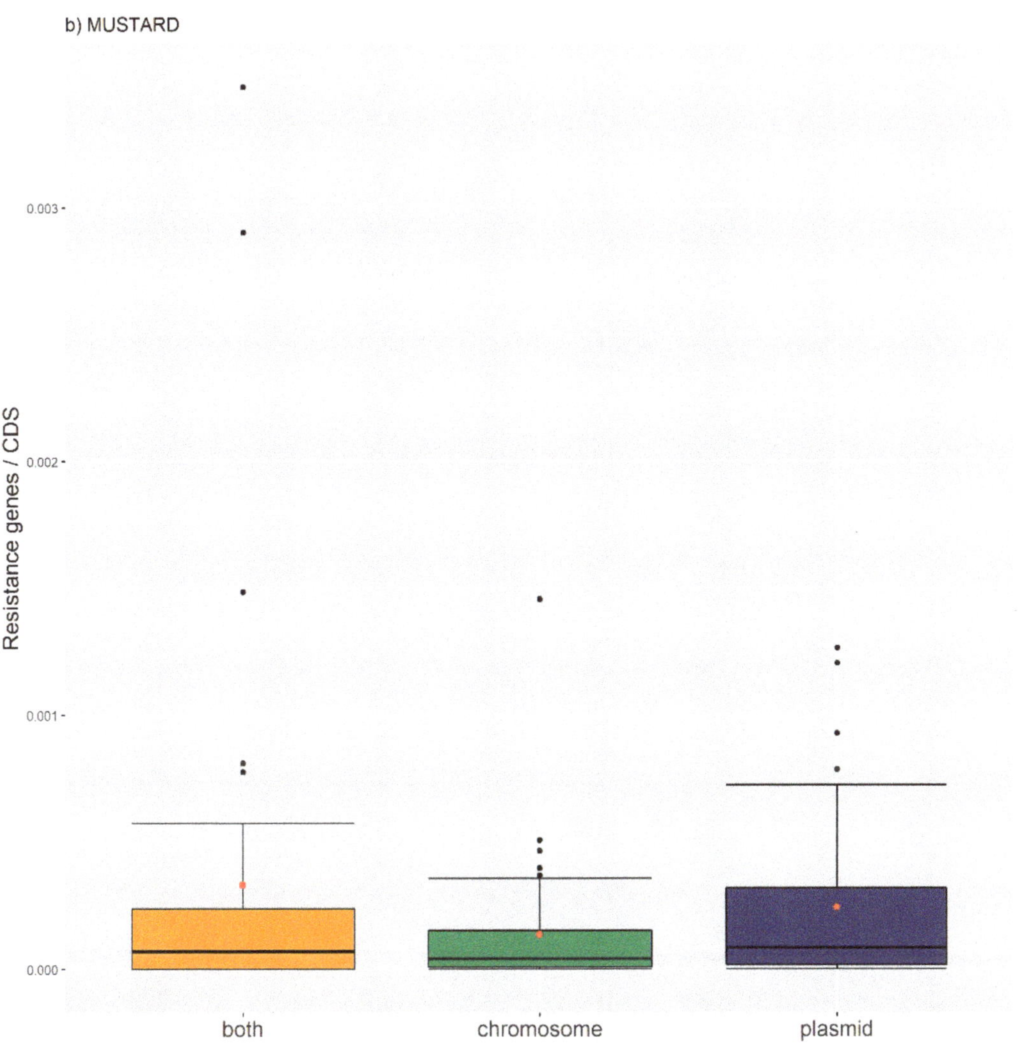

Figure 2. Boxplots of the antibiotic resistance orthologues per genomic location. (**a**) ResFinder ortho-logues. (**b**) MUSTARD orthologues. The red dot represents the mean.

2.2. Genomic Organization of Virulence Factor-Encoding Genes in Bacterial Genomes

We have also computed the number of virulence factor orthologues as described in the previous section. We calculated the number of orthologues of virulence factors (VF) grouped into various categories of the VFDB database, and according to their genomic location: only in the chromosome, only in a plasmid, or in both simultaneously. Once again, we may expect that larger plasmids or chromosomes would contain more VF orthologues. To remove this effect, we proceeded by dividing the counts by the total number of CDS in each of the genomic locations, as we did for the analysis of the ARs.

Although there are many virulence genes encoded on the plasmid, the chromosomal location is also very frequent in most bacteria (Figures 3 and 4). The results shown in Figure 3 show an uneven distribution of the various categories of virulence factors in each genome location. Pili, fimbriae, and flagella are the most representative virulence factors encoded on the chromosomes (Figure 3a). They all belong to the adhesion and invasion systems, which are essential to the colonization of the host or substrate, in the case of environmental species.

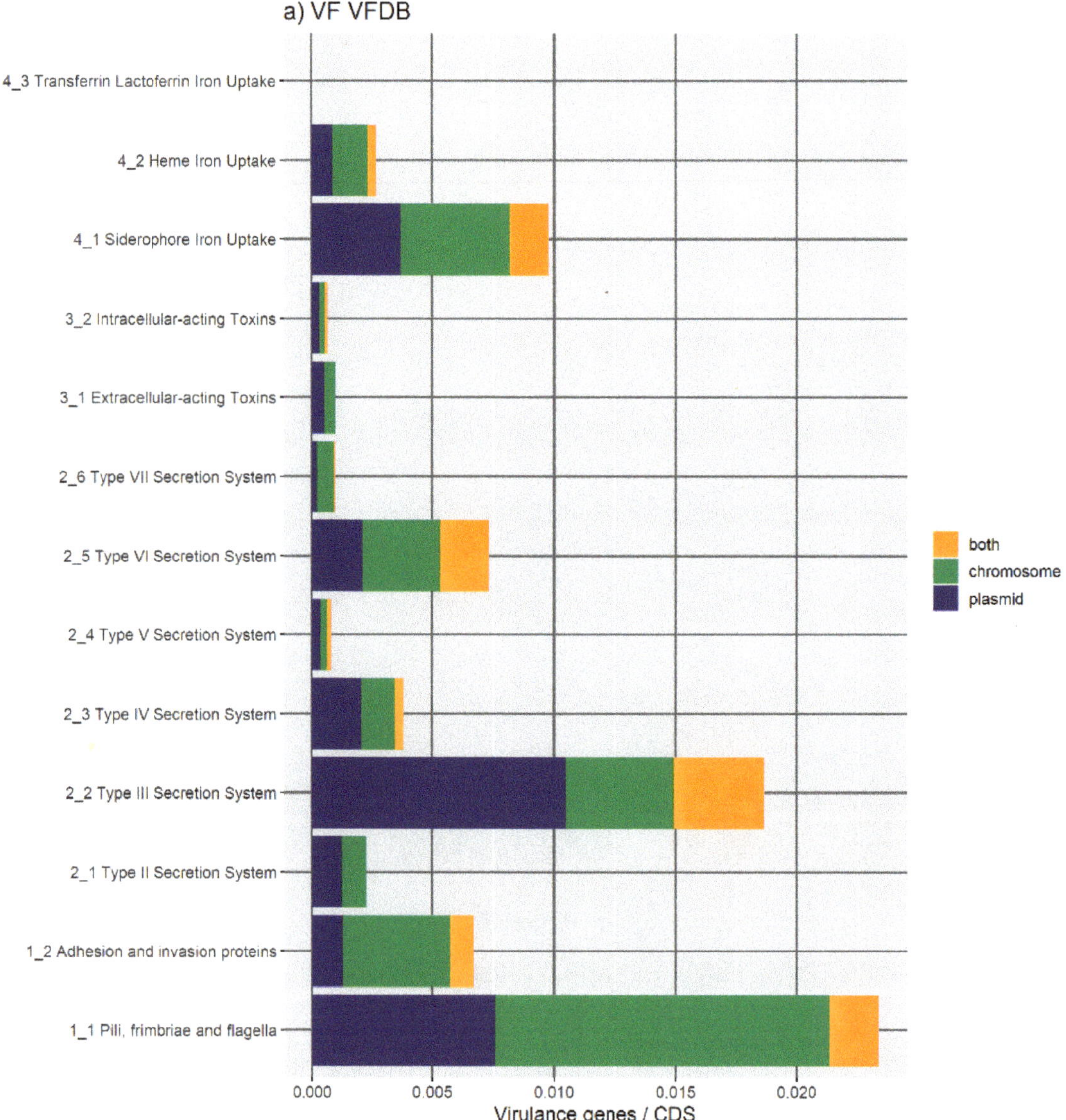

Figure 3. *Cont.*

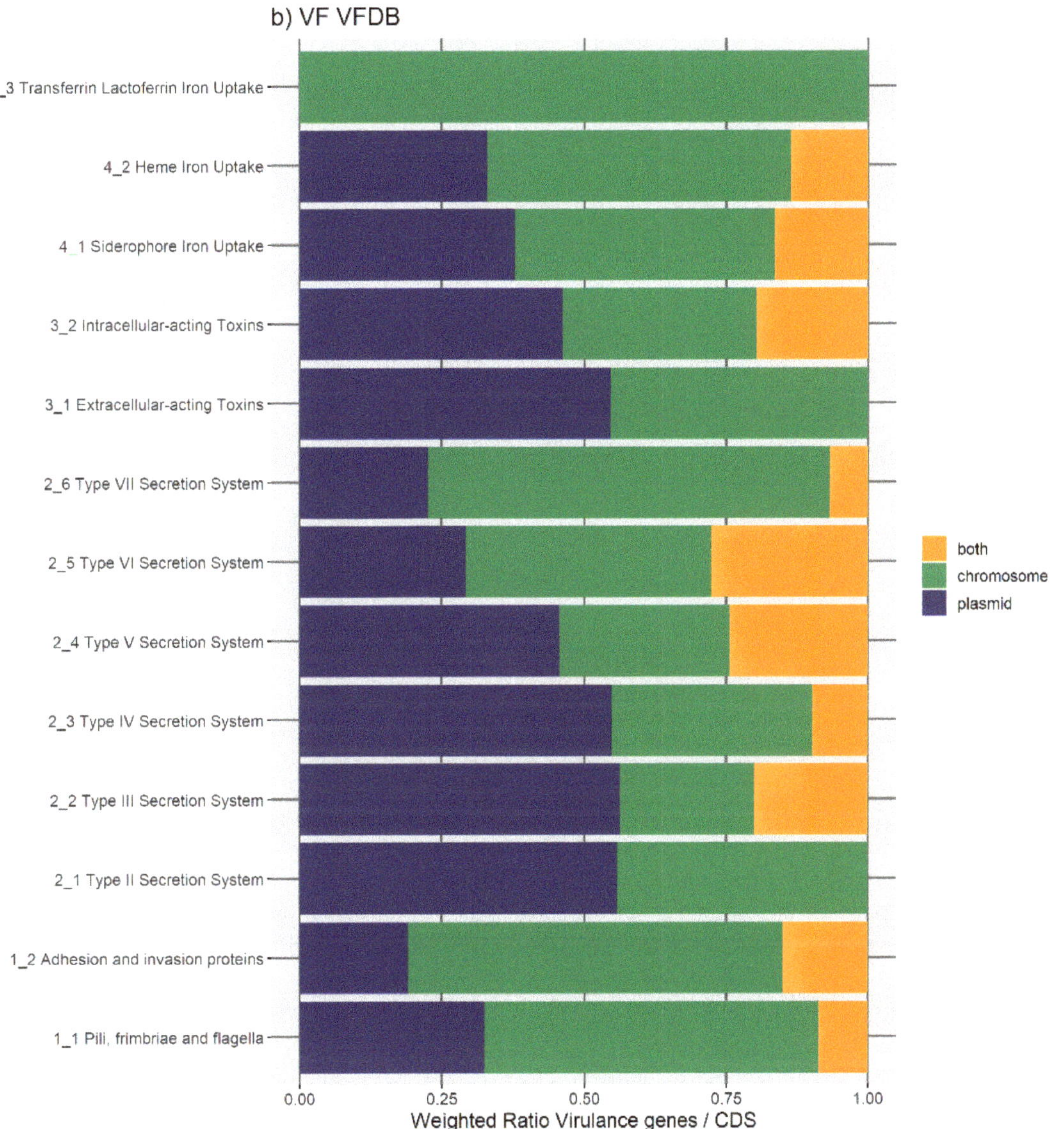

Figure 3. Bar plots (**a**) and weighted bar plots (**b**) of the number virulence factor orthologues in VFDB. Because replicons vary in size, the number of virulence orthologues in each genome was divided by the amount of coding DNA sequences (CDS) in that genome (in that plasmid, in that chromosome, or in both). (**a**) There is a higher number of virulence factor orthologues located in chromosomes (green) than only in a plasmid or in both a plasmid and the chromosome. There is a higher number of orthologues of Pili, fimbriae, and flagella and type III secretion systems than orthologues of the other virulence genes. (**b**) Orthologues of virulence genes are proportionally more present in the chromosome (green) only or in a plasmid only (blue). Transferrin and lactoferrin iron uptake orthologues are only located in chromosomes (green). Extracellular acting toxins and type III secretion system orthologues were either located in a plasmid (blue) or in the chromosome (green), not in both (yellow).

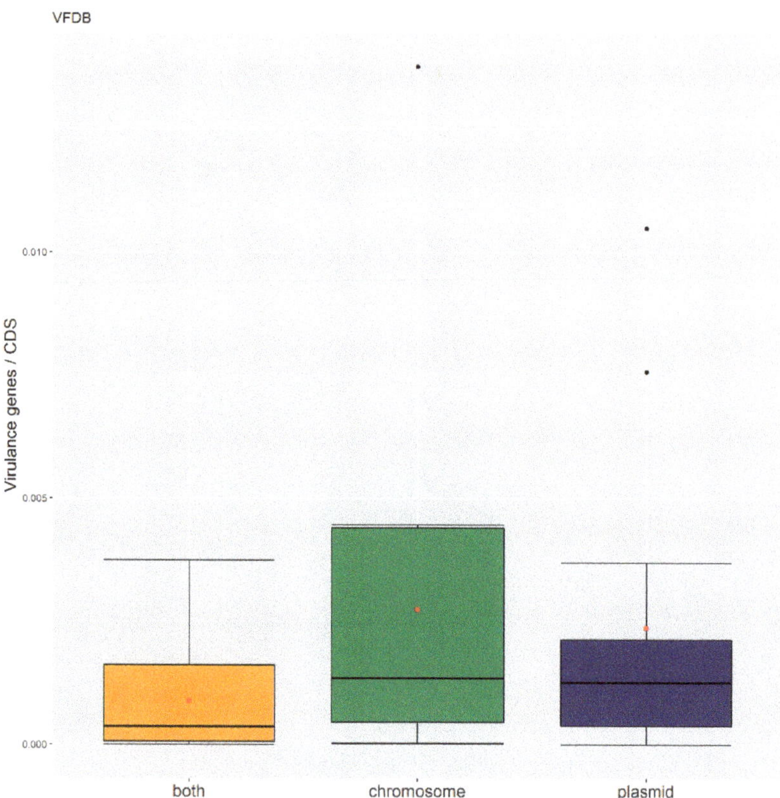

Figure 4. Boxplots of virulence factor orthologues per genomic location. The variance is higher in chromosomes (green). Orthologues found in plasmid (blue) and both (yellow) locations have similar variation. The red dot represents the mean.

Another important category of virulence factor is the type III secretion system (T3SS) [26], which, on the other hand, appears to be mainly encoded on plasmids (Figure 3a). They are important for the interaction of the cell with the host and are involved in the secretion of bacterial toxins directly into the host cell [27].

Regarding iron uptake systems, there is an uneven distribution according to the various categories: siderophore, heme, transferrin, and lactoferrin-mediated iron uptake systems. Siderophores are the most abundant, followed by heme-mediated iron uptake proteins, while orthologues of transferrin- and lactoferrin-mediated iron uptake proteins appear to be rare and chromosomally encoded (Figure 3a).

We can also see elevated levels of type IV and type VI secretion systems (T4SS and T6SS) [28,29] that, together with T3SS, are responsible for cell-to-cell interactions, such as the injection of toxins into host cells, or in conjugation and quorum sensing in bacteria. All these coding traits can be located on the chromosome, on plasmids, or present in both locations on the same genome (Figure 3a).

T4SS is part of the conjugation systems that are put in place to allow the transfer of mobilizable plasmids and conjugators as well as ICE from one cell to another [30]. The majority of the T4SS share a plasmid location.

We have also generated boxplots (Figure 4) and conducted the Kruskal–Wallis test to analyze the differences between the relative number of virulence gene orthologues according to the three types of genomic location. No significant differences were found (Chi square = 33.615, p-value = 0.4375, df = 12).

2.3. Correlation Analysis of Antibiotic Resistance- and Virulence Trait-Encoding Genes Per Genomic Location

The hypothesis that antibiotic resistance and virulence are genetically linked stems from the fact that, in human and veterinary medicine, the use of antibiotics, when not for prophylaxis, is adopted specifically to treat a bacterial infection, i.e., antibiotic therapy is targeted at pathogenic bacteria. If this assumption holds, we expect these two traits to be co-selected and mobilized together in plasmids—for example, in microbial communities [14]. Thus, first, we decided to assess the relationship between the number of ARs and VFs in each genome, at each genomic location, following a similar algorithm as in the previous sections. We assigned each orthologue to a genomic location: only on the chromosome, only on the plasmid, or both on the chromosome and plasmid. Then, we counted, for each genome, the number of ARs and VFs in each location. Finally, we divided these counts by the number of coding DNA sequences (CDS) in each location. These new values were used to generate the linear model in Figure 5. In general, there are no correlations between the number of ARs and VFs in bacterial genomes (Figure 5).

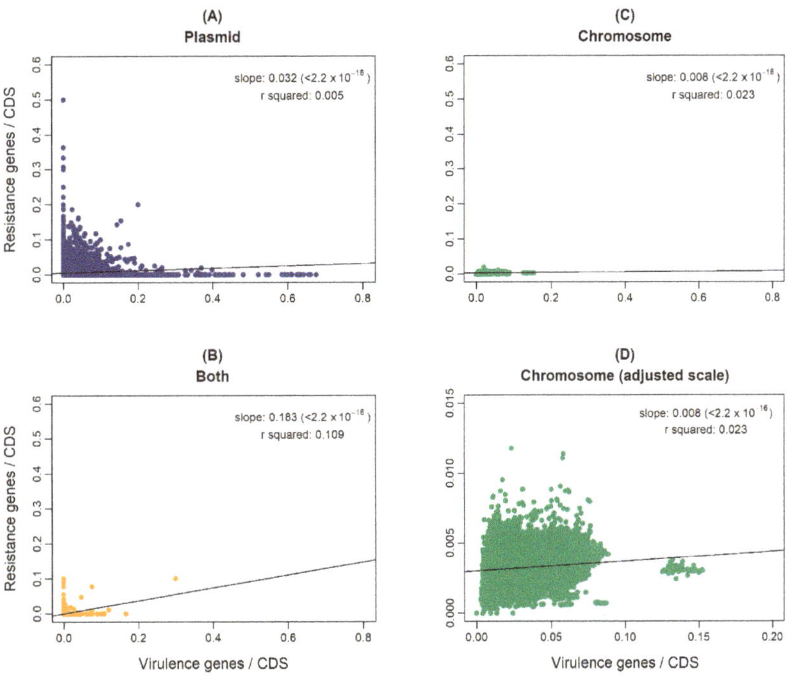

Figure 5. Relationship between antibiotic resistance gene orthologues of the ResFinder database (vertical axes) and virulence gene orthologues of the VFDB (horizontal axes). To correct for the size of the corresponding locations, the number of resistance and virulence genes in each genome was divided by the amount of coding DNA sequences (CDS) in that genomic location. Each point represents a genome. The colors indicate the genomic location: plasmid (blue), chromosome (green), or 'both' (yellow). (**C,D**) show the same data; however, (**D**) has the scale adjusted to better demonstrate the distribution of points. In (**A**), plasmid location (R = 0.005, slope = 0.032, *p*-value = ~0). In (**B**), both locations (R = 0.109, slope = 0.183, *p*-value = ~0). In (**C,D**), chromosome location (R = 0.023, slope = 0.008, *p*-value = ~0).

For the case of genes that are present both in the chromosome and in a plasmid, the sample size is much smaller than for the other locations. Furthermore, the size of a possible ICE-like mobile genetic element is expected to be even smaller than that of plasmids and chromosomes, which makes it risky to make direct comparisons. This smaller size makes the co-occurrence of ARs and VFs less likely and may explain the vertical and horizontal clustering of points seen in the 'both' panel of Figure 5B.

In order to evaluate the distribution of the co-occurrence of each of the antibiotic resistance gene categories and the virulence gene categories in the genomes used in this study, we constructed a contingency table. Each field of the contingency table contains the number of genomic elements containing orthologues from each of the AR and VF categories. These counts were made for all the AR and VF categories' combinations. Next, we generated a matrix containing the expected values of counts for each of the previous combinations. They are the product of the orthologous counts of each of the AR and VF categories in all genomes, divided by the total number of counts. Finally, we have computed the $\mathrm{Log_2}$ of the quotient of each observed value by the respective expected value. Heatmaps, as represented in Figures 6 and 7, were generated for the co-occurrences between all classes of ARs and VFs. Values in each position of the matrix that are greater than zero represent the genomes for which there are a higher number of co-occurrences between ARs from a given category and VFs from another category than expected at random and may mean that there is co-selection of these two genetic determinants. Conversely, a value below zero means that co-occurrence between a given AR category and another given category of VF is rarer than what would be expected to happen by chance, perhaps due to counter-selection. However, we have only considered as relevant those that are above 1 (represented in the yellow to red color shades) or below -1 (represented in blue shades). For these calculations, we have considered that each protein of the dataset can be classified into different AR or VF categories. With this choice, we are assuming that the same protein can belong to different antibiotic resistance mechanisms. An example of this is proteins that are part of the efflux pumps that perform the nonspecific extrusion of antibiotics belonging to different classes.

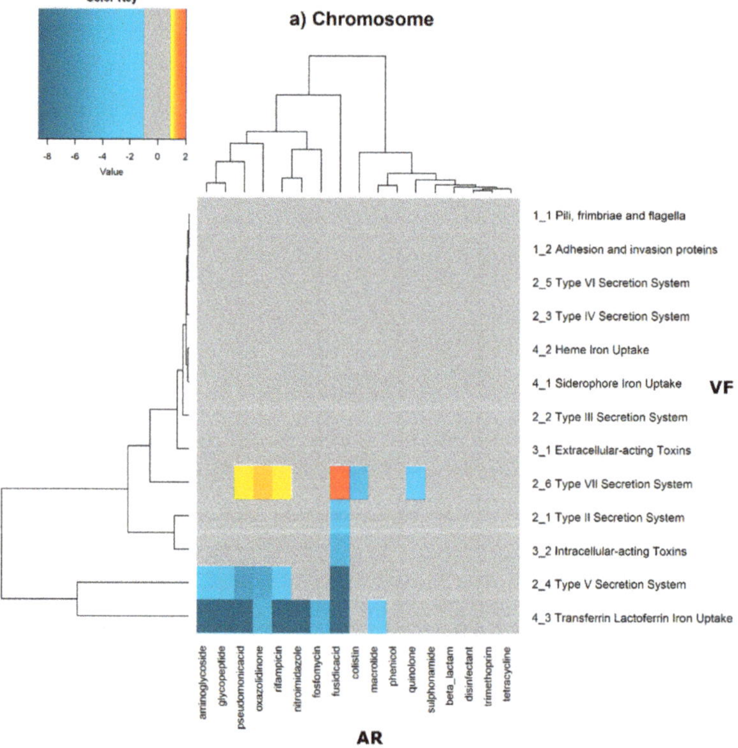

Figure 6. *Cont.*

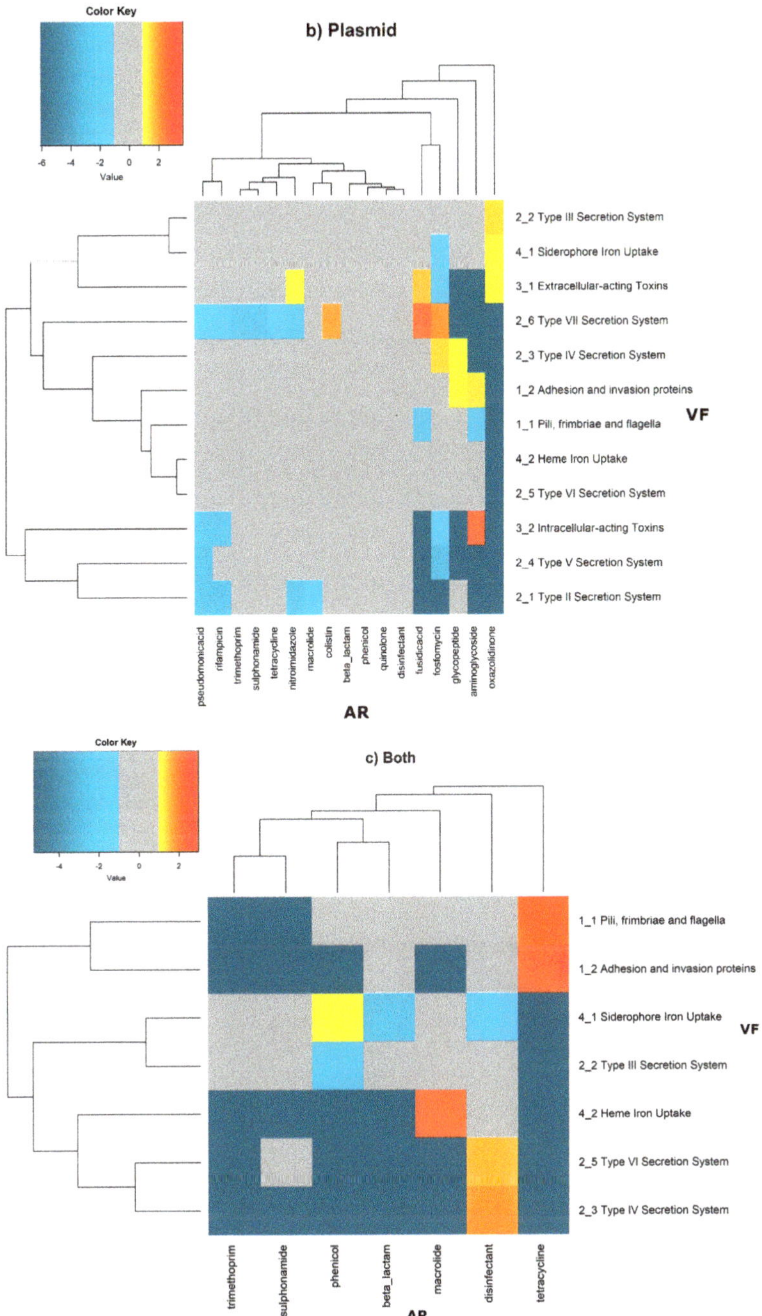

Figure 6. Heatmaps of the number of antibiotic resistance orthologues of the ResFinder database (*x* axis) and virulence factor orthologues of the VFDB database (*y* axis) in each genomic position. The values correspond to the Log$_2$ of the quotient of observed/expected values. The yellow to red shades represent a higher-than-expected number of co-occurrences; the blue shades represent a lower-than-expected number of co-occurrences; and the grey represents values close to the expected

values. (**a**) The higher number of ResFinder's orthologous proteins that co-occurred with VFDB's orthologous in the chromosome is between type VII secretion systems and fusidic acid. (**b**) In plasmids, the highest number of co-occurrences is between type VII secretion systems and fusidic acid and between intracellular acting proteins and aminoglycosides. (**c**) The number of co-occurrences in both the chromosome and plasmid is low. Consequently, we detected no co-occurrences involving six out of thirteen orthologues of virulence genes. There is a high number of co-occurrences of disinfectant orthologues with type IV and type VI secretion systems. The highest number of co-occurrences is between adhesion and invasion proteins and tetracyclines and between heme-mediated iron uptake and macrolides.

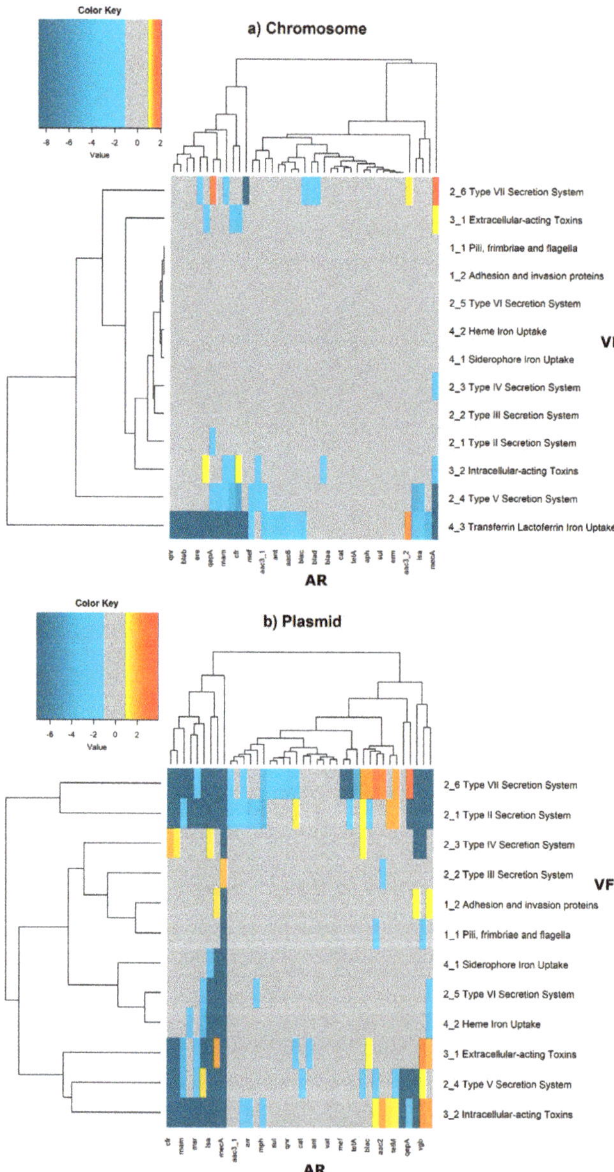

Figure 7. *Cont.*

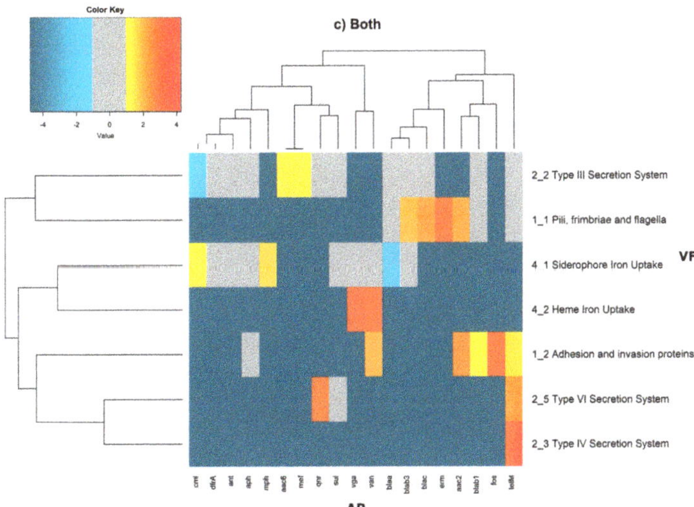

Figure 7. Heatmaps of the number of antibiotic resistance orthologues of the MUSTARD database (*x* axis) and virulence factor orthologues of the VFDB database (*y* axis) in each genomic position. The values correspond to the Log_2 of the quotient of observed/expected values. The yellow to red shades represent a higher-than-expected number of co-occurrences; the blue shades represent a lower-than-expected number of co-occurrences; and the grey represents values close to the expected values. (**a**) The highest number of co-occurrences of proteins orthologous to the MUSTARD database is between type VII secretion systems and the *mecA* antibiotic resistance gene, and between type VII secretion systems and *aac2*. (**b**) In plasmids, the highest number of co-occurrences appear to involve type VII secretion systems and toxins. (**c**) The 'both' genomic position has the lower number of co-occurrences; there is a more diverse pattern of co-occurrences. We observed no co-occurrence involving six out of thirteen orthologues of virulence genes.

The co-occurrences that vary most according to genomic location involve T7SS, which frequently co-occur with fusidic acid resistance genes, either on the chromosome or on the plasmid, but never on both simultaneously (Figures 6 and 7).

On the chromosome, fusidic acid resistance genes, on the other hand, do not occur with many different categories of other antibiotic resistance genes, such as colistin. At this location, apart from genes encoding transferrin-mediated iron uptake (for which there are no co-occurrences at any genomic locus), several other co-occurrences are exceedingly rare, as is the case for T5SS, with several distinct categories of antibiotic resistance genes.

In plasmids, there is a strong association between genes encoding for intracellularly acting toxins and those for aminoglycoside resistance, and oxazolidinone resistance genes co-occur frequently with some virulence factor genes, such as T3SS, the siderophore iron uptake system, or extracellularly acting proteins, while it co-occurs less than expected with all other virulence factors.

It is also worth noting that the heme-mediated iron uptake-encoding traits tend to co-occur with those for macrolide resistance at the 'both' loci. Transferrin-mediated iron uptake-coding genes only co-occur frequently with antibiotic resistance genes on the chromosome. The results in Figure 6 also highlight that adhesion and invasion virulence factor-encoding genes often co-occur with tetracyclines only at the 'both' loci, and with aminoglycosides and glycopeptides at the plasmid site, but to a lesser extent.

3. Discussion

This work aimed to study the evolutionary genomic dynamics of antibiotic resistance and virulence in bacteria. To have large-scale data and to perform a comprehensive study, we collected 16,632 reference genomes in the RefSeq database to address the correlation between the genomic locations of genes encoding for antibiotic resistance and bacterial virulence at two different scales. We first observed that the number of AR orthologues of all categories taken together does not differ significantly between the three locations (on the chromosome only, on a plasmid only, or on both). We reached the same conclusion for orthologues of VFs. In a second approach, which is novel, we analyzed the co-occurrences of each AR category with each VF category, by genomic location. This latter approach revealed co-occurrence patterns that are worth being addressed in more detail, and this will pave the way for further studies.

For this study, we have used the dataset of bacterial genomes from RefSeq, the NCBI Reference Sequence database, which consists of a non-redundant, well-annotated set of reference sequences including genomic and protein sequences. This dataset, however, does not represent a random sample of bacterial genomes. In fact, we expect this collection to be skewed towards bacterial genomes of human and medical interest. In this sense, an overrepresentation of human pathogens would explain the overrepresentation of VFs that we have found.

The asymmetrical distribution per replicon of the ARs and VFs can result from the fact that we have 37-fold more coding sequences (CDS) at the chromosomal location than the plasmid one, and 51-fold more CDS in plasmids than in 'both' locations. This is also because we have as twice as many chromosomes as plasmids on our dataset, but also because most of the plasmids are indicated to encode around 60 proteins and are thus much smaller than chromosomes [30], and ICE are even smaller than plasmids.

The fact that a particular gene is found on both the chromosome and a plasmid may indicate that this gene is in mobile regions of the chromosome, namely in ICE. According to Johnson and Grossman [23], antibiotic resistance genes mobilized from cell to cell by conjugative plasmids tend to move transiently or permanently onto the chromosome and do not remain in the cell as a stable plasmid. Therefore, one would expect that the antibiotic resistance genes found to have both a chromosome and plasmid location would be those clustered in the ICE and, therefore, also belong to the mobilome (the collection of mobile genetic elements).

3.1. The Antibiotic Resistance Genes Located in Plasmids Are Involved in Bacterial Local Adaptation

The normalization of the number of ARs by the number of CDS allowed the observation that there are many antibiotic resistance genes that are encoded in mobile regions of the genome, such as the 'both' category and in plasmids. This observation suggests that antibiotic resistance is a volatile and evolving trait, easily gained and lost, reinforcing the idea that these genetic elements are drivers of bacterial adaptation to the environment.

In this study, plasmids appear to be enriched with the following antibiotic resistance mechanisms: nucleotidyltransferase, beta-lactamase, phosphotransferase, and acetyltransferase. The production of hydrolyzing enzymes such as beta-lactamases encoded by *bla* genes is among the most important mechanisms for beta-lactams [31]. Bacteria encoding beta-lactamases are able to detoxify the local environment, hence rescuing nearby sensitive bacteria [32]. Other important enzymes include family members of nucleotidyltransferases (e.g., *ant* genes), phosphotransferase (e.g., *mph* and *fos* genes), and acetyltransferase (e.g., *aac* and *cat* genes), which confer resistance to aminoglycosides, macrolides, chloramphenicol, and fosfomycin, among others [33]. The link between antimicrobial use and resistance is complex, but the emergence of resistance is likely to be specific to each drug and to each microorganism [34]. Therefore, horizontal gene transfer and clonal expansion are giving rise to highly resistant microorganisms [5,35]. Nevertheless, other factors, such as cross-selection and co-selection, are highly pertinent while addressing antibiotic resistance [34].

We have also noticed that resistance to disinfectants is a very frequent trait, whose encoding genes preferentially belong to the 'both' category—that is, orthologues of disinfectant resistance are found in both the chromosome and a plasmid of the same cell. The link between disinfectants and antibiotics can be illustrated by the example of the use of quaternary ammonium compounds and sulphonamides since the 1930s, which has facilitated the spread of class 1 integrons and, thus, the evolution of antibiotic resistance in clinically relevant bacteria [36]. Given the fact that the use of disinfectants to clear pathogenic bacteria is recent, the results obtained here allow us to access the recent evolution of bacteria. The finding that the determinants of resistance to disinfectants are frequent and preferentially encoded in the mobile genome is a perfect example of the speed of the adaptive response and evolution of bacterial genomes in response to new environmental stresses. It has already been shown that biocidal agents used for disinfection can enhance antibiotic resistance in Gram-negative species [37]. Furthermore, in almost 90% of the outbreaks, the isolated pathogen turned out to be highly resistant to the disinfectant [36].

3.2. Mobile Genetic Elements Encode for Extracellular Traits and Cell-to-Cell Interactions

If we consider the considerable number of VFs that exist in the totality of genomes under study in this paper, we can observe that they are frequently encoded on the bacterial chromosome. This observation can reinforce the idea that many virulence traits are, in fact, essential to bacterial lifestyle and tend to be part of the core genome, rather than part of the arsenal for the colonization of new niches or environments. However, when we look at each of the VF categories individually, we can see that there are different patterns of genomic localization.

In this study, plasmids appear to be enriched with genes encoding extracellular traits, and for proteins targeted to the cell envelope, such as those pertaining to the bacterial secretion systems. Our previous results show that virulence factors that are targeted to the cell envelope or the outside medium are encoded in the mobilome [21]. This result seems coherent with the hypothesis that genes encoding for environmental adaptation and communication tend to be more mobile.

ICE are characterized by encoding T4SS, the conjugation pilus that enables them to 'jump' from the bacterial chromosome into an autonomous plasmid-like replicon able to synthetize the conjugative pilus, a T4SS, and be transferred from one cell into another [23], which may explain the elevated levels of co-occurrence between the T4SS and resistance to disinfectants.

3.3. Co-Selection between Antibiotic Resistance and Bacterial Virulence at Genomic Level

The natural selection exerted by antibiotics administered in the context of infection is not only exerted on bacterial pathogens. The entire host's microbiome will be under antibiotics' selective pressure, which weakens the association between antibiotic resistance and virulence in bacteria.

If, however, in modern industrialized societies, the onset of a bacterial infection triggers compliance to a standardized therapeutic protocol for that bacterial infection, then we are reinforcing the idea that there is an association between certain pathogenicity mechanisms, characteristics of certain bacterial infections, and resistance to the antibiotics most used to treat that type of infection. On the other hand, approaching the question from the point of view of bacterial evolution, we can imagine that resistance to the antibiotics most described for a given infection will confer an adaptive advantage to the pathogenic clones. From this point of view, antibiotic resistance is seen as a virulence factor, and is, therefore, in the limit, part of the infection mechanism since it will be under selective pressure.

This large-scale study did not detect a strong correlation between antibiotic resistance and virulence in bacterial genomes, even though bacterial reference genomes were biased towards human bacterial pathogens. The widespread transmission of antibiotic resistance traits, together with the permanent exposure to ubiquitous antibiotics, also undermines the hypothesis of the correlation between antibiotic resistance and bacterial virulence, thus explaining the results.

The novelty of this study, however, comes from the extremely broad and comprehensive analysis that highlights which combinations of distinct categories of AR and VF are in genetic linkage, and which could potentially be epidemic-prone due to their location in a mobile genomic locus.

On chromosomes and in plasmids, there are some strong co-occurrences between fusidic acid resistance and type VII secretion systems (T7SS), suggesting that resistance to this type of antibiotic is an important feature. Fusidic acid is bacteriostatic that is widely used to treat skin infections by Gram-positive bacteria, and that is also active against tuberculosis [27,38]. T7SS plays a significant role in mycobacteria, and in the human pathogen *Mycobacterium tuberculosis*, the main causative agent of tuberculosis [39]. These strong co-occurrences in many genomes, and especially on the chromosome, may highlight the evolution of the bacterial genome in response to antibiotic therapy for tuberculosis.

T5SS is a widely distributed autotransporter system involved in the secretion of bacterial toxins of *Escherichia coli* into the surrounding extracellular milieu [40]. *E. coli* is a particularly important human and animal pathogen, and, so, this differential co-occurrence with different antibiotic resistance categories of genes may result from differential selection pressure exerted, for example, due to the therapeutic protocols to cure *E. coli* infections.

Co-occurrence analysis leverages an understanding of the genomic evolution of human and animal pathogens and will therefore foster our understanding of the selective pressures favoring the antibiotic resistance of major bacterial pathogens.

3.4. Conclusions

With this work, we aimed to test whether there would be an association between antibiotic resistance and bacterial virulence at the genomic level. To do this, we analyzed an exceptionally large set of reference bacterial genomes. Although the dataset studied here was expected to be enriched with pathogenic bacterial genomes, we did not identify an overall trend towards the existence of a genetic linkage between virulence and antibiotic resistance at either the chromosomal or plasmid level. However, there are patterns of association between some virulence categories and antimicrobial resistance factors at different genomic locations that point to the evolution of bacterial genomes dictated by antibiotic use.

4. Methods

4.1. Data Gathering and Preparation

As a first step to begin this large-scale analysis, 16,632 complete bacterial genomes were retrieved from the National Center for Biotechnology Information (NCBI) Reference Sequence (RefSeq) database, which provides a non-redundant collection of sequences representing genomic data, transcripts, and proteins, in fasta format. Our dataset is composed of protein sequence fasta format files derived from every complete bacterial genome present in the repository, downloaded via command line, from https://ftp.ncbi.nlm.nih.gov/genomes/refseq/bacteria/ (accessed on 19 October 2020).

4.2. Blast against Selected Databases

The BLAST+ executables package (ncbi-blast-2.9.0+ version) was downloaded from the NCBI website (ftp://ftp.ncbi.nlm.nih.gov/blast/executables/blast+ (accessed on 11 November 2019)) [41]. MUSTARD is a database of antimicrobial resistance determinants that was generated using a 3-dimensional modeling-based approach, called pairwise comparative modeling (PCM), that accurately predicts functions of proteins that are distantly related to proteins with known functions from the human intestinal microbiota. This database is structured in 41 categories. MUSTARD was downloaded from http://mgps.eu/Mustard/db/all_ard.zip (accessed on 26 October 2020) [25].

ResFinder is a comprehensive and up-to-date database of acquired genes and chromosomal mutations mediating antimicrobial resistance in total or partial DNA sequences of bacteria of major public health relevance. This database is divided into 17 categories. The ResFinder database was downloaded from https://bitbucket.org/genomicepidemiology/resfinder_db.git, accessed on 30 October 2020 [24].

The Virulence Factor Database (VFDB) is a collection of bacterial virulence factor protein families with current and in-depth coverage of the major virulence factors of the best-characterized bacterial pathogens. This database contains a total of 13 functionally classified fasta files of bacterial virulence factor protein sub-families. Files were downloaded from http://www.mgc.ac.cn/VFs/Down/VFDB_setB_pro.fas.gz, accessed on 6 November 2020 [42].

Protein sequence fasta files pertaining to the databases were formatted in the bash shell, using the BLAST manual as a guideline. The BLAST program makeblastdb was run to set up the database. The blastdb_aliastool was then run to compile each database. The BLASTP analysis was performed using a small bash script, using non-default parameters, qcov 0.6 for the coverage, e-value was set at 1×10^5, and the output format was set to '6', which gives a tabular format to the BLAST results.

4.3. Post-BLAST Processing

Following the BLAST analysis, another filtering algorithm was applied to identify the hits with at least 40% identity. Of the alignments that passed this identity filter, we aimed to identify only the best alignment from each of the queries (alignment with the highest coverage). That is, even if a query aligns with more than one subject within the same family, or aligns with more than one family, only the best one is considered.

Each alignment was then divided into three categories based on its genomic location: (i) exclusively on the plasmid; (ii) exclusively on the chromosome; or (iii) both (on the plasmid and on the chromosome). There are no overlapping counts; this means that if a given genome has one or more copies of genes of a certain category on a plasmid and chromosome, the 'both' category count will be incremented by one and will not increase the counts on any of the individual plasmid or chromosome category. To perform this division, we used the feature tables for each genome that are available on RefSeq.

4.4. Normalization to the Size of the Genomic Location

Chromosomes and plasmids differ in size, which could impact the probability of encoding ARs or VFs. To avoid this effect, we also counted, for each genome, how many CDS exist for each genomic location. Subsequently, the counts of VF and AR orthologues per genomic location were divided by the number of CDS in that location.

4.5. Graphical and Statistical Analysis

The bar plots and the boxplots were created using the R ggplot2 package [43]; the linear regression models were created using R [44]; the heatmaps were created using the R gplots package [45], and the Kruskal–Wallis tests were performed with R [44].

4.6. Observed/Expected Ratio Calculation

To calculate the extent to which each AR orthologue and each VF orthologue co-localized more often or less often than expected, we proceeded as follows. First, we built a matrix where, in each position, we had the number of times each AR orthologue and each VF co-localizes (in the chromosome of the same bacterial cell, in plasmids of the same bacterial cell, or in both). Then, we calculated another matrix, but with the expected values. These expected values were calculated for RcoFinder, MUSTARD, and VFDB orthologues, from the observed count of orthologues per category multiplied by the total number of orthologues in that genomic location (of all categories) divided by the total number of orthologues found pertaining to the respective database. From this table, we constructed another matrix, for each database, with the observed/expected ratios of antibiotic resistance gene orthologues in each database category per genomic location.

The details to build the matrix with the expected values are as follows. Firstly, for each observation, the expected value was calculated using

$$E\left(AR_i,\, VF_j\right) = \frac{\sum_k Obs(AR_i, VF_k) \cdot \sum_m Obs(AR_m, VF_i)}{\sum_k \sum_m Obs(AR_m, VF_k)} \tag{1}$$

Secondly, each observation was divided by the expected value. The (i, j) position of the resulting matrix is

$$\frac{Obs(AR_i, VF_j)}{E(AR_i, VF_j)} \tag{2}$$

These calculations were performed for each genomic position and for the co-occurrences between Resfinder and VFDB and between MUSTARD and VFDB. The third step was to calculate the $\mathrm{Log}_2\left[\frac{Obs(AR_i,VF_j)}{E(AR_i,VF_j)}\right]$ for each (i, j) pair. The final tables were used to create the heatmaps. We considered relevant those cases where the number of observations was less than half of the expected value, so $Obs/E < 1/2$ (therefore, $\mathrm{Log}_2\left[\frac{Obs}{E}\right]$ is less than $\mathrm{Log}_2(1/2) = -1$) or where the number of observations was higher than double the expected value $Obs/E > 2$ (therefore, $\mathrm{Log}_2\left[\frac{Obs}{E}\right] > \mathrm{Log}_2[2] = 1$.

If $Obs(AR_i,VF_j) = 0$ and $E(AR_i,VF_j) > 0$, $Obs(AR_i,VF_j)/E(AR_i,VF_j) = 0$. For these cases, and because $\mathrm{Log}(0) = -\infty$, we changed the zero to the lowest number of the table divided by 10. Furthermore, we did not include the (AR_i,VF_j) cases in the heatmaps when $Obs(AR_i,VF_j) = 0$ and $E(AR_i,VF_j) = 0$.

Author Contributions: Conceptualization, J.P. and T.N.; methodology, F.D., J.P. and T.N.; software, H.D., C.P.F.D., J.S.R. and O.S.; formal analysis, H.D., C.P.F.D., J.S.R., A.A., F.D. and T.N.; investigation, H.D., C.P.F.D., J.S.R., A.A., F.D., A.A., J.P. and T.N.; writing—original draft preparation, H.D., C.P.F.D., J.S.R., A.A., F.D. and T.N.; writing—review and editing, H.D., C.P.F.D., J.S.R., A.A., F.D. and T.N.; supervision, T.N. All authors have read and agreed to the published version of the manuscript.

Funding: Fundação para a Ciência e a Tecnologia (FCT) supported this research by grant number PTDC/CVT-CVT/28469/2017, T.N. by contract PTDC/BIA-MIC/28824/2017, C.D. by the PhD grant UI/BD/153078/2022 and J.R. by the PhD grant SFRH/BD/04631/2021.

Institutional Review Board Statement: Not applicable.

Informed Consent Statement: Not applicable.

Data Availability Statement: Not applicable.

Acknowledgments: We would like to thank Fernanda Simões for the access to the server, Sérgio Crisóstomo for the all the interest during this work, and all the members of our group for the fruitful discussions during the work.

Conflicts of Interest: The authors declare no conflict of interest. The funders had no role in the design of the study; in the collection, analyses, or interpretation of data; in the writing of the manuscript, or in the decision to publish the results.

References

1. Woolhouse, M.; Ward, M.; van Bunnik, B.; Farrar, J. Antimicrobial Resistance in Humans, Livestock and the Wider Environment. *Philos. Trans. R. Soc. Lond. B. Biol. Sci.* **2015**, *370*, 20140083. [CrossRef]
2. David, P.H.C.; Sá-Pinto, X.; Nogueira, T. Using SimulATe to Model the Effects of Antibiotic Selective Pressure on the Dynamics of Pathogenic Bacterial Populations. *Biol. Methods Protoc.* **2019**, *4*, bpz004. [CrossRef]
3. Michelle, K.P.; Joan, M. Klebsiella Pneumoniae: Going on the Offense with a Strong Defense. *Microbiol. Mol. Biol. Rev.* **2016**, *80*, 629–661. [CrossRef]
4. Zhou, S.; Ren, G.; Liu, Y.; Liu, X.; Zhang, L.; Xu, S.; Wang, T. Challenge of Evolving Klebsiella Pneumoniae Infection in Patients on Hemodialysis: From the Classic Strain to the Carbapenem-Resistant Hypervirulent One. *Int. J. Med. Sci.* **2022**, *19*, 416–424. [CrossRef]
5. Lan, P.; Jiang, Y.; Zhou, J.; Yu, Y. A Global Perspective on the Convergence of Hypervirulence and Carbapenem Resistance in Klebsiella Pneumoniae. *J. Glob. Antimicrob. Resist.* **2021**, *25*, 26–34. [CrossRef]
6. Kiyaga, S.; Kyany'a, C.; Muraya, A.W.; Smith, H.J.; Mills, E.G.; Kibet, C.; Mboowa, G.; Musila, L. Genetic Diversity, Distribution, and Genomic Characterization of Antibiotic Resistance and Virulence of Clinical Pseudomonas Aeruginosa Strains in Kenya. *Front. Microbiol.* **2022**, *13*, 835403. [CrossRef]
7. Lam, M.M.C.; Wyres, K.L.; Duchêne, S.; Wick, R.R.; Judd, L.M.; Gan, Y.-H.; Hoh, C.-H.; Archuleta, S.; Molton, J.S.; Kalimuddin, S.; et al. Population Genomics of Hypervirulent Klebsiella Pneumoniae Clonal-Group 23 Reveals Early Emergence and Rapid Global Dissemination. *Nat. Commun.* **2018**, *9*, 2703. [CrossRef]

8. Risk Assessment: Emergence of Hypervirulent Klebsiella Pneumoniae ST23 Carrying Carbapenemase Genes in EU/EEA Countries. Available online: https://www.ecdc.europa.eu/en/publications-data/risk-assessment-emergence-hypervirulent-klebsiella-pneumoniae-eu-eea (accessed on 18 May 2022).

9. McEwen, S.A.; Collignon, P.J. Antimicrobial Resistance: A One Health Perspective. *Microbiol. Spectr.* **2018**, *6*, 2–6. [CrossRef]

10. Wall, B.A.; Mateus, A.; Marshall, L.; Pfeiffer, D.U.; Lubroth, J.; Ormel, H.J.; Otto, P.; Patriarchi, A. *Drivers, Dynamics and Epidemiology of Antimicrobial Resistance in Animal Production*; FAO: Roma, Italy, 2016.

11. Andersson, D.I.; Hughes, D. Microbiological Effects of Sublethal Levels of Antibiotics. *Nat. Rev. Microbiol.* **2014**, *12*, 465–478. [CrossRef]

12. Davies Julian; Davies Dorothy Origins and Evolution of Antibiotic Resistance. *Microbiol. Mol. Biol. Rev.* **2010**, *74*, 417–433. [CrossRef]

13. Czekalski, N., Sigdel, R., Birtel, J.; Matthews, B.; Bürgmann, H. Does Human Activity Impact the Natural Antibiotic Resistance Background? Abundance of Antibiotic Resistance Genes in 21 Swiss Lakes. *Environ. Int.* **2015**, *81*, 45–55. [CrossRef]

14. Escudeiro, P.; Pothier, J.; Dionisio, F.; Nogueira, T.; Fey, P.D. Antibiotic Resistance Gene Diversity and Virulence Gene Diversity Are Correlated in Human Gut and Environmental Microbiomes. *mSphere* **2019**, *4*, e00135-19. [CrossRef]

15. Domingues, C.P.F.; Rebelo, J.S.; Pothier, J.; Monteiro, F.; Nogueira, T.; Dionisio, F. The Perfect Condition for the Rising of Superbugs: Person-to-Person Contact and Antibiotic Use Are the Key Factors Responsible for the Positive Correlation between Antibiotic Resistance Gene Diversity and Virulence Gene Diversity in Human Metagenomes. *Antibiotics* **2021**, *10*, 605. [CrossRef]

16. Hall, J.P.J.; Wright, R.C.T.; Guymer, D.; Harrison, E.; Brockhurst, M.A. Extremely Fast Amelioration of Plasmid Fitness Costs by Multiple Functionally Diverse Pathways. *Microbiology* **2019**, *166*, 56–62. [CrossRef]

17. Loftie-Eaton, W.; Bashford, K.; Quinn, H.; Dong, K.; Millstein, J.; Hunter, S.; Thomason, M.K.; Merrikh, H.; Ponciano, J.M.; Top, E.M. Compensatory Mutations Improve General Permissiveness to Antibiotic Resistance Plasmids. *Nat. Ecol. Evol.* **2017**, *1*, 1354–1363. [CrossRef]

18. Andersson, D.I.; Hughes, D. Antibiotic Resistance and Its Cost: Is It Possible to Reverse Resistance? *Nat. Rev. Microbiol.* **2010**, *8*, 260–271. [CrossRef]

19. Geisinger, E.; Isberg, R.R. Interplay Between Antibiotic Resistance and Virulence During Disease Promoted by Multidrug-Resistant Bacteria. *J. Infect. Dis.* **2017**, *215*, S9–S17. [CrossRef]

20. Pan, Y.; Zeng, J.; Li, L.; Yang, J.; Tang, Z.; Xiong, W.; Li, Y.; Chen, S.; Zeng, Z.; Gilbert, J.A. Coexistence of Antibiotic Resistance Genes and Virulence Factors Deciphered by Large-Scale Complete Genome Analysis. *Msystems* **2020**, *5*, e00821-19. [CrossRef]

21. Nogueira, T.; Rankin, D.J.; Touchon, M.; Taddei, F.; Brown, S.P.; Rocha, E.P.C. Horizontal Gene Transfer of the Secretome Drives the Evolution of Bacterial Cooperation and Virulence. *Curr. Biol. CB* **2009**, *19*, 1683–1691. [CrossRef]

22. Guglielmini, J.; Quintais, L.; Garcillán-Barcia, M.P.; de la Cruz, F.; Rocha, E.P.C. The Repertoire of ICE in Prokaryotes Underscores the Unity, Diversity, and Ubiquity of Conjugation. *PLoS Genet.* **2011**, *7*, e1002222. [CrossRef]

23. Johnson, C.M.; Grossman, A.D. Integrative and Conjugative Elements (ICEs): What They Do and How They Work. *Annu. Rev. Genet.* **2015**, *49*, 577–601. [CrossRef]

24. Bortolaia, V.; Kaas, R.S.; Ruppe, E.; Roberts, M.C.; Schwarz, S.; Cattoir, V.; Philippon, A.; Allesoe, R.L.; Rebelo, A.R.; Florensa, A.F.; et al. ResFinder 4.0 for Predictions of Phenotypes from Genotypes. *J. Antimicrob. Chemother.* **2020**, *75*, 3491–3500. [CrossRef]

25. Ruppé, E.; Ghozlane, A.; Tap, J.; Pons, N.; Alvarez, A.-S.; Maziers, N.; Cuesta, T.; Hernando-Amado, S.; Clares, I.; Martínez, J.L.; et al. Prediction of the Intestinal Resistome by a Three-Dimensional Structure-Based Method. *Nat. Microbiol.* **2019**, *4*, 112–123. [CrossRef]

26. Burkinshaw, B.J.; Strynadka, N.C.J. Assembly and Structure of the T3SS. *Protein Traffick. Secret. Bact.* **2014**, *1843*, 1649–1663. [CrossRef]

27. Horna, G.; Ruiz, J. Type 3 Secretion System as an Anti-Pseudomonal Target. *Microb. Pathog.* **2021**, *155*, 104907. [CrossRef]

28. Monjarás Feria, J.; Valvano, M.A. An Overview of Anti-Eukaryotic T6SS Effectors. *Front. Cell. Infect. Microbiol.* **2020**, *10*, 584751. [CrossRef]

29. Cabezón, E.; Ripoll-Rozada, J.; Peña, A.; de la Cruz, F.; Arechaga, I. Towards an Integrated Model of Bacterial Conjugation. *FEMS Microbiol. Rev.* **2015**, *39*, 81–95. [CrossRef]

30. Smillie, C.; Garcillán-Barcia, M.P.; Francia, M.V.; Rocha, E.P.C.; de la Cruz, F. Mobility of Plasmids. *Microbiol. Mol. Biol. Rev. MMBR* **2010**, *74*, 434–452. [CrossRef]

31. Bush, K. Past and Present Perspectives on β-Lactamases. *Antimicrob. Agents Chemother.* **2018**, *62*, e01076-18. [CrossRef]

32. Domingues, I.L.; Gama, J.A.; Carvalho, L.M.; Dionisio, F. Social Behaviour Involving Drug Resistance: The Role of Initial Density, Initial Frequency and Population Structure in Shaping the Effect of Antibiotic Resistance as a Public Good. *Heredity* **2017**, *119*, 295–301. [CrossRef]

33. McArthur, A.G.; Waglechner, N.; Nizam, F.; Yan, A.; Azad, M.A.; Baylay, A.J.; Bhullar, K.; Canova, M.J.; De Pascale, G.; Ejim, L.; et al. The Comprehensive Antibiotic Resistance Database. *Antimicrob. Agents Chemother.* **2013**, *57*, 3348–3357. [CrossRef] [PubMed]

34. Holmes, A.H.; Moore, L.S.P.; Sundsfjord, A.; Steinbakk, M.; Regmi, S.; Karkey, A.; Guerin, P.J.; Piddock, L.J.V. Understanding the Mechanisms and Drivers of Antimicrobial Resistance. *Lancet Lond. Engl.* **2016**, *387*, 176–187. [CrossRef]

35. Leão, C.; Clemente, L.; Moura, L.; Seyfarth, A.M.; Hansen, I.M.; Hendriksen, R.S.; Amaro, A. Emergence and Clonal Spread of CTX-M-65-Producing Escherichia Coli From Retail Meat in Portugal. *Front. Microbiol.* **2021**, *12*, 653595. [CrossRef] [PubMed]

36. van Dijk, H.F.G.; Verbrugh, H.A.; Abee, T.; Andriessen, J.W.; van Dijk, H.F.G.; ter Kuile, B.H.; Mevius, D.J.; Montforts, M.H.M.M.; van Schaik, W.; Schmitt, H.; et al. Resisting Disinfectants. *Commun. Med.* **2022**, *2*, 6. [CrossRef]
37. Kampf, G. Biocidal Agents Used for Disinfection Can Enhance Antibiotic Resistance in Gram-Negative Species. *Antibiotics* **2018**, *7*, 110. [CrossRef]
38. Carr, W.; Kurbatova, E.; Starks, A.; Goswami, N.; Allen, L.; Winston, C. Interim Guidance: 4-Month Rifapentine-Moxifloxacin Regimen for the Treatment of Drug-Susceptible Pulmonary Tuberculosis-United States, 2022. *MMWR Morb. Mortal. Wkly. Rep.* **2022**, *71*, 285–289. [CrossRef]
39. Rivera-Calzada, A.; Famelis, N.; Llorca, O.; Geibel, S. Type VII Secretion Systems: Structure, Functions and Transport Models. *Nat. Rev. Microbiol.* **2021**, *19*, 567–584. [CrossRef]
40. Wells, T.J.; Henderson, I.R. Chapter 16-Type 1 and 5 Secretion Systems and Associated Toxins. In *Escherichia coli*, 2nd ed.; Donnenberg, M.S., Ed.; Academic Press: Cambridge, MA, USA, 2013; pp. 499–532. ISBN 978-0-12-397048-0.
41. Johnson, M.; Zaretskaya, I.; Raytselis, Y.; Merezhuk, Y.; McGinnis, S.; Madden, T.L. NCBI BLAST: A Better Web Interface. *Nucleic Acids Res.* **2008**, *36*, W5–W9. [CrossRef]
42. Chen, L.; Yang, J.; Yu, J.; Yao, Z.; Sun, L.; Shen, Y.; Jin, Q. VFDB: A Reference Database for Bacterial Virulence Factors. *Nucleic Acids Res.* **2005**, *33*, D325–D328. [CrossRef]
43. Create Elegant Data Visualisations Using the Grammar of Graphics • Ggplot2. Available online: https://ggplot2.tidyverse.org/ (accessed on 29 April 2022).
44. R: The R Project for Statistical Computing. Available online: https://www.r-project.org/ (accessed on 29 April 2022).
45. Warnes, G.R.; Bolker, B.; Bonebakker, L.; Gentleman, R.; Huber, W.; Liaw, A.; Lumley, T.; Maechler, M.; Magnusson, A.; Moeller, S.; et al. Gplots: Various R Programming Tools for Plotting Data. 2005. Available online: https://CRAN.R-project.org/package=gplots (accessed on 29 April 2022).

Article

Genomic Analysis of Multidrug-Resistant Hypervirulent (Hypermucoviscous) *Klebsiella pneumoniae* Strain Lacking the Hypermucoviscous Regulators (*rmpA/rmpA2*)

Hisham N. Altayb [1,2,*], Hana S. Elbadawi [3], Othman Baothman [1], Imran Kazmi [1], Faisal A. Alzahrani [1,2,4], Muhammad Shahid Nadeem [1], Salman Hosawi [1,2] and Kamel Chaieb [1,5]

1 Department of Biochemistry, Faculty of Science, King Abdulaziz University, Jeddah 21589, Saudi Arabia; oabaothman@kau.edu.sa (O.B.); ikazmi@kau.edu.sa (I.K.); faisalzh@gmail.com (F.A.A.); mhalim@kau.edu.sa (M.S.N.); shosawi@kau.edu.sa (S.H.); kalshaib@kau.edu.sa (K.C.)
2 Centre for Artificial Intelligence in Precision Medicine, King Abdulaziz University, Jeddah 21589, Saudi Arabia
3 Microbiology and Parasitology Department, Soba University Hospital, University of Khartoum, Khartoum 11115, Sudan; hanasalah200@gmail.com
4 King Fahd Medical Research Center, Embryonic Stem Cells Unit, Department of Biochemistry, Faculty of Science, King Abdulaziz University, Jeddah 21589, Saudi Arabia
5 Laboratory of Analysis, Treatment and Valorization of Pollutants of the Environmental and Products, Faculty of Pharmacy, University of Monastir, Monastir 5000, Tunisia
* Correspondence: hdemmahom@kau.edu.sa; Tel.: +0096-6549087515

Citation: Altayb, H.N.; Elbadawi, H.S.; Baothman, O.; Kazmi, I.; Alzahrani, F.A.; Nadeem, M.S.; Hosawi, S.; Chaieb, K. Genomic Analysis of Multidrug-Resistant Hypervirulent (Hypermucoviscous) *Klebsiella pneumoniae* Strain Lacking the Hypermucoviscous Regulators (*rmpA/rmpA2*). *Antibiotics* **2022**, *11*, 596. https://doi.org/10.3390/antibiotics11050596

Academic Editor: Teresa V. Nogueira

Received: 21 March 2022
Accepted: 26 April 2022
Published: 28 April 2022

Publisher's Note: MDPI stays neutral with regard to jurisdictional claims in published maps and institutional affiliations.

Abstract: Hypervirulent *K. pneumoniae* (hvKP) strains possess distinct characteristics such as hypermucoviscosity, unique serotypes, and virulence factors associated with high pathogenicity. To better understand the genomic characteristics and virulence profile of the isolated hvKP strain, genomic data were compared to the genomes of the hypervirulent and typical *K. pneumoniae* strains. The *K. pneumoniae* strain was isolated from a patient with a recurrent urinary tract infection, and then the string test was used for the detection of the hypermucoviscosity phenotype. Whole-genome sequencing was conducted using Illumina, and bioinformatics analysis was performed for the prediction of the isolate resistome, virulome, and phylogenetic analysis. The isolate was identified as hypermucovisrous, type 2 (K2) capsular polysaccharide, ST14, and multidrug-resistant (MDR), showing resistance to ciprofloxacin, ceftazidime, cefotaxime, trimethoprim-sulfamethoxazole, cephalexin, and nitrofurantoin. The isolate possessed four antimicrobial resistance plasmids (*pKPN3-307_type B*, *pECW602*, *pMDR*, and *p3K157*) that carried antimicrobial resistance genes (ARGs) (*bla*$_{OXA-1}$, *bla*$_{CTX-M-15}$, *sul2*, *APH(3″)-Ib*, *APH(6)-Id*, and *AAC(6′)-Ib-cr6*). Moreover, two chromosomally mediated ARGs (*fosA6* and *SHV-28*) were identified. Virulome prediction revealed the presence of 19 fimbrial proteins, one aerobactin (*iutA*) and two salmochelin (*iroE* and *iroN*). Four secretion systems (T6SS-I (13), T6SS-II (9), T6SS-III (12), and Sci-I T6SS (1)) were identified. Interestingly, the isolate lacked the known hypermucoviscous regulators (*rmpA/rmpA2*) but showed the presence of other *RcsAB* capsule regulators (*rcsA* and *rcsB*). This study documented the presence of a rare MDR hvKP with hypermucoviscous regulators and lacking the common capsule regulators, which needs more focus to highlight their epidemiological role.

Keywords: antimicrobial resistance; hvKP; K2 capsule; ST14; fimbrial proteins; aerobactin

1. Introduction

Klebsiella pneumoniae is a Gram-negative bacterium associated with invasive hospital-acquired infections [1]. Hypervirulent *K. pneumoniae* (hvKP) overproduces a polysaccharide capsule and is an important clinical pathogen responsible for several infections in healthy and immunosuppressed patients [2,3]. The presence of capsular polysaccharides (CPS) and lipopolysaccharides (LPS) are associated with organism dissemination and virulence [4].

This pathotype with hypermucoviscosity has acquired antimicrobial resistance capable of causing serious invasive disease, unlike the old drug-susceptible strains [3]. The presence of hvKP has been linked to endophthalmitis, pneumonia, liver abscesses, and meningitis [5]. The hvKP phenotype, which contributes to the hypermucoviscous phenotype, is related to the presence of a virulence plasmid containing two capsular polysaccharide regulator genes (*rmpA* and *rmpA2*) as well as multiple siderophore gene clusters and capsular K antigens (*K1, K2, K5, K20, K54,* and *K57*) [6,7]. Most of the hvKPs belong to a small collection of clonal groups; the more dominant groups are CG23 and include ST23, 26, 57, and 1633 [8].

Capsules, siderophores, lipopolysaccharides (LPS), fimbriae, outer membrane proteins, and type 6 secretion systems (T6SS) are among the virulence components that contribute to hvKP strains [9]. Most of the hypermucoviscous and hypervirulent strains of *K. pneumoniae* are characterized by the presence of the *rmpA* and *rmpA2* (transcriptional activators, which regulate the mucoid phenotype) regulatory genes [10], but in a few cases, these strains could lack the *rmpA* and *rmpA2* regulators [8,11].

Aerobactin is considered one of the most critical virulence factors in hvKP and is used for the definition of hypermucoviscous strains such as hvKP [6]. Aerobactin-producing isolates are more likely to cause a severe immune response in the host and more invasive infections [6]. In Taiwan, hypermucoviscosity was seen in 88.8% of *K. pneumoniae* isolates from individuals with pyogenic liver abscesses [12]. A purulent liver abscess caused by a very invasive community-acquired *K. pneumoniae* has recently been reported [3]. Furthermore, an outbreak of ST11-type carbapenem-resistant hvKP was reported in a Chinese hospital in 2016 [13].

Most of the hvKPs have remained susceptible to a variety of routinely used antimicrobial agents with the exception of ampicillin, but recently MDR isolates have been increasingly reported worldwide [14–16]. Carbapenem-resistant *K. pneumoniae* strains from the clonal group (CG) 258 are the most prevalent, with ST258 and ST11 being the most common multilocus sequence types globally [17]. The acquisition of virulence plasmids by *K. pneumoniae* harboring the insertion of the drug resistance genes *bla*$_{KPC-2}$ and *catA1* has been reported [18,19]. According to Hao et al. [3] the rates of the virulence-associated genes *rmp*A, *iro*B, *fib*, and *hib* were considerably greater in hvKP than in non-hvKP. Furthermore, plasmids carrying two replicons (IncHI1B–IncFIB and IncFIIK–IncFIBK) coding for drug-resistant and virulence genes were discovered [20,21]. The presence of a wide range of β-lactamases, aminoglycoside, and carbapenem-resistant genes could result in the increasing difficulty of treatment and long hospital stays [16,22]. More recently, hvKP belonging to ST147 in COVID-19 patients has been reported in Italy with three plasmid replicons of the IncFIB (Mar), IncR, and IncHI1B types as well as different resistance genes [23]. Additionally, fourteen colistin-resistant *K. pneumoniae* (CoRKp) strains were screened retrospectively in China between 2017 and 2018 [24]. Among them, six CoRKp strains belonging to ST11 were MDR [24].

Khartoum is one of the most crowded cities in Africa [25,26] which facilitates the horizontal transfer of antimicrobial-resistant bacteria. Additionally, Sudan suffers from the inappropriate use of antibiotics; most of the antibiotics are frequently sold over the counter and even without a medical prescription [27,28]. In a recent study conducted in Khartoum state, strains positive for β-lactamase and carbapenemase genes have been reported in hvKP isolates [29]. To better understand the genomic characteristics and virulence profile of the newly isolated hvKP strain (named 9KP), this comparative genomic study was conducted.

2. Results

2.1. Patient Details and Phenotypic Characterization of the Isolate

The isolate was obtained from a patient with CKD in Soba University Hospital in Sudan, and it was identified with a hypermucoviscous phenotype using the string test, in which mucus is measured more than 9 cm by lop (Supplementary file 1, Figure S1). The isolate was classified according to CLSI breakpoints as MDR when showing resistance to ciprofloxacin, ceftazidime, cefotaxime, trimethoprim-sulfamethoxazole, cephalexin, nitrofurantoin, amoxicillin-clavulanic acid, and ampicillin, while it was susceptible to meropenem, imipenem, amikacin, and gentamicin. A high resistance level was observed for cephalosporins and penicillin, in which a no inhibition zone (0 mm) was observed for amoxicillin-clavulanic acid and ampicillin. Additionally, for the first-generation and third-generation cephalosporins, a small zone of inhibition (10 mm) was observed. Among non-β-lactams, a high resistance level was observed for trimethoprim-sulfamethoxazole (0 mm) and a small zone of inhibition (10 mm) was observed with ciprofloxacin (Table 1).

Table 1. Antimicrobial susceptibility testing of selected antimicrobial agents used against 9KP strain.

Antibiotic	Inhibition Zone (mm)	MIC (µg/mL)	Susceptibility [a]
ciprofloxacin	12	128	R
ceftazidime	10	-	R
cefotaxime	10	128	R
trimethoprim-sulfamethoxazole	No inhibition	-	R
cephalexin	10	-	R
nitrofurantoin	10	-	R
amoxicillin-clavulanic acid	No inhibition	-	R
ampicillin	No inhibition	1024	R
tetracycline	-	256	R
meropenem	32	-	S
imipenem	30	-	S
amikacin	20	-	S
gentamicin	20	4	S
chloramphenicol	-	4	S

Abbreviation: R = Resistant, S = Sensitive, - = Not tested, mm = millimeter; [a] Antimicrobial susceptibility testing determined according to CLSI guidelines [30].

For the determination of the minimum inhibitory concentrations (MIC) of the antibiotics, we used the microtitre broth dilution method, which revealed that the isolate possessed a high resistance level against ampicillin (MIC = 1024 µg/mL), tetracycline (MIC = 256 µg/mL), cefotaxime (MIC = 128 µg/mL), and ciprofloxacin (MIC = 128 µg/mL), while two antimicrobial (gentamicin and chloramphenicol) scored a very low MIC (4 µg/mL), falling within the susceptibility range according to CLSI guidelines [30] (Table 1).

2.2. Genome Characteristics and Typing

The total genome was assembled into 5364730 bp, with 83 contigs and an average contig length of 64635, while N50 was 220979, L50 7, the average coverage was 100X, and the GC content was 57.3%. The total number of predicted genes was 5248, 76 tRNA, and 202 genes associated with stress response, defense, and virulence (Supplementary file 1, Figure S2). The isolate was identified as *K. pneumoniae* with sequence type (ST) 14 by the Institut Pasteur MLST and MLST 2.0 databases. The global platform for genomic surveillance, Pathogenwatch, was used for the prediction of the capsule (K) and O serotypes; the isolate was identified with the K2 (wzi2 genotype) capsule and O1 serotype. The 9KP strain harbored ten antimicrobial resistance genes including β-lactam resistance genes (*bla*$_{OXA-1}$, *bla*$_{CTX-M-15}$, and *bla*$_{SHV-28}$), sulfonamide resistance (*sul2*), fosfomycin resistance (*fosA6*), aminoglycoside resistance (*APH(3″)-Ib*, *APH(6)-Id*, and *AAC(6′)-Ib-cr6*), and the gene causing resistance to tetracycline (*tet(A)*). The chloramphenicol O-acetyltransferase (*CatB3*) gene was detected in the 9KP strain with 70% coverage and 100% identity (Supplementary file 1, Table S1). Additionally, three efflux pumps were identified, including *K. pneumoniae KpnF, LptD*, and *oqxA*. Two chromosomal mutations conferring resistance to fosfomycin (E350Q) and elfamycin EF-Tu (R234F) were also identified. The PlasmidFinder tool revealed the presence of four plasmid replicons (Col440II, IncFII, IncFIB(K), and IncFII(K)) in the 9KP strain with 100% identity and coverage. Additionally, the use of a BLASTn search against the PLSDB database revealed the presence of four plasmids in the 9KP strain, carrying different ARGs, *p*KPN3-307_type B, *p*ECW602, *p*MDR, and *p*3K157, which showed a matching of 99.56%, 99.75%, 100%, and 100%, respectively. The *p*KPN3-307_type B plasmid of the *K. pneumoniae* strain H151440672 was identified in our strain as carrying genes corresponding to *bla*$_{CTX-M-15}$, RND efflux, and IS1 sequences (Supplementary file 1, Figures S3 and S4). The *Escherichia coli* plasmid *p*ECW602 was detected in the 9KP strain carrying different mobile elements and ARGs-encoding genes, which included sulfonamide (*sul2*) and aminoglycoside resistance genes (*APH(3″)-Ib* and *APH(6)-Id*) (Figure 1). *K. pneumoniae* *p*MDR was identified with two transposases capturing *tet(A)* MFS-family efflux-pump-encoding genes (Supplementary file 1, Figure S5). Moreover, we detected the chloramphenicol O-acetyltransferase (*CatB3*) gene, class D beta-lactamase (*bla*$_{OXA-1}$), and aminoglycoside N(6′)-acetyltransferase (*aac(6′)-Ib-cr*) genes in the 9KP plasmid (*p*3K157) (Supplementary file 1, Figure S6) while *SHV-28* and fosfomycin resistance (*fosA6*) genes were detected only in chromosomal sequences and were absent in the assembled plasmid, indicating their possible chromosomal association.

One plasmid belonging to the IncFIB(K) type was identified by a BLASTn search against PLSDB and showed 99.7% identity to the *K. pneumoniae* strain SCKP020143 plasmid *p*1_020143, and it was negative for ARGs (Supplementary file 1, Figure S7).

The virulence factor database was used for the prediction and comparison of the virulence genes of the 9KP strain with others. Different types of fimbrial proteins were discovered including type I (10), type 3 (8), and type IV pili (*pilW*) (Table 2) (Supplementary file 1, Table S2). A total of 15 iron uptake proteins were identified, including 1 aerobactin (*iutA*), 12 Ent siderophores, and 2 salmochelin, while it lacked the other aerobactin (*iucA*, *iucB*, *iucC*, and *iucD*) reported in the hvKP strains (NTUH-K2044 and KCTC 2242). The most closely related strains (kkp066 and kkp0e7) were positive for the hvKP marker, the *RmpA* gene, and lacked aerobactin (*iucA*, *iucB*, *iucC*, and *iucD*), similar to our strain. High similarity in the iron uptake system of 9KP and the other Sudanese strain (23KE) was observed, including the complete absence of genes related to yersiniabactin and the presence of two salmochelin and one aerobactin. Four secretion systems that are crucial virulence factors of pathogenic bacteria were identified in the 9KP strain, including T6SS-I (13), T6SS-II (9), T6SS-III (12), and one Sci-I T6SS exclusively detected in our strain. The isolate was positive for two *RcsAB* (*rcsA* and *rcsB*) regulatory proteins and one serum resistance LPS protein. The mediator of the hyper adherence *YidE* in enterobacteria and its conserved region were predicted in the isolate.

Table 2. Comparison of virulence factors of K. pneumoniae 9KP with other control strains (K. pneumoniae 342, MGH 78578, NTUH-K2044, 1084, HS11286, JM45, KCTC 2242, SB3432) and the most related strains (kkp066, kkp0e6, and 23KE).

Virulence Factor	Related Genes	9KP	342	MGH78578	NTUH-K2044	1084	HS11286	JM45	KCTC 2242	SB3432	kkp066	kkp0e6	kkp0e7	23KE
Adherence														
Type 3 fimbriae	8	+	+	7	+	+	+	+	+	+	+	+	7	+
Type I fimbriae	10	+	+	+	+	+	+	+	+	+	9	9	8	+
Type IV pili	12	1	-	-	-	-	-	-	-	-	-	-	-	-
Antiphagocytosis														
Capsule	1	+	+	+	+	+	+	+	+	+	+	+	+	+
Efflux pump														
AcrAB	2	+	+	+	+	+	+	+	+	+	1	+	+	+
Iron uptake														
Aerobactin	5	1	1	1	+	1	1	1	+	+	1	1	1	1
Ent siderophore	13	12	+	+	+	+	+	12	+	-	12	10	11	+
Salmochelin	5	2	2	2	+	4	2	2	2	4	2	2	2	2
Yersiniabactin	11	-	-	-	+	+	+	-	-	-	-	+	+	-
Nutritional factor														
Allantoin utilization	6	-	-	-	+	+	-	-	-	-	+	-	-	-
Regulation														
RcsAB	2	+	+	+	+	+	+	+	+	+	+	+	+	+
RmpA	1	-	-	-	+	-	-	-	+	-	1	-	1	-
Secretion system														
T6SS-I	18	13	11	11	13	13	+	+	+	10	16	15	15	12
T6SS-II	10	9	+	8	1	1	1	1	-	4	1	1	-	1
T6SS-III	18	12	+	11	14	13	14	13	14	11	10	8	5	12
Sci-I T6SS	27	1	-	-	-	-	-	-	-	-	-	-	-	-
Serum resistance														
LPS rfb locus	1	+	+	+	+	+	+	+	+	+	+	+	+	-
Toxin														
Colibactin	18	-	-	-	-	+	-	-	-	-	-	-	-	+

Key: + means the presence of the same number of genes, - means gene absent, numbers in tables indicate numbers of virulence-factors-related genes.

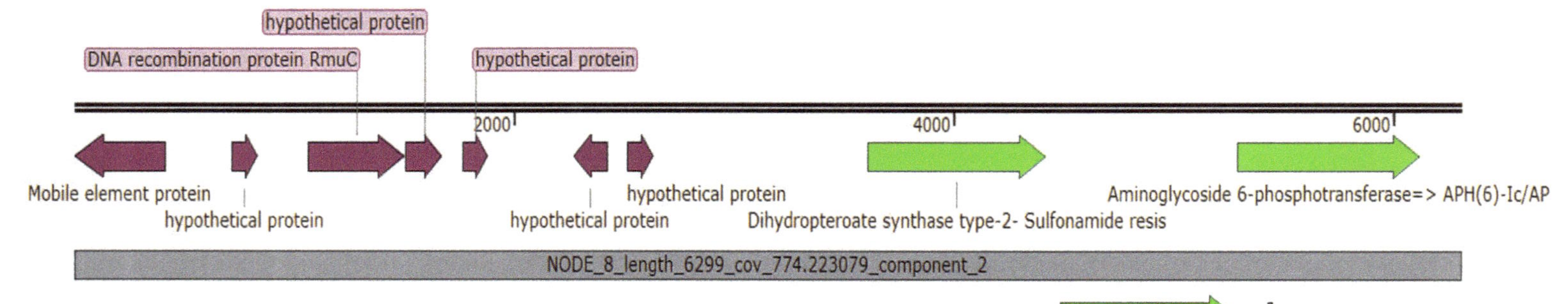

Figure 1. Linear map of *E. coli* plasmid *p*ECW602 which was detected in 9KP strain; the horizontal black lines indicate the length of the plasmid, the middle gray line contains information about plasmid length and coverage. In addition, the purple arrows indicate mobile elements and hypothetical proteins. The green arrows indicate ARGs.

2.3. Comparative Genomics and Phylogenomics Analysis

After the genome comparison, the species formed 6142 protein clusters, 3185 orthologous, and 2957 single-copy gene clusters. 9KP showed 192 single-copy genes and 4843 proteins clustered with others (Supplementary file 1, Table S3). A high degree of variability was observed at different chromosomal regions of 9KP, which contains ARGs, incF plasmid proteins, IS, and other mobile elements.

Comparative genomics revealed that the strains TCC BAA-2146, 23KE, kkp066, kkp0e6, and NTUH-K2044 exhibited a high similarity to 9KP, in which different virulent regions were similar, such as the outer membrane protein OmpN, LysR-*type* transcriptional regulators, kinase, and fimbrial proteins (Figure 2) detected at a region located between the chromosomal range 1.5–1.6 Mb. Ferric enterobactin-related proteins and phage-related proteins were clustered in *K. pneumoniae* 9KP similarly to the strains ATCC BAA-2146 and NTUH-K2044 (Figure 3), while the secretion systems T6SS were located in a region adjacent to the VgrG protein, transposases, putative kinase, mobile elements, transcriptional regulator, LysR family, and phage proteins. The PTS system in the 9KP strain was most similar to the PTS system of the 23KE strain from Sudan others (Supplementary file 1, Figure S3, and Supplementary file 2).

A phylogenetic tree was generated among the African strains by the iTOL—Interactive Tree of Life—*Klebsiella* Pasteur MLST database. The 9KP strain was clustered in a clade containing three strains from Kenya, one was isolated from a patient with a soft tissue infection (kkp066) and the others (kkp0e6 and kkp0e7) were isolated from hospital environment. And it was also clustered to one MDR Sudanese strain (K23) isolated from drinking water in Khartoum state (Figure 4).

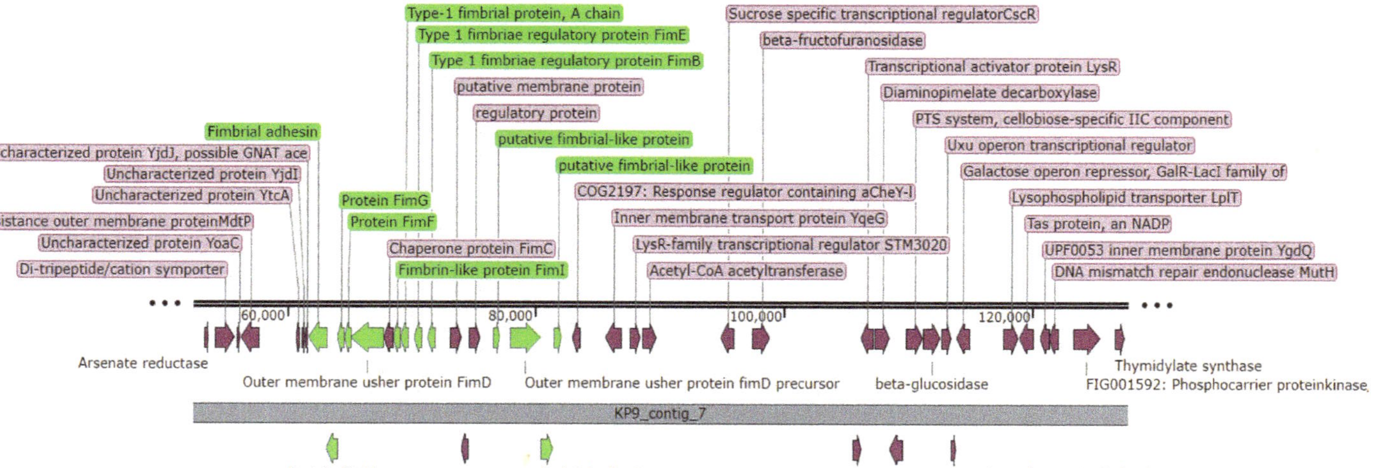

Figure 2. Clustering of fimbrial proteins in contig 7 of *K. pneumoniae* 9KP; the horizontal black lines indicate the contig length, the green arrows indicate genes encoding fimbrial proteins, and the purple arrows indicate other genes located at the same contig.

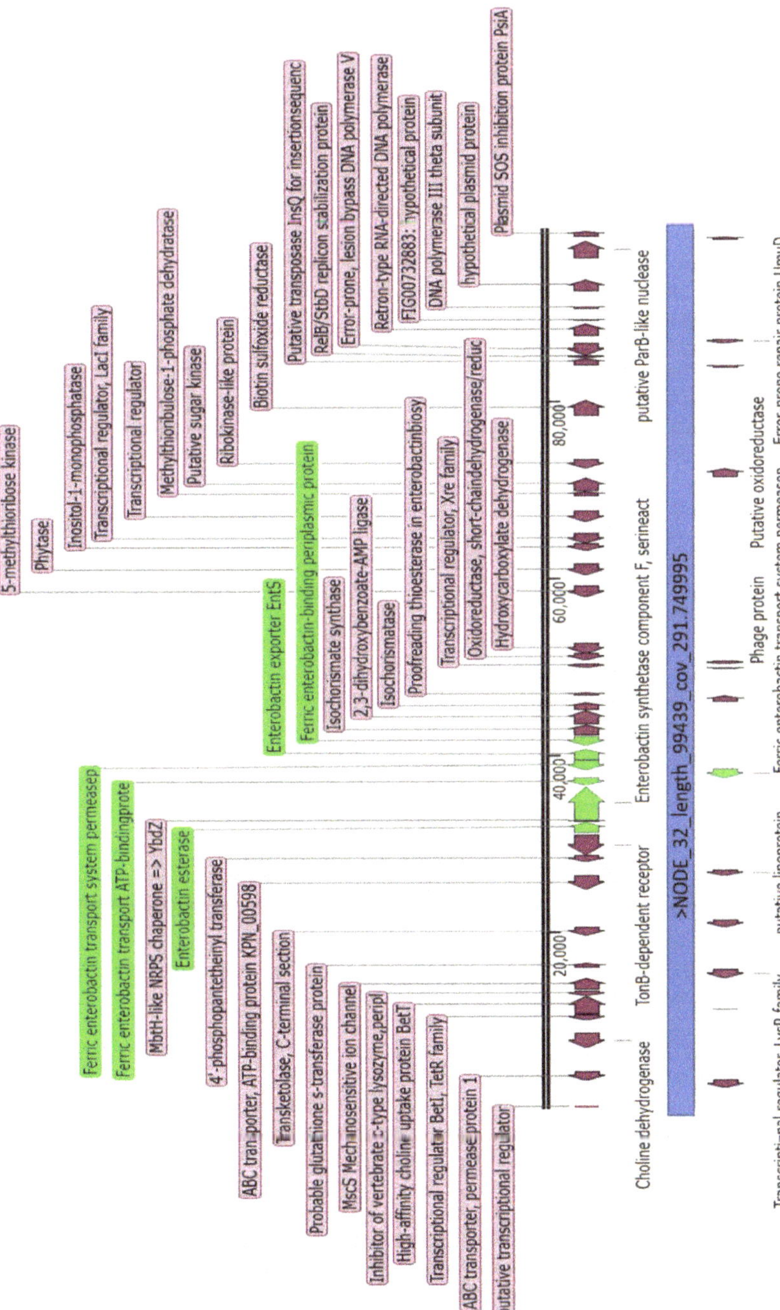

Figure 3 Clustering enterobactin proteins in contig 32 of *K. pneumoniae* 9KP; the horizontal black lines indicate the contig length, the green arrows indicate genes encoding enterobactin proteins, and the purple arrows indicate other genes located at the same contig.

The isolate lacked the common regulators of the hypermucoviscous phenotype (*rmpA/rmpA2*) [49] and yersiniabactin system but showed the presence of aerobactin-(*iutA*) and salmochelin-(*iroE* and *iroN*) encoding genes, which are clear markers for hvKP identification [50]. Additionally, the strain was predicted with the K2 capsule type and hypermucoviscosity, which are common virulence factors in hvKP [51]. Similarly, strains belonging to hvKP and lacking the *rmpA* and *rmpA2* genes were previously reported without knowledge of the mechanisms of capsule overexpression [52,53]. One possible explanation of the mucoviscosity in *K. pneumoniae* 9KP is the presence of the *RcsA* and *RcsB* genes; the *RcsA* gene binds with *RcsB* to activate the genes responsible for capsular polysaccharide production in *E. coli* [54]. Another explanation for the presence of the siderophore receptors without biosynthetic genes in hvKP is that these strains can acquire the siderophores from other bacteria found in the same environment [8]. Similar to our finding, a highly virulent and invasive *K. pneumoniae* strain possessing genes such as aerobactin (*iutA*), hypermucoviscosity, salmochelin, and lacking *rmpA/rmpA2* was reported in a patient suffering from necrotizing soft tissue infection at Northwestern Memorial Hospital, USA [51].

In this study, four T6SS systems were detected. The type VI secretion system (T6SS) is usually located at the chromosomes or pathogenicity islands of virulent bacteria, and they have a role in host infection and colonization [55]. Additionally, eight type 3 fimbrial proteins were reported. Usually, isolates that express type 3 fimbriae are more biofilm-producing compared to other strains [56]. Biofilm-producing isolates can cause community or hospital infections and are associated with 65% of microbial infections and 80% of chronic infections globally [57]. Furthermore, the genomic analysis of the *K. pneumoniae* 9KP strain demonstrated a large abundance of LysR-family transcriptional regulators in the genomic regions containing a cluster of virulence and antimicrobial resistance genes. *LysR* is found in different bacterial species and has a role in the regulation of virulence factors in pathogenic bacteria [58]. A novel type of the LysR family has been demonstrated to have a pleiotropic role in mediating the resistance and increasing the virulence of the hvKP NTUH-K2044 strain [59].

The phylogenetic analysis showed that the 9KP strain is more related to strains from Kenya and Sudan. This could be due to the fact that Kenya is a neighboring country to Sudan, and the Sudanese clustered isolate was from the same location (Khartoum) of the sample collection in our study. Two of the Kenyan strains (kkp066 and kkp0e7) were hvKPs possessing the *RmpA* gene and lacked aerobactins (*iucA, iucB, iucC* and *iucD*), similar to our strain. Additionally, the 9KP strain showed a high similarity in the PTS system to the 23KE strain from Sudan. This could be one of the reasons behind their high similarity to our strain.

MDR and hvKP strains previously developed in distinct clonal groups [60] but the recent emergence of hvKP isolates carrying MDR genes needs more attention. Such a strain has the potential to produce fatal hospital outbreaks, so more focus is needed to highlight its epidemiological role.

4. Methods

4.1. Bacterial Isolation, Identification, Susceptibility Testing, and DNA Extraction

Klebsiella spp. was isolated from the urine sample of a 40-year-old male patient with a history of recurrent UTI, hypertension, and chronic kidney disease (CKD) admitted for hemodialysis in Soba Hospital, Khartoum in July 2021. The patient was visiting the dialysis unit regularly 2 times in a week; the patient received a course of ciprofloxacin twice daily for 3 days without a response. The bacterium was isolated using a MacConkey and blood agar (HiMedia, Mumbai, India), then was identified using routine conventional microbiology methods [61] and Chromogenic UTI media (bioMérieux, Lyon, France). The isolate was identified as a hypermucoviscous strain using the string test [62]. Antimicrobial susceptibility testing was performed using the disk diffusion method to test the activity of amoxicillin-clavulanate (30 µg), cefuroxime (30 µg), ceftriaxone (30 µg), ceftazidime (30 µg), cephalexin (30 µg), meropenem (10 µg), imipenem (10 µg), amikacin (30 µg), gentamicin

(10 µg), ciprofloxacin (5 µg), trimethoprim-sulfamethoxazole (25 µg), and nitrofurantoin (300 µg). *K. pneumoniae* ATCC 700603 was used for testing the quality of the culture media, antibiotic disc, and MIC. CLSI guidelines [30] were used for the susceptibility test results interpretation. DNA was extracted using the quinidine chloride protocol [63]. The gel electrophoresis and Nanodrop, Qubit (Thermo Scientific TM, Carlsbad, CA, USA), were used for the estimation of the integrity and quantification of the extracted DNA.

4.2. Minimum Inhibitory Concentration (MIC)

The microtitre broth dilution method [64] was used to determine the minimum inhibitory concentration of ciprofloxacin, gentamicin, cefotaxime, ampicillin, chloramphenicol, and tetracycline. A two-fold serial dilution of the antibiotics was prepared in Muller–Hinton (MH) broth, and 100 µL of overnight-grown bacteria adjusted to 5–10^5 CFU/mL was poured into each well. The antibiotics concentration used was in the range of 2 to 1024 µg/mL [65]. MIC results were interpreted according to CLSI guidelines [30].

4.3. Genome Sequencing and Assembly

Whole-genome sequencing was conducted by Novogene Company (Beijing, China) using HiSeq 2500 platform (Illumina, San Diego, CA, USA). The generated short reads (2×150 bp) were assembled into contigs using a de novo assembly of Velvet v. 1.2.10 [66]; then, reads with low quality and less than 200 bp were removed. The assembled sequences were submitted to GenBank under bioproject (PRJNA767482), biosample (SAMN26332310), and accession number JAKWFM000000000, and were assigned the 9KP strain. The isolate was identified using MLST 2.0 and the Pasteur MLST. The PATRIC web server and the NCBI Prokaryotic Genome Annotation Pipeline (PGAP) [67] were used for genome annotation.

4.4. Plasmid Assembly and Identification

The plasmidSPAdes tool v3.15.4 [68] was used for the assembly of the putative plasmids sequences from the illumine short read, using different k-mer sizes (21, 33, and 55). The generated plasmids were further evaluated by the Plasmid Finder 2.1 tool using 95% identity and 60% coverage. Additionally, the generated plasmids were aligned using BLASTn against the plasmid sequences obtained from the plasmid database (PLSDB); then, a local database of the obtained plasmids was generated at OmicsBox v2.1, and a local blast search was used for the identification of the plasmids. A plasmid circular map was generated by the SnapGene Viewer 6.0.2 software.

4.5. Identification of Antimicrobial-Resistant Genes (ARGs) and Mobile Elements

To identify plasmid-mediated ARGs, the generated plasmids were submitted to the Resistance Gene Identifier (RGI) 5.2.1 and ResFinder 4.0 [69] databases; hits with $\geq$95% identity and $\geq$98% coverage were accepted. Furthermore, ResFinder 4.0 was used to detect chromosomal mutations conferring resistance to antibiotics; this tool contains a hit that can be flagged to indicate whether the hit is a plasmid or chromosomally mediated. Insertion sequences (IS) were identified by an IS Finder.

4.6. Prediction and Comparison of Virulence Genes

The virulence factors of the hvKP strain were screened using RAST 2.0 and the virulence factor database (VFDB) [70]. The capsule-type genes were identified using the Kleborate v2.2.0 [71] and Pathogenwatch database. The isolate (9KP) virulence profile was compared to a list of *K. pneumoniae* strains including the most closely related strains (23KE, kkp066, kkp0e6, and kkp0e7) and those found in the VFDB database which includes *K. pneumoniae* 342, MGH78578, NTUH-K2044, 1084, HS11286, KCTC 2242, and SB3432; among these strains, two (NTUH-K2044 and KCTC 2242) were hvKP [72]. SnapGene Viewer v.6.0.2 (GSL Biotech; available at snapgene.com, accessed on 20 March 2022) was used for the visualization of the virulence genes cassettes.

4.7. Comparative Genomics and Phylogenetic Analysis

The PATRIC v3.6.12 proteome comparison tool [73] was used to perform a protein-sequence-based genome comparison using bidirectional BLASTp. The OrthoVenn2 server [74] was used for protein orthologous clustering analysis. The most closely related genomes (23KE, kkp066, and kkp0e7) and the commonly used strains (*K. pneumoniae* BAA2146, HS11286, MGH78578, NTUH-K2044, NUHL24835, and PittNDM01) for *K. pneumoniae* genome comparison [31,75,76] were used as references. The phylogenetic tree was generated and visualized by the online Interactive Tree of Life (iTOL v6) tool available at Pasteur MLST. This tool generates neighbor-joining trees from concatenated nucleotide sequences; we considered all loci that contained allele sequence identifiers and cgMLST schemes for tree generation. The tree was generated against the most similar African strains of *K. pneumoniae* submitted to the Pasteur MLST database.

5. Conclusions

This study documented the presence of a rare MDR hvKP, *K. pneumoniae* 9KP, belonging to *K2* and ST14 with hypermucoviscous; it lacked the yersiniabactin system and the common regulators (*rmpA*/*rmpA2*) of the hypermucoviscous but showed the presence of other capsule regulators, such as *RcsAB* (*rcsA* and *rcsB*) and aerobactin (*iutA*), as well as the presence of salmochelin-(*iroE*, *iroN*) encoding genes, which are clear markers for hvKP identification.

The MIC revealed that the isolate possessed a high resistance level against ampicillin (1024 µg/mL), tetracycline (256 µg/mL), cefotaxime (128 µg/mL), and ciprofloxacin (128 µg/mL).

The isolate possessed four antimicrobial resistance plasmids (*pKPN3-307_type B*, *pECW602*, *pMDR*, and *p3K157*) that carried different ARGs and transposases, indicating their possible horizontal transfer and the clonal spread. The *pECW602* plasmid is a novel plasmid reported recently in an extensively drug-resistant (XDR) *E. coli* isolate in China [41]; here, for the first time, we reported it in a *K. pneumoniae* (9KP) isolate with high identity (99.75%).

Supplementary Materials: The following are available online at https://www.mdpi.com/article/10.3390/antibiotics11050596/s1, Supplementary file 1: contains Tables S1–S3, representing ARGs (Table S1), virulence factors (Table S2), and numbers of the clustered and singletons proteins in the 9KP strain compared to others (Table S3). Additionally, it contains figures from Figures S1–S7 representing the string test photograph (Figure S1), a pie chart of the annotated subsystem and genes of *K. pneumoniae* 9KP (Figure S2), a circular map of the whole-genome comparison of the 9KP strain to different *K. pneumoniae* strains (Figure S3), and a map of the *K. pneumoniae* strain 9PK plasmids (Figures S4–S7)). Supplementary file 2: contains the complete data of the whole-genome comparison of the 9KP strain to different *K. pneumoniae* strains.

Author Contributions: H.N.A.: conceptualization, supervision, bioinformatics analysis, writing—review and editing, and funding acquisition. H.S.E.: data acquisition, carried out the microbiological analysis, writing—review and editing. O.B. and I.K.: methodology, software, data curation, formal analysis, writing—review and editing, F.A.A.: methodology, software, data curation, writing—review and editing. M.S.N.: methodology, analysis, writing—review and editing. S.H.: investigations, resources, writing—review and editing. K.C.: validation, supervision, visualization, writing—review and editing. All authors have read and agreed to the published version of the manuscript.

Funding: The authors extend their appreciation to the Deputyship for Research & Innovation; Ministry of Education in Saudi Arabia, for funding this research work through the project number IFPRC-072-130-2020; and King Abdulaziz University DSR, Jeddah, Saudi Arabia.

Institutional Review Board Statement: This study was approved by the Ethics Committee of the Khartoum State Ministry of Health (REF: 2/2021).

Informed Consent Statement: Not applicable because we were collecting sample remnants without the patient's identifiable information.

Data Availability Statement: The data for this project was submitted to GenBank under the Bioproject PRJNA767482 and in the additional files.

Acknowledgments: The authors extend their appreciation to the Deputyship for Research & Innovation; Ministry of Education in Saudi Arabia, for funding this research work through the project number IFPRC-072-130-2020; and King Abdulaziz University DSR, Jeddah, Saudi Arabia.

Conflicts of Interest: The authors declare no conflict of interest.

References

1. Meatherall, B.L.; Gregson, D.; Ross, T.; Pitout, J.D.; Laupland, K.B. Incidence, risk factors, and outcomes of Klebsiella pneumoniae bacteremia. *Am. J. Med.* **2009**, *122*, 866–873. [CrossRef] [PubMed]
2. Li, P.; Liang, Q.; Liu, W.; Zheng, B.; Liu, L.; Wang, W.; Xu, Z.; Huang, M.; Feng, Y. Convergence of carbapenem resistance and hypervirulence in a highly-transmissible ST11 clone of K. pneumoniae: An epidemiological, genomic and functional study. *Virulence* **2021**, *12*, 377–388. [CrossRef] [PubMed]
3. Hao, Z.; Duan, J.; Liu, L.; Shen, X.; Yu, J.; Guo, Y.; Wang, L.; Yu, F. Prevalence of community-acquired, hypervirulent Klebsiella pneumoniae isolates in Wenzhou, China. *Microb. Drug Resist.* **2020**, *26*, 21–27. [CrossRef] [PubMed]
4. Hsieh, P.-F.; Lin, T.-L.; Yang, F.-L.; Wu, M.-C.; Pan, Y.-J.; Wu, S.-H.; Wang, J.-T. Lipopolysaccharide O1 antigen contributes to the virulence in Klebsiella pneumoniae causing pyogenic liver abscess. *PLoS ONE* **2012**, *7*, e33155.
5. Shon, A.S.; Bajwa, R.P.; Russo, T.A. Hypervirulent (hypermucoviscous) Klebsiella pneumoniae: A new and dangerous breed. *Virulence* **2013**, *4*, 107–118. [CrossRef] [PubMed]
6. Liu, C.; Guo, J. Hypervirulent Klebsiella pneumoniae (hypermucoviscous and aerobactin positive) infection over 6 years in the elderly in China: Antimicrobial resistance patterns, molecular epidemiology and risk factor. *Ann. Clin. Microbiol. Antimicrob.* **2019**, *18*, 1–11. [CrossRef]
7. Alcántar-Curiel, M.D.; Girón, J.A. Klebsiella Pneumoniae and the Pyogenic Liver Abscess: Implications and Association of the Presence of rpmA Genes and Expression of Hypermucoviscosity. *Virulence* **2015**, *6*, 407–409. [CrossRef]
8. Choby, J.; Howard-Anderson, J.; Weiss, D. Hypervirulent Klebsiella pneumoniae—Clinical and molecular perspectives. *J. Intern. Med.* **2020**, *287*, 283–300. [CrossRef]
9. Lan, Y.; Zhou, M.; Li, X.; Liu, X.; Li, J.; Liu, W. Preliminary Investigation of Iron Acquisition in Hypervirulent Klebsiella pneumoniae Mediated by Outer Membrane Vesicles. *Infect. Drug Resist.* **2022**, *15*, 311. [CrossRef]
10. Shankar, C.; Basu, S.; Lal, B.; Shanmugam, S.; Vasudevan, K.; Mathur, P.; Ramaiah, S.; Anbarasu, A.; Veeraraghavan, B. Aerobactin Seems To Be a Promising Marker Compared With Unstable RmpA2 for the Identification of Hypervirulent Carbapenem-Resistant Klebsiella pneumoniae: In Silico and In Vitro Evidence. *Front. Cell. Infect. Microbiol.* **2021**, *11*, 709681. [CrossRef]
11. Yu, W.-L.; Lee, M.-F.; Tang, H.-J.; Chang, M.-C.; Chuang, Y.-C. Low prevalence of rmpA and high tendency of rmpA mutation correspond to low virulence of extended spectrum β-lactamase-producing Klebsiella pneumoniae isolates. *Virulence* **2015**, *6*, 162–172. [CrossRef] [PubMed]
12. Ku, Y.-H.; Chuang, Y.-C.; Yu, W.-L. Clinical spectrum and molecular characteristics of Klebsiella pneumoniae causing community-acquired extrahepatic abscess. *J. Microbiol. Immunol. Infect.* **2008**, *41*, 311–317. [PubMed]
13. Gu, D.; Dong, N.; Zheng, Z.; Lin, D.; Huang, M.; Wang, L.; Chan, E.W.-C.; Shu, L.; Yu, J.; Zhang, R. A fatal outbreak of ST11 carbapenem-resistant hypervirulent Klebsiella pneumoniae in a Chinese hospital: A molecular epidemiological study. *Lancet Infect. Dis.* **2018**, *18*, 37–46. [CrossRef]
14. Tang, M.; Kong, X.; Hao, J.; Liu, J. Epidemiological characteristics and formation mechanisms of multidrug-resistant hypervirulent Klebsiella pneumoniae. *Front. Microbiol.* **2020**, *11*, 2774. [CrossRef] [PubMed]
15. Shankar, C.; Santhanam, S.; Kumar, M.; Gupta, V.; Devanga Ragupathi, N.K.; Veeraraghavan, B. Draft genome sequence of an extended-spectrum-β-lactamase-positive hypervirulent Klebsiella pneumoniae strain with novel sequence type 2318 isolated from a neonate. *Genome Announc.* **2016**, *4*, e01273-16. [CrossRef]
16. Hao, M.; Shi, X.; Lv, J.; Niu, S.; Cheng, S.; Du, H.; Yu, F.; Tang, Y.-W.; Kreiswirth, B.N.; Zhang, H. In vitro activity of apramycin against carbapenem-resistant and hypervirulent Klebsiella pneumoniae isolates. *Front. Microbiol.* **2020**, *11*, 425. [CrossRef]
17. Chen, L.; Mathema, B.; Chavda, K.D.; DeLeo, F.R.; Bonomo, R.A.; Kreiswirth, B.N. Carbapenemase-producing Klebsiella pneumoniae: Molecular and genetic decoding. *Trends Microbiol.* **2014**, *22*, 686–696. [CrossRef]
18. Chen, Y.; Marimuthu, K.; Teo, J.; Venkatachalam, I.; Cherng, B.P.Z.; De Wang, L.; Prakki, S.R.S.; Xu, W.; Tan, Y.H.; Nguyen, L.C. Acquisition of plasmid with carbapenem-resistance gene blaKPC2 in hypervirulent Klebsiella pneumoniae, Singapore. *Emerg. Infect. Dis.* **2020**, *26*, 549. [CrossRef]
19. Shankar, C.; Jacob, J.J.; Vasudevan, K.; Biswas, R.; Manesh, A.; Sethuvel, D.P.M.; Varughese, S.; Biswas, I.; Veeraraghavan, B. Emergence of multidrug resistant hypervirulent ST23 Klebsiella pneumoniae: Multidrug resistant plasmid acquisition drives evolution. *Front. Cell. Infect. Microbiol.* **2020**, *10*, 575289. [CrossRef]
20. Lam, M.M.; Wyres, K.L.; Wick, R.R.; Judd, L.M.; Fostervold, A.; Holt, K.E.; Löhr, I.H. Convergence of virulence and MDR in a single plasmid vector in MDR Klebsiella pneumoniae ST15. *J. Antimicrob. Chemother.* **2019**, *74*, 1218–1222. [CrossRef]
21. Turton, J.; Davies, F.; Turton, J.; Perry, C.; Payne, Z.; Pike, R. Hybrid resistance and virulence plasmids in "high-risk" clones of Klebsiella pneumoniae, including those carrying blaNDM-5. *Microorganisms* **2019**, *7*, 326. [CrossRef] [PubMed]

22. Navon-Venezia, S.; Kondratyeva, K.; Carattoli, A. Klebsiella pneumoniae: A major worldwide source and shuttle for antibiotic resistance. *FEMS Microbiol. Rev.* **2017**, *41*, 252–275. [CrossRef] [PubMed]

23. Falcone, M.; Tiseo, G.; Arcari, G.; Leonildi, A.; Giordano, C.; Tempini, S.; Bibbolino, G.; Mozzo, R.; Barnini, S.; Carattoli, A.; et al. Spread of hypervirulent multidrug-resistant ST147 Klebsiella pneumoniae in patients with severe COVID-19: An observational study from Italy, 2020–2021. *J. Antimicrob. Chemother.* **2022**, *77*, 1140–1145. [CrossRef] [PubMed]

24. Liu, X.; Wu, Y.; Zhu, Y.; Jia, P.; Li, X.; Jia, X.; Yu, W.; Cui, Y.; Yang, R.; Xia, W. Emergence of colistin-resistant hypervirulent Klebsiella pneumoniae (CoR-HvKp) in China. *Emerg. Microbes Infect.* **2022**, *11*, 648–661. [CrossRef] [PubMed]

25. Elhadary, Y.; Ali, S. A new trend in urban housing: Gated communities in Khartoum, Sudan. *Am. J. Sociol. Res.* **2017**, *7*, 45–55.

26. Gatari, M. Air pollution over East Africa. In Proceedings of the Oral Presentation, First International Workshop on Climate Variability over Africa, Alexandria, Egypt, 15–26 May 2005.

27. Alamin, A.S.A.; Kheder, S.I. Knowledge, Attitudes and Practices of Prescribers towards Antimicrobial Stewardship at Hospitals in Khartoum State—Sudan. *J. Med. Inform. Decis. Mak.* **2020**, *1*, 12–25. [CrossRef]

28. Musa, M.M.Y.A. *Medicine Prices, Availability and Affordability in Sudan*; Partial fulfillment of MSc of Health Economics and Health Care Management; Chulalongkorn University: Bangkok, Thailand, 2013.

29. Albasha, A.M.; Abd-Alhalim, S.; Alshaib, E.F.; Al-Hassan, L.; Altayb, H.N. Detection of several carbapenems resistant and virulence genes in classical and hyper-virulent strains of Klebsiella pneumoniae isolated from hospitalized neonates and adults in Khartoum. *BMC Res. Notes* **2020**, *13*, 312. [CrossRef]

30. Clinical and Laboratory Standards Institute. *Performance Standards for Antimicrobial Susceptibility Testing: Thirtieth Informational Supplement*; Document M100-S130 CLSI; Clinical and Laboratory Standards Institute: Wayne, PA, USA, 2020.

31. Du, L.; Zhang, J.; Liu, P.; Li, X.; Su, K.; Yuan, L.; Zhang, Z.; Peng, D.; Li, Y.; Qiu, J. Genome sequencing and comparative genome analysis of 6 hypervirulent Klebsiella pneumoniae strains isolated in China. *Arch. Microbiol.* **2021**, *203*, 3125–3133. [CrossRef]

32. Bouza, E.; Cercenado, E. Klebsiella and enterobacter: Antibiotic resistance and treatment implications. *Semin. Respir. Infect.* **2002**, *17*, 215–230. [CrossRef]

33. Soge, O.O.; Queenan, A.M.; Ojo, K.K.; Adeniyi, B.A.; Roberts, M.C. CTX-M-15 extended-spectrum β-lactamase from Nigerian Klebsiella pneumoniae. *J. Antimicrob. Chemother.* **2006**, *57*, 24–30. [CrossRef]

34. Huang, Q.-S.; Liao, W.; Xiong, Z.; Li, D.; Du, F.-L.; Xiang, T.-X.; Wei, D.; Wan, L.-G.; Liu, Y.; Zhang, W. Prevalence of the NTEKPC-I on IncF plasmids among Hypervirulent Klebsiella pneumoniae isolates in Jiangxi Province, South China. *Front. Microbiol.* **2021**, *12*, 622280. [CrossRef] [PubMed]

35. Du, P.; Liu, C.; Fan, S.; Baker, S.; Guo, J. The Role of Plasmid and Resistance Gene Acquisition in the Emergence of ST23 Multi-Drug Resistant, Hypervirulent Klebsiella pneumoniae. *Microbiol. Spectr.* **2022**, *21*, e0192921. [CrossRef] [PubMed]

36. Marr, C.M.; Russo, T.A. Hypervirulent Klebsiella pneumoniae: A new public health threat. *Expert Rev. Anti-Infect. Ther.* **2019**, *17*, 71–73. [CrossRef] [PubMed]

37. Ludden, C.; Moradigaravand, D.; Jamrozy, D.; Gouliouris, T.; Blane, B.; Naydenova, P.; Hernandez-Garcia, J.; Wood, P.; Hadjirin, N.; Radakovic, M. A One Health study of the genetic relatedness of Klebsiella pneumoniae and their mobile elements in the East of England. *Clin. Infect. Dis.* **2020**, *70*, 219–226. [CrossRef]

38. Saeed Mohammed, S.A.; Musa, A.; Ahmed Mohammed, A.; Mohammed, H. CTX-M B-lactamase–producing *Escherichia coli* in sudan tertiary hospitals: Detection genotypes variants and bioinformatics analysis. *Int. J. Med. Biomed. Stud.* **2019**, *3*, 146–157.

39. Altayb, H.N.; Salih, E.K.; Moglad, E.H. Molecular detection of beta-lactamase blaCTX-M group 1 in Escherichia coli isolated from drinking water in Khartoum State. *J. Water Health* **2020**, *18*, 1091–1097. [CrossRef] [PubMed]

40. Sütterlin, S. *Aspects of Bacterial Resistance to Silver*; Acta Universitatis Upsaliensis: Uppsala, Switzerland, 2015.

41. Wang, M.; Wang, W.; Niu, Y.; Liu, T.; Li, L.; Zhang, M.; Li, Z.; Su, W.; Liu, F.; Zhang, X. A clinical extensively-drug resistant (XDR) Escherichia coli and role of Its β-Lactamase Genes. *Front. Microbiol.* **2020**, *11*, 590357. [CrossRef]

42. Xu, J.; Zhu, Z.; Chen, Y.; Wang, W.; He, F. The plasmid-borne tet (A) gene is an important factor causing tigecycline resistance in ST11 carbapenem-resistant Klebsiella pneumoniae under selective pressure. *Front. Microbiol.* **2021**, *12*, 328. [CrossRef]

43. Foong, W.E.; Wilhelm, J.; Tam, H.-K.; Pos, K.M. Tigecycline efflux in Acinetobacter baumannii is mediated by TetA in synergy with RND-type efflux transporters. *J. Antimicrob. Chemother.* **2020**, *75*, 1135–1139. [CrossRef]

44. Enany, S.; Zakeer, S.; Diab, A.A.; Bakry, U.; Sayed, A.A. Whole genome sequencing of Klebsiella pneumoniae clinical isolates sequence type 627 isolated from Egyptian patients. *PLoS ONE* **2022**, *17*, e0265884. [CrossRef]

45. Guo, Q.; Tomich, A.D.; McElheny, C.L.; Cooper, V.S.; Stoesser, N.; Wang, M.; Sluis-Cremer, N.; Doi, Y. Glutathione-S-transferase FosA6 of Klebsiella pneumoniae origin conferring fosfomycin resistance in ESBL-producing Escherichia coli. *J. Antimicrob. Chemother.* **2016**, *71*, 2460–2465. [CrossRef]

46. Kieffer, N.; Poirel, L.; Mueller, L.; Mancini, S.; Nordmann, P. IS Ecp1-mediated transposition leads to fosfomycin and broad-spectrum cephalosporin resistance in Klebsiella pneumoniae. *Antimicrob. Agents Chemother.* **2020**, *64*, e00150-20. [CrossRef] [PubMed]

47. de Man, T.J.; Lutgring, J.D.; Lonsway, D.R.; Anderson, K.F.; Kiehlbauch, J.A.; Chen, L.; Walters, M.S.; Sjölund-Karlsson, M.; Rasheed, J.K.; Kallen, A. Genomic analysis of a pan-resistant isolate of Klebsiella pneumoniae, United States 2016. *mBio* **2018**, *9*, e00440-18. [CrossRef]

48. Liakopoulos, A.; Mevius, D.; Ceccarelli, D. A review of SHV extended-spectrum β-lactamases: Neglected yet ubiquitous. *Front. Microbiol.* **2016**, *7*, 1374. [CrossRef] [PubMed]

49. Piazza, A.; Perini, M.; Mauri, C.; Comandatore, F.; Meroni, E.; Luzzaro, F.; Principe, L. Antimicrobial Susceptibility, Virulence, and Genomic Features of a Hypervirulent Serotype K2, ST65 Klebsiella pneumoniae Causing Meningitis in Italy. *Antibiotics* **2022**, *11*, 261. [CrossRef]
50. Li, Q.; Zhu, J.; Kang, J.; Song, Y.; Yin, D.; Guo, Q.; Song, J.; Zhang, Y.; Wang, S.; Duan, J. Emergence of NDM-5-producing carbapenem-resistant Klebsiella pneumoniae and SIM-producing hypervirulent Klebsiella pneumoniae Isolated from aseptic body fluid in a large tertiary hospital, 2017–2018: Genetic traits of blaNDM-like and blaSIM-like genes as determined by NGS. *Infect. Drug Resist.* **2020**, *13*, 3075. [PubMed]
51. Krapp, F.; Morris, A.R.; Ozer, E.A.; Hauser, A.R. Virulence characteristics of carbapenem-resistant Klebsiella pneumoniae strains from patients with necrotizing skin and soft tissue infections. *Sci. Rep.* **2017**, *7*, 13533. [CrossRef]
52. Cubero, M.; Grau, I.; Tubau, F.; Pallarés, R.; Dominguez, M.; Linares, J.; Ardanuy, C. Hypervirulent Klebsiella pneumoniae clones causing bacteraemia in adults in a teaching hospital in Barcelona, Spain (2007–2013). *Clin. Microbiol. Infect.* **2016**, *22*, 154–160. [CrossRef]
53. Fang, C.-T.; Chuang, Y.-P.; Shun, C.-T.; Chang, S.-C.; Wang, J.-T. A novel virulence gene in Klebsiella pneumoniae strains causing primary liver abscess and septic metastatic complications. *J. Exp. Med.* **2004**, *199*, 697–705. [CrossRef]
54. Majdalani, N.; Gottesman, S. The Rcs phosphorelay: A complex signal transduction system. *Annu. Rev. Microbiol.* **2005**, *59*, 379–405. [CrossRef]
55. Wan, B.; Zhang, Q.; Ni, J.; Li, S.; Wen, D.; Li, J.; Xiao, H.; He, P.; Ou, H.-y.; Tao, J. Type VI secretion system contributes to Enterohemorrhagic Escherichia coli virulence by secreting catalase against host reactive oxygen species (ROS). *PLoS Pathog.* **2017**, *13*, e1006246. [CrossRef] [PubMed]
56. Schroll, C.; Barken, K.B.; Krogfelt, K.A.; Struve, C. Role of type 1 and type 3 fimbriae in Klebsiella pneumoniae biofilm formation. *BMC Microbiol.* **2010**, *10*, 179. [CrossRef] [PubMed]
57. Jamal, M.; Ahmad, W.; Andleeb, S.; Jalil, F.; Imran, M.; Nawaz, M.A.; Hussain, T.; Ali, M.; Rafiq, M.; Kamil, M.A. Bacterial biofilm and associated infections. *J. Chin. Med. Assoc.* **2018**, *81*, 7–11. [CrossRef]
58. Islam, M.M.; Kim, K.; Lee, J.C.; Shin, M. LeuO, a LysR-Type Transcriptional Regulator, Is Involved in Biofilm Formation and Virulence of Acinetobacter baumannii. *Front. Cell. Infect. Microbiol.* **2021**, *11*, 738706. [CrossRef] [PubMed]
59. Srinivasan, V.B.; Mondal, A.; Venkataramaiah, M.; Chauhan, N.K.; Rajamohan, G. Role of oxyRKP, a novel LysR-family transcriptional regulator, in antimicrobial resistance and virulence in Klebsiella pneumoniae. *Microbiology* **2013**, *159*, 1301–1314. [CrossRef]
60. Surgers, L.; Boyd, A.; Girard, P.-M.; Arlet, G.; Decré, D. ESBL-producing strain of hypervirulent Klebsiella pneumoniae K2, France. *Emerg. Infect. Dis.* **2016**, *22*, 1687. [CrossRef] [PubMed]
61. Osman, E.A.; El-Amin, N.; Adrees, E.A.; Al-Hassan, L.; Mukhtar, M. Comparing conventional, biochemical and genotypic methods for accurate identification of Klebsiella pneumoniae in Sudan. *Access Microbiol.* **2020**, *2*, acmi000096. [CrossRef] [PubMed]
62. Kumabe, A.; Kenzaka, T. String test of hypervirulent Klebsiella pneumonia. *QJM Int. J. Med.* **2014**, *107*, 1053. [CrossRef]
63. Sabeel, S.; Salih, M.A.; Ali, M.; El-Zaki, S.-E.; Abuzeid, N.; Elgadi, Z.A.M.; Altayb, H.N.; Elegail, A.; Ibrahim, N.Y.; Elamin, B.K. Phenotypic and genotypic analysis of multidrug-resistant Mycobacterium tuberculosis isolates from Sudanese patients. *Tuberc. Res. Treat.* **2017**, *2017*, 8340746.
64. Mogana, R.; Adhikari, A.; Tzar, M.; Ramliza, R.; Wiart, C. Antibacterial activities of the extracts, fractions and isolated compounds from Canarium patentinervium Miq. against bacterial clinical isolates. *BMC Complement. Med. Ther.* **2020**, *20*, 55. [CrossRef]
65. Kouidhi, B.; Zmantar, T.; Jrah, H.; Souiden, Y.; Chaieb, K.; Mahdouani, K.; Bakhrouf, A. Antibacterial and resistance-modifying activities of thymoquinone against oral pathogens. *Ann. Clin. Microbiol. Antimicrob.* **2011**, *10*, 29. [CrossRef] [PubMed]
66. Zerbino, D.R.; Birney, E. Velvet: Algorithms for de novo short read assembly using de Bruijn graphs. *Genome Res.* **2008**, *18*, 821–829. [CrossRef] [PubMed]
67. Tatusova, T.; DiCuccio, M.; Badretdin, A.; Chetvernin, V.; Nawrocki, E.P.; Zaslavsky, L.; Lomsadze, A.; Pruitt, K.D.; Borodovsky, M.; Ostell, J. NCBI prokaryotic genome annotation pipeline. *Nucleic Acids Res.* **2016**, *44*, 6614–6624. [CrossRef] [PubMed]
68. Antipov, D.; Hartwick, N.; Shen, M.; Raiko, M.; Lapidus, A.; Pevzner, P.A. plasmidSPAdes: Assembling plasmids from whole genome sequencing data. *Bioinformatics* **2016**, *32*, 3380–3387. [CrossRef]
69. Bortolaia, V.; Kaas, R.S.; Ruppe, E.; Roberts, M.C.; Schwarz, S.; Cattoir, V.; Philippon, A.; Allesoe, R.L.; Rebelo, A.R.; Florensa, A.F. ResFinder 4.0 for predictions of phenotypes from genotypes. *J. Antimicrob. Chemother.* **2020**, *75*, 3491–3500. [CrossRef]
70. Liu, B.; Zheng, D.; Jin, Q.; Chen, L.; Yang, J. VFDB 2019: A comparative pathogenomic platform with an interactive web interface. *Nucleic Acids Res.* **2019**, *47*, D687–D692. [CrossRef]
71. Lam, M.; Wick, R.R.; Watts, S.C.; Cerdeira, L.T.; Wyres, K.L.; Holt, K.E. A genomic surveillance framework and genotyping tool for Klebsiella pneumoniae and its related species complex. *Nat. Commun.* **2021**, *12*, 4188. [CrossRef]
72. Struve, C.; Roe, C.C.; Stegger, M.; Stahlhut, S.G.; Hansen, D.S.; Engelthaler, D.M.; Andersen, P.S.; Driebe, E.M.; Keim, P.; Krogfelt, K.A. Mapping the evolution of hypervirulent Klebsiella pneumoniae. *mBio* **2015**, *6*, e00630-15. [CrossRef]
73. Davis, J.J.; Wattam, A.R.; Aziz, R.K.; Brettin, T.; Butler, R.; Butler, R.M.; Chlenski, P.; Conrad, N.; Dickerman, A.; Dietrich, E.M. The PATRIC Bioinformatics Resource Center: Expanding data and analysis capabilities. *Nucleic Acids Res.* **2020**, *48*, D606–D612. [CrossRef]

74. Xu, L.; Dong, Z.; Fang, L.; Luo, Y.; Wei, Z.; Guo, H.; Zhang, G.; Gu, Y.Q.; Coleman-Derr, D.; Xia, Q. OrthoVenn2: A web server for whole-genome comparison and annotation of orthologous clusters across multiple species. *Nucleic Acids Res.* **2019**, *47*, W52–W58. [CrossRef]

75. Kwon, T.; Jung, Y.-H.; Lee, S.; Yun, M.-r.; Kim, W.; Kim, D.-W. Comparative genomic analysis of Klebsiella pneumoniae subsp. pneumoniae KP617 and PittNDM01, NUHL24835, and ATCC BAA-2146 reveals unique evolutionary history of this strain. *Gut Pathog.* **2016**, *8*, 34. [CrossRef] [PubMed]

76. Ramos, P.I.P.; Picão, R.C.; de Almeida, L.G.P.; Lima, N.C.B.; Girardello, R.; Vivan, A.C.P.; Xavier, D.E.; Barcellos, F.G.; Pelisson, M.; Vespero, E.C. Comparative analysis of the complete genome of KPC-2-producing Klebsiella pneumoniae Kp13 reveals remarkable genome plasticity and a wide repertoire of virulence and resistance mechanisms. *BMC Genom.* **2014**, *15*, 54. [CrossRef] [PubMed]

Review

Genetic Determinants of Tigecycline Resistance in *Mycobacteroides abscessus*

Hien Fuh Ng and Yun Fong Ngeow *

Centre for Research on Communicable Diseases, Faculty of Medicine and Health Sciences, Universiti Tunku Abdul Rahman, Kajang 43000, Malaysia; hfng@utar.edu.my
* Correspondence: ngeowyf@utar.edu.my; Tel.: +60-3-9086-0288 (ext. 158)

Abstract: *Mycobacteroides abscessus* (formerly *Mycobacterium abscessus*) is a clinically important, rapid-growing non-tuberculous mycobacterium notoriously known for its multidrug-resistance phenotype. The intrinsic resistance of *M. abscessus* towards first- and second-generation tetracyclines is mainly due to the over-expression of a tetracycline-degrading enzyme known as MabTetX (*MAB_1496c*). Tigecycline, a third-generation tetracycline, is a poor substrate for the MabTetX and does not induce the expression of this enzyme. Although tigecycline-resistant strains of *M. abscessus* have been documented in different parts of the world, their resistance determinants remain largely elusive. Recent work on tigecycline resistance or reduced susceptibility in *M. abscessus* revealed the involvement of the gene *MAB_3508c* which encodes the transcriptional activator WhiB7, as well as mutations in the *sigH-rshA* genes which control heat shock and oxidative-stress responses. The deletion of *whiB7* has been observed to cause a 4-fold decrease in the minimum inhibitory concentration of tigecycline. In the absence of environmental stress, the SigH sigma factor (*MAB_3543c*) interacts with and is inhibited by the anti-sigma factor RshA (*MAB_3542c*). The disruption of the SigH-RshA interaction resulting from mutations and the subsequent up-regulation of SigH have been hypothesized to lead to tigecycline resistance in *M. abscessus*. In this review, the evidence for different genetic determinants reported to be linked to tigecycline resistance in *M. abscessus* was examined and discussed.

Keywords: *Mycobacteroides abscessus*; tigecycline; resistance; genetic determinants; WhiB7; SigH; RshA

Citation: Ng, H.F.; Ngeow, Y.F. Genetic Determinants of Tigecycline Resistance in *Mycobacteroides abscessus*. *Antibiotics* **2022**, *11*, 572. https://doi.org/10.3390/antibiotics11050572

Academic Editor: Teresa V. Nogueira

Received: 27 February 2022
Accepted: 18 April 2022
Published: 25 April 2022

Publisher's Note: MDPI stays neutral with regard to jurisdictional claims in published maps and institutional affiliations.

1. Introduction

1.1. Tigecycline

Tigecycline is the first and only clinically available glycylcycline (a new class of tetracycline). It is a minocycline derivative, with an N,N-dimethyglycylamido moiety attached to the 9′ carbon on the tetracycline four-ringed skeleton [1]. Like other tetracyclines, tigecycline is a bacteriostatic antibiotic which inhibits translation by binding to the A site of the 30S ribosomal subunit (made up of the 16S rRNA and ribosomal proteins) [2]. The protein-synthesis inhibitory activity of tigecycline is 3- and 20-fold more potent than that of minocycline and tetracycline, respectively [3]. The ability of tigecycline to escape two common mechanisms of tetracycline resistance, active efflux and ribosomal protection [2], is attributed to its bulky side chain [4]. Furthermore, a molecular modelling study demonstrated that tigecycline has additional interaction with H34 and H18 nucleotides of ribosomes, in comparison to tetracycline and minocycline [3]. These characteristics are believed to help tigecycline to bind in a different orientation and with greater affinity than tetracycline [5], thus preventing recognition by ribosomal protection proteins and Tet efflux transporters [6,7].

Tigecycline is a broad-spectrum antibiotic. It is also active against important drug-resistant pathogens, such as methicillin-resistant *Staphylococcus aureus*, penicillin-resistant *Streptococcus pneumoniae*, vancomycin-resistant enterococci, and extended-spectrum beta-lactamase producers [2]. Furthermore, tigecycline is one of the rescue antibiotics, alongside colistin, to treat infections caused by pathogens expressing the New Delhi metallo-beta-lactamase-1 (a carbapenemase) that confers resistance to multiple antibiotics [8]. Fast-growing non-tuberculous mycobacteria are highly tigecycline-susceptible [9]. Specifically, this antibiotic has shown good in vitro and in vivo activities against *Mycobacteroides abscessus* complex (formerly known as *Mycobacterium abscessus* complex) [10,11]. On the other hand, slow-growing non-tuberculous mycobacteria and *Mycobacterium tuberculosis* complex are largely resistant to tigecycline [9,12].

1.2. The M. abscessus Complex

M. abscessus complex is a species complex, consisting of *M. abscessus* subspecies *abscessus*, *M. abscessus* subspecies *massiliense* and *M. abscessus* subspecies *bolletii* (hereafter referred to as *M. abscessus*, *M. massiliense* and *M. bolletii*, respectively), that causes a wide spectrum of infections in humans, including but not limited to pulmonary and soft-tissue infections, and disseminated infections [13]. It is also one of the most important pathogens in cystic fibrosis patients [14]. More importantly, this species complex is notorious for its resistance to multiple antibiotics, mediated through its intrinsic features or through chromosomal mutations that arise under the selective pressure of antibiotic use [15]. Thus, the *M. abscessus* complex poses a major threat to clinical management and public health as treatment options for the infections caused by it are limited.

The intrinsic resistance of the *M. abscessus* complex towards first- and second-generation tetracyclines is mainly due to the over-expression of a tetracycline-degrading enzyme known as MabTetX (*MAB_1496c*) [16]. Tigecycline is a poor substrate for the MabTetX and does not induce the expression of this enzyme [16], which could explain its potency against *M. abscessus* complex. Interestingly, tigecycline has shown synergistic activities with other antibiotics (clarithromycin, linezolid and teicoplanin) against the *M. abscessus* complex in vitro and in vivo [11,17,18]. In 2014, Wallace et al. reported that, after receiving tigecycline-containing salvage regimens for more than a month, approximately 66% of patients with *M. abscessus* complex or *M. chelonae* infections (n = 38) showed clinical improvement [19]. This led the authors to conclude that tigecycline might be a useful addition to other clinically available drugs in patients with these difficult-to-treat infections.

1.3. Genetic Determinants of Tigecycline Resistance or Reduced Susceptibility in Other Bacteria

Tigecycline resistance has emerged in the past 10 years and is most commonly observed among Gram-negative bacteria, mainly *Acinetobacter baumannii* and members of the Enterobacteriaceae [7]. The decreased susceptibility or resistance to tigecycline in these clinically important microorganisms has mostly been attributed to the over-expression of resistance-nodulation-cell division-type transporters, including the AcrAB efflux pumps [7]. Moreover, mutations in genes encoding the ribosomal protein S10 [20], a SAM-dependent methyltransferase [21], the acyl-sn-glycerol-3-phosphate acyltransferase [22], and proteins involved in the lipopolysaccharide core biosynthesis [23] have also been linked to tigecycline resistance in Gram-negative organisms. Another mechanism of tigecycline resistance is the TetX-mediated modification of the drug [24]. Tigecycline resistance has also been documented, albeit less frequently, in Gram-positive bacteria [7]. Through the characterization of laboratory-derived mutants, over-expression of MepA (a multidrug and toxic compound extrusion family efflux pump) and mutations in ribosomal genes (16S rRNA, ribosomal proteins and a 16S rRNA methyltransferase) were associated with resistance or decreased susceptibility to tigecycline in *S. aureus* and *S. pneumoniae*, respectively [25,26].

2. Genetic Determinants of Resistance or Reduced Susceptibility to Tigecycline in *M. abscessus*

Although tigecycline-resistant strains of *M. abscessus* complex have been documented in different parts of the world [27,28], their resistance determinants remain largely elusive. In this review, the evidence for different genetic determinants reported to be linked to tigecycline resistance or reduced tigecycline susceptibility in the subspecies *M. abscessus* was examined and discussed. These reported genetic determinants were identified from mutants generated from *M. abscessus* ATCC 19977, the type strain of *M. abscessus*.

2.1. An Intrinsic Feature Associated with Reduced Tigecycline Susceptibility: WhiB7

In mycobacteria, WhiB7 is a transcriptional activator of intrinsic antibiotic resistance that can be induced by exposure to stresses, such as heat shock, iron deficiency and redox imbalance, and many antibiotics, including aminoglycosides, lincosamides, macrolides, pleuromutilins and tetracyclines [29–32]. In 2017, Pryjma et al. found *whiB7* (*MAB_3508c*) to be associated with reduced tigecycline susceptibility in *M. abscessus* [33]. The deletion of the WhiB7-encoding gene caused a 4-fold decrease in the minimum inhibitory concentration (MIC—minimum inhibitory concentration) of tigecycline. Unfortunately, this group of authors did not identify the downstream effector gene(s) of WhiB7 that is linked to the reduced tigecycline susceptibility. To the best of our knowledge, this constitutes the earliest report on the genetic determinant associated with reduced tigecycline susceptibility in *M. abscessus*.

2.2. Acquired Tigecycline Resistance: RshA Mutations

In *M. abscessus*, the *sigH* gene (*MAB_3543c*) for the sigma factor SigH and *rshA* gene (*MAB_3542c*) for the anti-sigma factor RshA control heat shock and oxidative-stress responses. In the absence of environmental stress, RshA interacts with and inhibits SigH. In response to stress, however, the interaction between RshA and SigH is disrupted, leading to the release of SigH which would then form the RNA polymerase holoenzyme (with the core RNA polymerase) and initiate the transcription of *sigH* and other genes involved in stress response [34]. Other than heat and redox stress signals, the RshA-SigH interaction can also be disrupted by mutations in the HXXXCXXC motif of RshA [34].

Through the characterization of a tigecycline-resistant, spontaneous mutant of *M. abscessus* ATCC 19977 (MIC: 0.25 mg/L), designated as 7C (MIC: 2 mg/L), Ng et al. (2018) found the C51R mutation in the RshA to be associated with tigecycline resistance [35]. The non-species related breakpoints (sensitive $\leq$ 0.25 mg/L, resistant > 0.5 mg/L) proposed by the EUCAST (2018) [36] was used in this study. The C51R mutation changed the first cysteine residue in the HXXXCXXC motif to arginine. As a result, there was an up-regulation of *sigH* and other stress-response genes in 7C that was confirmed by transcriptome profiling [37]. The causal relationship between the mutation, identified by whole-genome sequencing, and the resistance phenotype was established using the complementation of 7C with the wild-type *MAB_3542c* gene. The *whiB7* gene was not differentially expressed in 7C. In a follow-up study, Lee et al. (2021) showed that the over-expression of the *sigH* gene alone was capable of inducing tigecycline resistance in the wild-type *M. abscessus* ATCC 19977 [38]. This is supported by a recent study by Schildkraut et al. (2021) which showed an increased expression of *sigH* following an exposure of *M. abscessus* to tigecycline at a sub-inhibitory concentration, suggesting that this gene is needed for the tigecycline adaptation [39]. Although it has been well-documented that dysregulated stress response can lead to antibiotic resistance in bacteria [40], the exact mechanism or downstream gene(s) through which the RshA mutation and the *sigH* up-regulation caused a tigecycline resistance phenotype remains unclear.

2.3. SigH Mutation

SigH is known to play two functions, which are to interact with and be inhibited by the RshA anti-sigma factor under normal circumstances and to initiate transcription in response to stressful conditions [34]. Lee et al. (2021) isolated a tigecycline-resistant mutant, designated as CL7 (MIC: 2 mg/L), which carried a stop-gain mutation (E229×) in SigH (*MAB_3543c*) [38]. The stop-gain mutation led to a seven-amino-acid truncation in the SigH protein. Interestingly, by transforming an expression plasmid carrying the mutant *sigH* gene, the previously sensitive ATCC 19977 developed resistance towards tigecycline, suggesting that truncated SigH might retain its capability to cause tigecycline resistance.

RT-qPCR analyses of CL7 showed an over-expression of *sigH* along with stress-response genes encoding the thioredoxin and heat-shock proteins, which are the known regulon of SigH [34]. As such, these findings suggested that the SigH mutation might not be a completely loss-of-function mutation, as it only disrupted the interaction of mutated SigH with RshA but retained the SigH ability to auto-up-regulate itself and key stress genes, ultimately leading to the development of tigecycline resistance.

2.4. rshA-Knockout Mutant

The demonstration of tigecycline resistance in *M. abscessus* following the disruption of the SigH-RshA interaction and subsequent up-regulation of *sigH* led to the prediction that knocking out the *rshA* gene should also result in the development of tigecycline resistance, owing to a decreased inhibition of SigH. Unexpectedly, a recent study by Schildkraut et al. (2021) suggested otherwise [39]. Their *rshA*-knockout mutant (ΔMAB_3542c), derived from ATCC 19977, had neither an increase in tigecycline MIC nor a *sigH* up-regulation. A possible explanation could be that *sigH* and *rshA* are co-transcribed in a polycistronic mRNA (Figure S1A) as the genome of ATCC 19977 shows a four-bp overlap (the final four bps of the *sigH* gene are the first four bps of the *rshA* gene) (Figure S1B). As such, the deletion of *rshA* could likely result in an unwanted polar effect on the neighboring *sigH* gene. One example of such a polar effect is the introduction of synonymous mutations in the final two codons of the *sigH* gene (the alanine and stop codons) (Figure S1C). Synonymous mutations are known to alter the target gene expression [41]. In addition, the tag stop codon, introduced after the deletion of *rshA*, has been associated with a higher read-through error rate than tga (the original stop codon) during the translation [42]. Thus, the unexpected findings by Schildkraut et al. were likely an outcome of the longer-than-usual, non-functional SigH which failed to induce tigecycline resistance and its auto-up-regulation or the altered gene expression of *sigH* due to the synonymous mutations.

3. Future Perspectives and Research Areas

Thus far, the reported genetic determinants of resistance or reduced susceptibility to tigecycline in *M. abscessus*, including WhiB7, RshA and SigH, are transcriptional regulators which respond to physiological stresses. Ribosome disruption via antibiotic exposure or mutation can lead to the production of aberrant polypeptides that are prone to oxidative modification/damage [43]. Although this aspect (tigecycline-induced oxidative damage) of tigecycline killing/inhibition has not been described before in bacteria, tigecycline has been shown to be able to induce oxidative stress in eukaryotic mitochondria [44] which have a bacterial origin [45,46]. If oxidative damage were indeed a part of the tigecycline killing/inhibition, it would be convenient for *M. abscessus* strains with WhiB7 or SigH over-expression, or RshA and SigH mutations to resist the antibiotic onslaught in clinical therapy. As oxidative damage is one of the human immune defense functions against microbes [47], and both WhiB7 and SigH are potential virulence factors in mycobacteria [48,49], it may also be interesting to investigate the pathogenicity of the WhiB7, SigH and RshA mutants in animal models.

With the emergence of tigecycline resistance in the past decade, it can be foreseen that molecular assays, such as those based on the PCR, line immunoassay and next-generation sequencing technologies, will be increasingly used for the rapid resistotyping of clinical isolates. Among the *M. abscessus* complex, studies on tigecycline resistance determinants have thus far been focused solely on *M. abscessus*. Since there is evidence suggesting a differential tigecycline susceptibility pattern among the subspecies of the *M. abscessus* complex [28], future studies in this area should focus more on the other two subspecies of *M. massiliense* and *M. bolletii*. In general, a thorough understanding of resistance determinants would help to determine the best way to utilize tigecycline for the treatment of *M. abscessus* complex infections, to prevent further escalation of tigecycline resistance in these pathogens.

Supplementary Materials: The following supporting information can be downloaded at: https://www.mdpi.com/article/10.3390/antibiotics11050572/s1, Figure S1: (A) The *sigH* (*MAB_3543c*) and *rshA* (*MAB_3542c*) genes are transcribed as an operon. RT-PCR analysis with the forward primer annealed to the *MAB_3543c* gene and the reverse primer annealed to the *MAB_3542c* gene. cDNA was prepared from the RNA of ATCC 19977. NoRT: no-reverse transcription control. (B) Both genes are neighbor genes in the ATCC 19977 genome with a 4-base overlap. (C) Partial DNA sequences of *MAB_3543c* from ATCC 19977 and ΔMAB_3542c.

Author Contributions: H.F.N. and Y.F.N. conceptualized and wrote the manuscript. All authors have read and agreed to the published version of the manuscript.

Funding: Y.F.N.'s research was supported by grant 4486/U0U from Universiti Tunku Abdul Rahman, Malaysia.

Institutional Review Board Statement: Not applicable.

Informed Consent Statement: Not applicable.

Data Availability Statement: Not applicable.

Acknowledgments: We thank Col Lin Lee and Kar Men Aw for their kind assistance in the preparation of this manuscript.

Conflicts of Interest: The authors declare that the research was conducted in the absence of any commercial or financial relationships that could be construed as a potential conflict of interest.

References

1. Townsend, M.L.; Pound, M.W.; Drew, R.H. Tigecycline: A new glycylcycline antimicrobial. *Int. J. Clin. Pract.* **2006**, *60*, 1662–1672. [CrossRef] [PubMed]
2. Noskin, G.A. Tigecycline: A new glycylcycline for treatment of serious infections. *Clin. Infect. Dis.* **2005**, *41*, S303–S314. [CrossRef] [PubMed]
3. Olson, M.W.; Ruzin, A.; Feyfant, E.; Rush, T.S.; O'Connell, J.; Bradford, P.A. Functional, biophysical, and structural bases for antibacterial activity of tigecycline. *Antimicrob. Agents Chemother.* **2006**, *50*, 2156–2166. [CrossRef]
4. Schedlbauer, A.; Kaminishi, T.; Ochoa-Lizarralde, B.; Dhimole, N.; Zhou, S.; López-Alonso, J.P.; Connell, S.R.; Fucini, P. Structural characterization of an alternative mode of tigecycline binding to the bacterial ribosome. *Antimicrob. Agents Chemother.* **2015**, *59*, 2849–2854. [CrossRef] [PubMed]
5. Bauer, G.; Berens, C.; Projan, S.J.; Hillen, W. Comparison of tetracycline and tigecycline binding to ribosomes mapped by dimethylsulphate and drug-directed Fe^{2+} cleavage of 16S rRNA. *J. Antimicrob. Chemother.* **2004**, *53*, 592–599. [CrossRef]
6. Rasmussen, B.A.; Gluzman, Y.; Tally, F.P. Inhibition of protein synthesis occurring on tetracycline-resistant, TetM-protected ribosomes by a novel class of tetracyclines, the glycylcyclines. *Antimicrob. Agents Chemother.* **1994**, *38*, 1658–1660. [CrossRef]
7. Sun, Y.; Cai, Y.; Liu, X.; Bai, N.; Liang, B.; Wang, R. The emergence of clinical resistance to tigecycline. *Int. J. Antimicrob. Agents* **2013**, *41*, 110–116. [CrossRef]
8. Kumarasamy, K.K.; Toleman, M.A.; Walsh, T.R.; Bagaria, J.; Butt, F.; Balakrishnan, R.; Chaudhary, U.; Doumith, M.; Giske, C.G.; Irfan, S.; et al. Emergence of a new antibiotic resistance mechanism in India, Pakistan, and the UK: A molecular, biological, and epidemiological study. *Lancet Infect. Dis.* **2010**, *10*, 597–602. [CrossRef]
9. Wallace, R.J.; Brown-elliott, B.A.; Crist, C.J.; Mann, L.; Wilson, R.W. Comparison of the In Vitro Activity of the Glycylcycline Tigecycline (Formerly GAR-936) with Those of Tetracycline, Minocycline, and Doxycycline against Isolates of Nontuberculous Mycobacteria. *Antimicrob. Agents Chemother.* **2002**, *46*, 3164–3167. [CrossRef]
10. Lerat, I.; Cambau, E.; dit Bettoni, R.R.; Gaillard, J.-L.; Jarlier, V.; Truffot, C.; Veziris, N. In Vivo Evaluation of Antibiotic Activity Against Mycobacterium abscessus. *J. Infect. Dis.* **2014**, *209*, 905–912. [CrossRef]
11. Oh, C.-T.; Moon, C.; Park, O.K.; Kwon, S.-H.; Jang, J. Novel drug combination for Mycobacterium abscessus disease therapy identified in a Drosophila infection model. *J. Antimicrob. Chemother.* **2014**, *69*, 1599–1607. [CrossRef] [PubMed]
12. Coban, A.Y.; Deveci, A.; Cayci, Y.T.; Uzun, M.; Akgunes, A.; Durupinar, B. In vitro effect of tigecycline against Mycobacterium tuberculosis and a review of the available drugs for tuberculosis. *Afr. J. Microbiol. Res.* **2011**, *5*, 311–315.
13. Griffith, D.E.; Aksamit, T.; Brown-Elliott, B.A.; Catanzaro, A.; Daley, C.; Gordin, F.; Holland, S.M.; Horsburgh, R.; Huitt, G.; Iademarco, M.F.; et al. An official ATS/IDSA statement: Diagnosis, treatment, and prevention of nontuberculous mycobacterial diseases. *Am. J. Respir. Crit. Care Med.* **2007**, *175*, 367–416. [CrossRef]
14. Skolnik, K.; Kirkpatrick, G.; Quon, B.S. Nontuberculous Mycobacteria in Cystic Fibrosis. *Curr. Treat. Options Infect. Dis.* **2016**, *8*, 259–274. [CrossRef] [PubMed]
15. Nessar, R.; Cambau, E.; Reyrat, J.M.; Murray, A.; Gicquel, B. Mycobacterium abscessus: A new antibiotic nightmare. *J. Antimicrob. Chemother.* **2012**, *67*, 810–818. [CrossRef] [PubMed]

16. Rudra, P.; Hurst-Hess, K.; Lappierre, P.; Ghosh, P. High Levels of Intrinsic Tetracycline Resistance in *Mycobacterium abscessus* Are Conferred by a Tetracycline-Modifying Monooxygenase. *Antimicrob. Agents Chemother.* **2018**, *62*, e00119-18. [CrossRef]
17. Huang, C.-W.; Chen, J.-H.; Hu, S.-T.; Huang, W.-C.; Lee, Y.-C.; Huang, C.-C.; Shen, G.-H. Synergistic activities of tigecycline with clarithromycin or amikacin against rapidly growing mycobacteria in Taiwan. *Int. J. Antimicrob. Agents* **2013**, *41*, 218–223. [CrossRef]
18. Aziz, D.B.; Teo, J.W.P.; Dartois, V.; Dick, T. Teicoplanin–Tigecycline Combination Shows Synergy Against Mycobacterium abscessus. *Front. Microbiol.* **2018**, *9*, 932. [CrossRef]
19. Wallace, R.J.; Dukart, G.; Brown-Elliott, B.A.; Griffith, D.E.; Scerpella, E.G.; Marshall, B. Clinical experience in 52 patients with tigecycline-containing regimens for salvage treatment of Mycobacterium abscessus and Mycobacterium chelonae infections. *J. Antimicrob. Chemother.* **2014**, *69*, 1945–1953. [CrossRef]
20. Villa, L.; Feudi, C.; Fortini, D.; García-Fernández, A.; Carattoli, A. Genomics of KPC-producing Klebsiella pneumoniae sequence type 512 clone highlights the role of RamR and ribosomal S10 protein mutations in conferring tigecycline resistance. *Antimicrob. Agents Chemother.* **2014**, *58*, 1707–1712. [CrossRef]
21. Chen, Q.; Li, X.; Zhou, H.; Jiang, Y.; Chen, Y.; Hua, X.; Yu, Y. Decreased susceptibility to tigecycline in Acinetobacter baumannii mediated by a mutation in trm encoding SAM-dependent methyltransferase. *J. Antimicrob. Chemother.* **2014**, *69*, 72–76. [CrossRef] [PubMed]
22. Li, X.; Liu, L.; Ji, J.; Chen, Q.; Hua, X.; Jiang, Y.; Feng, Y.; Yu, Y. Tigecycline resistance in Acinetobacter baumannii mediated by frameshift mutation in plsC, encoding 1-acyl-sn-glycerol-3-phosphate acyltransferase. *Eur. J. Clin. Microbiol. Infect. Dis.* **2015**, *34*, 625–631. [CrossRef]
23. Linkevicius, M.; Sandegren, L.; Andersson, D.I. Mechanisms and fitness costs of tigecycline resistance in Escherichia coli. *J. Antimicrob. Chemother.* **2013**, *68*, 2809–2819. [CrossRef] [PubMed]
24. Moore, I.F.; Hughes, D.W.; Wright, G.D. Tigecycline is modified by the flavin-dependent monooxygenase TetX. *Biochemistry* **2005**, *44*, 11829–11835. [CrossRef]
25. McAleese, F.; Petersen, P.; Ruzin, A.; Dunman, P.M.; Murphy, E.; Projan, S.J.; Bradford, P.A. A novel MATE family efflux pump contributes to the reduced susceptibility of laboratory-derived Staphylococcus aureus mutants to tigecycline. *Antimicrob. Agents Chemother.* **2005**, *49*, 1865–1871. [CrossRef]
26. Lupien, A.; Gingras, H.; Leprohon, P.; Ouellette, M. Induced tigecycline resistance in Streptococcus pneumoniae mutants reveals mutations in ribosomal proteins and rRNA. *J. Antimicrob. Chemother.* **2015**, *70*, 2973–2980. [CrossRef]
27. Broda, A.; Jebbari, H.; Beaton, K.; Mitchell, S.; Drobniewski, F. Comparative drug resistance of Mycobacterium abscessus and M. chelonae isolates from patients with and without cystic fibrosis in the United Kingdom. *J. Clin. Microbiol.* **2013**, *51*, 217–223. [CrossRef]
28. Ananta, P.; Kham-ngam, I.; Chetchotisakd, P.; Chaimanee, P.; Reechaipichitkul, W.; Namwat, W.; Lulitanond, V.; Faksri, K. Analysis of drug-susceptibility patterns and gene sequences associated with clarithromycin and amikacin resistance in serial Mycobacterium abscessus isolates from clinical specimens from Northeast Thailand. *PLoS ONE* **2018**, *13*, e0208053. [CrossRef] [PubMed]
29. Burian, J.; Ramón-García, S.; Sweet, G.; Gómez-Velasco, A.; Av-Gay, Y.; Thompson, C.J. The mycobacterial transcriptional regulator whiB7 gene links redox homeostasis and intrinsic antibiotic resistance. *J. Biol. Chem.* **2012**, *287*, 299–310. [CrossRef]
30. Burian, J.; Yim, G.; Hsing, M.; Axerio-Cilies, P.; Cherkasov, A.; Spiegelman, G.B.; Thompson, C.J. The mycobacterial antibiotic resistance determinant WhiB7 acts as a transcriptional activator by binding the primary sigma factor SigA (RpoV). *Nucleic Acids Res.* **2013**, *41*, 10062–10076. [CrossRef]
31. Morris, R.P.; Nguyen, L.; Gatfield, J.; Visconti, K.; Nguyen, K.; Schnappinger, D.; Ehrt, S.; Liu, Y.; Heifets, L.; Pieters, J.; et al. Ancestral antibiotic resistance in Mycobacterium tuberculosis. *Proc. Natl. Acad. Sci. USA* **2005**, *102*, 12200–12205. [CrossRef] [PubMed]
32. Geiman, D.E.; Raghunand, T.R.; Agarwal, N.; Bishai, W.R. Differential gene expression in response to exposure to antimycobacterial agents and other stress conditions among seven Mycobacterium tuberculosis whiB-like genes. *Antimicrob. Agents Chemother.* **2006**, *50*, 2836–2841. [CrossRef] [PubMed]
33. Pryjma, M.; Burian, J.; Kuchinski, K.; Thompson, C.J. Antagonism between Front-Line Antibiotics Clarithromycin and Amikacin in the Treatment of Mycobacterium abscessus Infections Is Mediated by the *whiB7* Gene. *Antimicrob. Agents Chemother.* **2017**, *61*, 61. [CrossRef] [PubMed]
34. Song, T.; Dove, S.L.; Lee, K.H.; Husson, R.N. RshA, an anti-sigma factor that regulates the activity of the mycobacterial stress response sigma factor SigH. *Mol. Microbiol.* **2003**, *50*, 949–959. [CrossRef]
35. Ng, H.F.; Tan, J.L.; Zin, T.; Yap, S.F.; Ngeow, Y.F. A mutation in anti-sigma factor MAB_3542c may be responsible for tigecycline resistance in Mycobacterium abscessus. *J. Med. Microbiol.* **2018**, *67*, 1676–1681. [CrossRef]
36. EUCAST. Breakpoint Tables for Interpretation of MICs and Zone Diameters. Version 8. 2018. Available online: http://www. eucast.org (accessed on 7 August 2018).
37. Ng, H.F.; Ngeow, Y.F.; Yap, S.F.; Zin, T.; Tan, J.L. Tigecycline resistance may be associated with dysregulated response to stress in Mycobacterium abscessus. *Int. J. Med. Microbiol.* **2020**, *310*, 151380. [CrossRef]
38. Lee, C.L.; Ng, H.F.; Ngeow, Y.F.; Thaw, Z. A stop-gain mutation in sigma factor SigH (MAB_3543c) may be associated with tigecycline resistance in Mycobacteroides abscessus. *J. Med. Microbiol.* **2021**, *70*, 001378. [CrossRef]

39. Schildkraut, J.A.; Coolen, J.P.M.; Burbaud, S.; Sangen, J.J.N.; Kwint, M.P.; Floto, R.A.; Op den Camp, H.J.M.; Te Brake, L.H.M.; Wertheim, H.F.L.; Neveling, K.; et al. RNA-sequencing elucidates drug-specific mechanisms of antibiotic tolerance and resistance in M. abscessus. *Antimicrob. Agents Chemother.* **2021**, *66*, e0150921. [CrossRef]
40. Poole, K. Bacterial stress responses as determinants of antimicrobial resistance. *J. Antimicrob. Chemother.* **2012**, *67*, 2069–2089. [CrossRef]
41. Bailey, S.F.; Hinz, A.; Kassen, R. ARTICLE Adaptive synonymous mutations in an experimentally evolved Pseudomonas fluorescens population. *Nat. Commun.* **2014**, *5*, 4076. [CrossRef]
42. Korkmaz, G.; Holm, M.; Wiens, T.; Sanyal, S. Comprehensive Analysis of Stop Codon Usage in Bacteria and Its Correlation with Release Factor Abundance. *J. Biol. Chem.* **2014**, *289*, 30334–30342. [CrossRef] [PubMed]
43. Dukan, S.; Farewell, A.; Ballesteros, M.; Taddei, F.; Radman, M.; Nyström, T. Protein oxidation in response to increased transcriptional or translational errors. *Proc. Natl. Acad. Sci. USA* **2000**, *97*, 5746–5749. [CrossRef] [PubMed]
44. Tan, J.; Song, M.; Zhou, M.; Hu, Y. Antibiotic tigecycline enhances cisplatin activity against human hepatocellular carcinoma through inducing mitochondrial dysfunction and oxidative damage. *Biochem. Biophys. Res. Commun.* **2017**, *483*, 17–23. [CrossRef] [PubMed]
45. Margulis, L. Symbiotic theory of the origin of eukaryotic organelles; criteria for proof. *Symp. Soc. Exp. Biol.* **1975**, *29*, 21–38.
46. Suárez-Rivero, J.M.; Pastor-Maldonado, C.J.; Povea-Cabello, S.; Álvarez-Córdoba, M.; Villalón-García, I.; Talaverón-Rey, M.; Suárez-Carrillo, A.; Munuera-Cabeza, M.; Sánchez-Alcázar, J.A. Mitochondria and Antibiotics: For Good or for Evil? *Biomolecules* **2021**, *11*, 1050. [CrossRef]
47. Spooner, R.; Yilmaz, Ö. The Role of Reactive-Oxygen-Species in Microbial Persistence and Inflammation. *Int. J. Mol. Sci.* **2011**, *12*, 334. [CrossRef]
48. Rohde, K.H.; Veiga, D.F.T.; Caldwell, S.; Balázsi, G.; Russell, D.G. Linking the Transcriptional Profiles and the Physiological States of Mycobacterium tuberculosis during an Extended Intracellular Infection. *PLoS Pathog.* **2012**, *8*, e1002769. [CrossRef]
49. Kaushal, D.; Schroeder, B.G.; Tyagi, S.; Yoshimatsu, T.; Scott, C.; Ko, C.; Carpenter, L.; Mehrotra, J.; Manabe, Y.C.; Fleischmann, R.D.; et al. Reduced immunopathology and mortality despite tissue persistence in a Mycobacterium tuberculosis mutant lacking alternative sigma factor, SigH. *Proc. Natl. Acad. Sci. USA* **2002**, *99*, 8330–8335. [CrossRef]

 *antibiotics*

Article

Comparative Genomic Analysis of a Panton–Valentine Leukocidin-Positive ST22 Community-Acquired Methicillin-Resistant *Staphylococcus aureus* from Pakistan

Nimat Ullah [1,2], Samavi Nasir [1], Zaara Ishaq [1], Farha Anwer [1], Tanzeela Raza [1], Moazur Rahman [3], Abdulrahman Alshammari [4,*], Metab Alharbi [4], Taeok Bae [2], Abdur Rahman [1] and Amjad Ali [1,*]

1　Atta-ur-Rahman School of Applied Biosciences (ASAB), National University of Sciences and Technology (NUST), Sector H-12, Islamabad 44000, Pakistan; nullah.phdabs12asab@asab.nust.edu.pk (N.U.); snasir.msib08asab@student.nust.edu.pk (S.N.); zishaq.msib07asab@student.nust.edu.pk (Z.I.); fanwar.phdabs17asab@student.nust.edu.pk (F.A.); tanzeela.raza12@gmail.com (T.R.); a.rahman@asab.nust.edu.pk (A.R.)

2　Department of Microbiology and Immunology, Indiana University School of Medicine-Northwest, 3400 Broadway, Gary, IN 46408, USA; tbae@iun.edu

3　School of Biological Sciences, University of the Punjab, Quaid-i-Azam Campus, Lahore 54590, Pakistan; moaz.sbs@pu.edu.pk

4　Department of Pharmacology and Toxicology, College of Pharmacy, King Saud University, P.O. Box 2455, Riyadh 11451, Saudi Arabia; mesalharbi@ksu.edu.sa

*　Correspondence: abdalshammari@ksu.edu.sa (A.A.); amjad.ali@asab.nust.edu.pk (A.A.)

Citation: Ullah, N.; Nasir, S.; Ishaq, Z.; Anwer, F.; Raza, T.; Rahman, M.; Alshammari, A.; Alharbi, M.; Bae, T.; Rahman, A.; et al. Comparative Genomic Analysis of a Panton–Valentine-Positive ST22 Community-Acquired Methicillin-Resistant *Staphylococcus aureus* from Pakistan. *Antibiotics* **2022**, *11*, 496. https://doi.org/10.3390/antibiotics11040496

Academic Editor: Teresa V. Nogueira

Received: 2 March 2022
Accepted: 6 April 2022
Published: 8 April 2022

Publisher's Note: MDPI stays neutral with regard to jurisdictional claims in published maps and institutional affiliations.

Abstract: *Staphylococcus aureus* (*S. aureus*) ST22 is considered a clinically important clone because an epidemic strain EMRSA-15 belongs to ST22, and several outbreaks of this clone have been documented worldwide. We performed genomic analysis of an *S. aureus* strain Lr2 ST22 from Pakistan and determined comparative analysis with other ST22 strains. The genomic data show that Lr2 belongs to *spa*-type t2986 and harbors staphylococcal cassette chromosome *mec* (SCC*mec*) type IVa(2B), one complete plasmid, and seven prophages or prophage-like elements. The strain harbors several prophage-associated virulence factors, including Panton–Valentine leukocidin (PVL) and toxic shock syndrome toxin (TSST). The single nucleotide polymorphism (SNPs)-based phylogenetic relationship inferred from whole genome and core genome revealed that strain Lr2 exhibits the nearest identities to a South African community-acquired methicillin-resistant *S. aureus* (CA-MRSA) ST22 strain and makes a separate clade with an Indian CA-MRSA ST22 strain. Although most ST22 strains carry *blaZ*, *mecA*, and mutations in *gyrA*, the Lr2 strain does not have the *blaZ* gene but, unlike other ST22 strains, carries the antibiotic resistance genes *erm(C)* and *aac(6′)-Ie-aph(2″)-Ia*. Among ST22 strains analyzed, only the strain Lr2 possesses both PVL and TSST genes. The functional annotation of genes unique to Lr2 revealed that mobilome is the third-largest Cluster of Orthologous Genes (COGs) category, which encodes genes associated with prophages and transposons. This possibly makes methicillin-resistant *S. aureus* (MRSA) Lr2 ST22 strain highly virulent, and this study would improve the knowledge of MRSA ST22 strains in Pakistan. However, further studies are needed on a large collection of MRSA to comprehend the genomic epidemiology and evolution of this clone in Pakistan.

Keywords: comparative genomic analysis; CA-MRSA; EMRSA-15; PVL-positive; ST22

1. Introduction

Staphylococcus aureus (*S. aureus*) has long been considered an important human pathogen that causes both hospital- and community-acquired (HA and CA) infections. This pathogen is also well-known for acquiring resistance to a variety of antibiotics [1,2]. Particularly, methicillin resistance is becoming more common in *S. aureus*, posing a growing public health threat with substantial mortality and morbidity. The methicillin-resistant *S. aureus*

(MRSA) produces a low-affinity penicillin-binding protein (PBP2a), which enables it to confer resistance to almost all beta-lactam antibiotics [3,4]. The PBP2a is encoded by the gene *mecA*, which is carried by a large mobile genetic element known as staphylococcal cassette chromosome *mec* (SCC*mec*) [5]. In the past, MRSA caused mostly hospital-acquired infections; however, now, it also causes community-acquired infections, including necrotizing pneumonia and skin and soft tissue infection (SSTI) [6]. Apart from humans, MRSA can also colonize other animal species, particularly livestock, and CA-MRSA infections can also be caused by livestock-associated MRSA (LA-MRSA) [7].

Compared with HA-MRSA, the CA-MRSA strains belong to a diverse lineage with smaller SCC*mec*, for example, SCC*mec* IV, and harbors distinct virulence factors, particularly Panton–Valentine leukocidin (PVL) [8]. PVL is a two-component toxin encoded by prophage-associated genes *lukF-PV* and *lukS-PV* [9,10]. The toxin targets phagocytic leucocytes and triggers leukocyte lysis and/or apoptosis by forming pores. Therefore, PVL-positive CA-MRSA strains cause various infections, particularly necrotizing pneumonia and SSTI [11]. Although not all CA-MRSA contain the *pvl* gene, it can be considered as a molecular marker for CA-MRSA along with the SCC*mec* type IV because PVL is associated with the global spread of CA-MRSA [12]. The well-known clones of PVL-positive CA-MRSA include ST8, frequently found in the United States [13]; ST80, a Europe clone [10]; ST59, reported in Taiwan [14]; ST30, a global strain [15]; and ST22 (SCC*mec* IV/PVL-positive), which emerged in the United Kingdom and is now spreading worldwide [16]. *S. aureus* ST22 is considered a clinically important clone because an epidemic strain EMRSA-15 belongs to ST22. The EMRSA-15 emerged in the United Kingdom in the 1990s [17] and very quickly expanded all over Europe, Australia, and Asia [18]. EMRSA-15 has SCC*mec* type IV, a deletion of 2268 bp region in fibronectin-binding protein (FnBP) locus, and a point mutation in the *ureC* gene, and it is resistant to fluoroquinolone and macrolide. The genomic analysis of ST22 strains revealed that the epidemic clone EMRSA-15 makes a distinct clade (ST22-A clade) [18].

Although there is a wealth of literature available on comprehensive epidemiological and molecular characterization of MRSA in the USA, Europe, and Asia, there is a scarcity of detailed genomic characterization of MRSA strains in Pakistan. According to some recent studies, Pakistan has a greater rate of MRSA infection, and the data suggest that MRSA clones are becoming more diverse [19–21]. The whole-genome analysis provides high resolution in both global and local outbreak investigations as well as further exploring genes associated with pathogenicity and antibiotic resistance [22]. Therefore, whole-genome-based approaches have become an indispensable tool for the real-time comparative genomic study of a variety of pathogens in terms of antibiotic resistance, the emergence of new virulent clones, and niche adaptation [23]. Furthermore, investigating virulence- and antibiotic resistance-associated genes can help identify risk factors for MRSA infection and develop efficient infection control programs. Given the lack of genome-based surveillance and epidemiological studies of MRSA in Pakistan, we aimed to study a CA-MRSA ST22 strain from Pakistan at the genomic level and compare the virulence and antibiotic resistance genes and evolutionary relationships to other ST22 strains. To our knowledge, this is the first genome-based analysis of PVL-positive CA-MRSA ST22 from Pakistan, and we found many genomic differences compared to other complete genomes of MRSA ST22 strains.

2. Results

2.1. Preliminary Identification and Phenotypic Antibiotic Resistance Pattern

The isolate Lr2 was found positive for coagulase and catalase while negative for oxidase. The isolate was resistant to oxacillin, methicillin, ampicillin, erythromycin, gentamicin, streptomycin, clindamycin, ciprofloxacin, linezolid, and tetracycline but susceptible to vancomycin, chloramphenicol, and rifampicin (Table 1). The amplification of the 500 bp *mecA* gene confirmed the nature of methicillin resistance.

Table 1. The phenotypic and genotypic resistance profile of Lr2.

Antibiotic	Sensitivity (Zone of Growth Inhibition in mm)	Related Gene(s)
Oxacillin	R (no zone)	
Ampicillin	R (no zone)	*mecA*
Methicillin	R (no zone)	
Gentamicin	R (4 mm)	*aac(6′)-Ie-aph(2″)-Ia*
Streptomycin	R (7 mm)	
Erythromycin	R (4 mm)	*erm(C)*
Clindamycin	R (9 mm)	
Ciprofloxacin	R (11 mm)	*gyrA*
Vancomycin	S (23 mm)	ND
Chloramphenicol	S (17 mm)	ND
Linezolid	R (13 mm)	ND
Rifampicin	S (21 mm)	ND
Tetracycline	R (9 mm)	ND
Fusidic acid	R (11 mm)	ND

R—resistant; S—susceptible; ND—not determined.

2.2. Genomic Features and Epidemiological Typing

The genomic DNA (gDNA) was successfully sequenced, and a total of 1,363,009 reads with a mean length of 565 bp (272-fold sequence coverage) were obtained. The kmerFinder 2.0 identified the isolate as *S. aureus*. The N_{50}, N_{75}, and L_{50} values are 131,746, 94,139, and 7, respectively; de novo assembly resulted in 52 contigs (>500 bp), and the size of the largest contig is 425,597 bp. The genome size of Lr2 is 2,831,239 bp with 32.7% GC content. Genome annotation predicted 2835 total genes, of which 2768 are CDSs. The number of predicted tRNA are 56 and rRNA are 7 (5S = 4, 16S = 2, 23S = 1). In silico genome-based epidemiological typing revealed that strain Lr2 belongs to ST22, *spa*-type t2986, and carries SCC*mec* type IVa(2B) (Table 2).

Table 2. The genomic features and characteristics of the MRSA strain Lr2 ST22.

Characteristics	Lr2
Genome size (bp)	2,831,239
Contigs	52
GC content %	32.7
N_{50}	131,746
N_{75}	94,139
L_{50}	7
Largest contig (bp)	425,597
Genes (total)	2835
CDSs (total)	2768
No. of tRNA	56
No. of rRNA	4, 2, 1 (5S, 16S, 23S)
ST	22
SCC*mec* type	IVa(2B)
spa-type	t2986
NCBI Accession number	JAIGOF000000000

2.3. Predicted Antibiotic Resistance Determinants and Virulence Factors

The Lr2 genome contains a methicillin resistance gene *mecA*, an aminoglycoside resistance gene *aac(6′)-Ie-aph(2″)-Ia*, a point mutation in *gyrA* conferring resistance to fluoroquinolones, and *erm(C)*—a plasmid-associated gene conferring resistance to lincosamide, macrolide (erythromycin), and streptogramin. However, the *blaZ* gene conferring penicillin resistance was found absent in Lr2 (Table 1 and Figure 1).

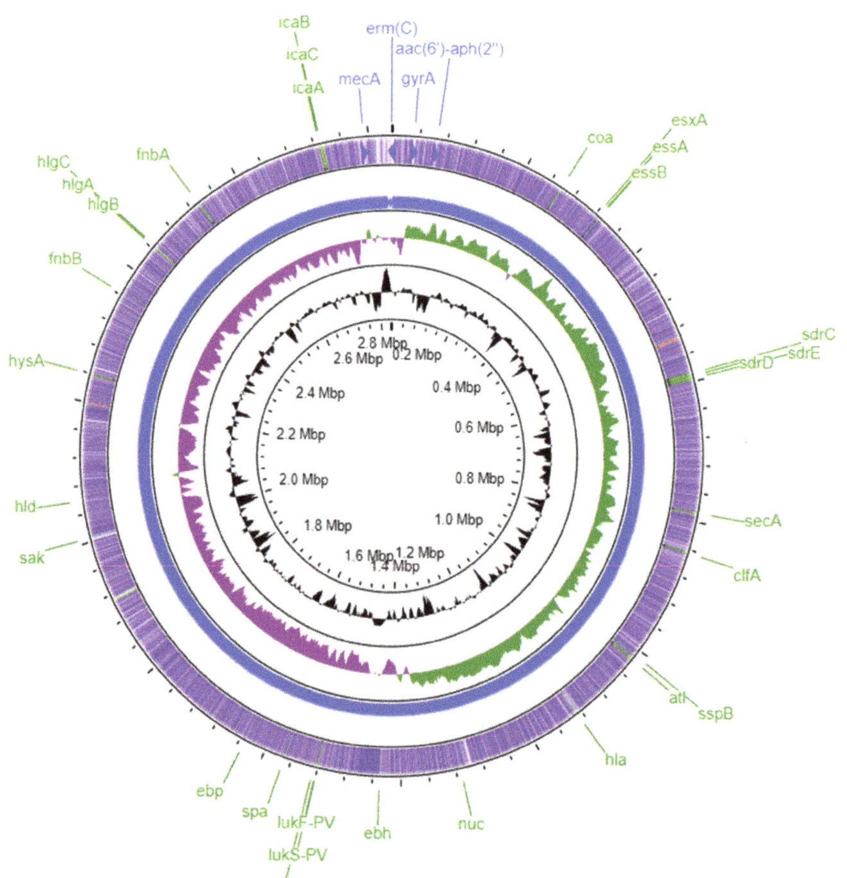

Figure 1. Circular visualization of Lr2 genome via CGViewer, showing virulence factors (light green) and antibiotic resistance determinants (light blue).

The Virulence Factor Database (VFDB) predicated 62 virulence factors responsible for adherence, protein secretion, immune evasion, and toxin production. Adherence-associated genes are autolysin (*atl*); clumping factor A (*clfA*) and B (*clfB*); an elastin binding protein (*ebp*); cell-wall-associated fibronectin-binding protein (*ebh*); collagen adhesion (*cna*); fibrinogen-binding protein (*efb*); fibronectin-binding protein A (*fnbA*) and B (*fnbB*); Ser-Asp rich fibrinogen-binding protein C (*sdrC*), D (*sdrD*), and E (*sdrE*); intercellular adhesin (*icaA*, *icaB*, and *icaC*); and staphylococcal protein A (*spa*) (Figure 1). The virulome also consists of several exoenzymes genes, including hyaluronate lyase (*hysA*), staphylokinase (*sak*), cysteine protease (*sspB* and *sspC*), serine V8 protease (*sspa*), lipase (*geh* and *lip*), staphylocoagulase (*coa*), and thermonuclease (*nuc*). However, a six gene-cluster (*splA*, *splB*, *splC*, *splD*, *splE*, and *splF*) of serine was absent in Lr2 (Figure 1). The Lr2 genome also contains virulence factors associated with the host immune evasion, such as staphylococcal complement inhibitor (*scn*), IgG-binding protein (*sbi*), and chemotaxis inhibiting protein (*chp*) (Figure 1). It also carries type VII secretion system associated virulence factors, including soluble cytosolic (*esaB* and *esaG*) and membrane-associated protein A (*essA*), B (*essB*), and C (*essC*), and A (*esxA*) but no *esxB*, *esxC*, *esxD*, *esaD*, and *esaE* were found in Lr2 genome (Figure 1).

The genome of Lr2 carries toxin associated genes, particularly alpha-hemolysin gene (*hla*), gamma hemolysin (*hlgA*, *hlgB*, and *hlgC*), delta hemolysin gene (*hld*), exotoxins (*set10*, *set13*, *set15*, and *set16*), enterotoxin-like O (*selo*), exotoxins (*set1*, *set2*, *set4*, *set7*, *set10*, *set13*, *set15*, and *set16*), enterotoxin-like M (*selm*), enterotoxin-like N (*seln*), enterotoxin B (*seg*), PVL (*lukF-PV* and *lukS-PV*), and TSST (tsst-1). However, beta-hemolysin gene (*hlb*), exfoliative toxin type A (*eta*), enterotoxin A (*sea*), enterotoxin Yent1 (*yent1*), enterotoxin-like K (*selk*), exotoxins (*set34*, *set37*, and *set39*), and leukotoxins (*lukM*, *lukD*, and *lukE*) were found absent (Figure 1).

2.4. Predicted Plasmids and Prophages

PlasmidFinder identified one plasmid of 2402 bp length with 99.75% sequence similarity to *S. aureus* strain E14 plasmid pDLK1 (GU562624.1). The pDLK1 plasmid consists of an erythromycin resistance gene (*emrC*) and a replication gene (*repL*), and it encodes no other factors.

The genome of Lr2 strain presents seven putative prophages including a complete prophage (PHAGE_Staphy_tp310_3), five incomplete prophages (PHAGE_Bacill_IEBH, PHAGE_Staphy_PT1028, PHAGE_Clostr_phiC2, PHAGE_Staphy_phiPVL_CN125, and PHAGE_Staphy_96), and a questionable prophage (PHAGE_Staphy_phiN315) (Table 3). Several virulence factors were predicted in the identified prophages, e.g., *sec*, *sell*, and *tsst* in the prophage PHAGE_Bacill_IEBH; the *ebp*, *lukF-PV*, and *lukS-PV* in PHAGE_Staphy_phiPVL_CN125; the *seg*, *sei*, *yent2*, *selm*, *seln*, and *selo* in PHAGE_Staphy_96; *sak*, *chp*, and *scn* in PHAGE_Staphy_tp310_3; and *cna* in PHAGE_Staphy_phiN315 (Table 3). However, no antibiotic resistance gene was found in any of the identified prophages.

Table 3. Characteristics of prophages present in MRSA Lr2 ST22.

Prophage	Length (Kb)	Total Proteins	Phage Hit Proteins	GC %	Annotation	Most Common Phage	Virulence Factors
1 (incomplete)	9	15	7	32.8	Transposase, tail	PHAGE_Bacill_IEBH	None
2 (incomplete)	19	24	20	31.6	Integrase	PHAGE_Staphy_PT1028	*sec, sell, tsst*
3 (incomplete)	9.1	20	10	29.0	Head	PHAGE_Clostr_phiC2	None
4 (incomplete)	42.8	42	35	31.9	Integrase	PHAGE_Staphy_phiPVL_CN125	*ebp, lukF-PV, lukS-PV*
5 (incomplete)	12.5	24	21	28.8	Portal, transposase	PHAGE_Staphy_96	*seg, sei, yent2, selm, seln, selo*
6 (complete)	55.1	79	68	32.6	Tail, head, portal, terminase, integrase	PHAGE_Staphy_tp310_3	*sak, chp, scn*
7 (questionable)	46	37	26	33.2	Tail, transposase, integrase	PHAGE_Staphy_phiN315	*cna*

2.5. Comparative Phylogenetic Analysis

The whole-genome-based single nucleotide polymorphisms (SNPs) phylogenetic analysis of the Lr2 ST22 strain with all other ST22 strains indicated that the Lr2 strain exhibits the nearest identities to South African CA-MRSA ST22 strain 71A_S11 (CP010940) and Indian strain CA-MRSA ST22 VB31683 (CP035671) (Figure 2).

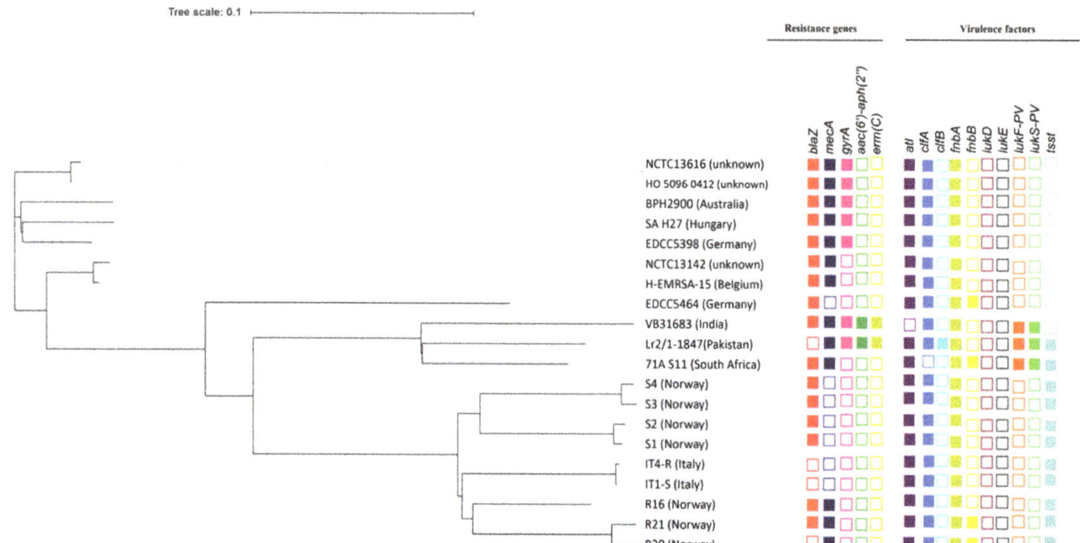

Figure 2. Whole-genome-based SNPs phylogenetic tree of CA-MRSA ST22 strain Lr2 and other ST22 strains. The heatmap shows presence and absence of antimicrobial resistance determinants and virulence factors.

2.6. Comparative Analysis of Antibiotic Resistance Determinants and Virulence Factors

The heatmap shows that aminoglycoside resistance gene *aac(6′)-Ie-aph(2″)-Ia* and a plasmid-associated gene *erm(C)* conferring resistance to lincosamide, macrolide (erythromycin), and streptogramin are only present in Lr2 and Indian strain VB31683. Both Lr2 and Indian strain VB31683 also have a point mutation in *gyrA*, conferring fluoroquinolones resistance, which is absent in a majority of ST22 strains. However, the penicillin-resistance gene (*blaZ*) is absent in Lr2, while it is present in almost all ST22 strains. In contrast, the South African strain 71A_S11, a sister strain to Lr2, carries only *blaZ* and *mecA* genes (Figure 2). However, South African strain 71A_S11 has a very similar profile of virulence factors to that of the Lr2 strain since both TSST (*tsst*) and PVL (*lukF-PV* and *lukS-PV*) genes are only present in these two ST22 strains. In addition, Lr2 strains also harbor clumping factor B (*clfB*), which is absent in South African strain 71A_S11 and all other ST22 strains. Interestingly, leukotoxin D (*lukD*) and E (*lukE*) were absent in all ST22 strains, including the Lr2 strain (Figure 2).

2.7. Pan-Genome Analysis and Functional Annotation

The pan genome of MRSA ST22 strains consists of 2941 genes, of which 2408 (81.8%) genes are part of the core genome, 321 (10.9%) genes are accessory, and 212 (7.2%) genes are unique (Figure 3A). The plot of the pan and core genome shows fluctuation in the number of pan and core genes, which suggests that pan- and core-genome sizes are not stable (Figure 3B).

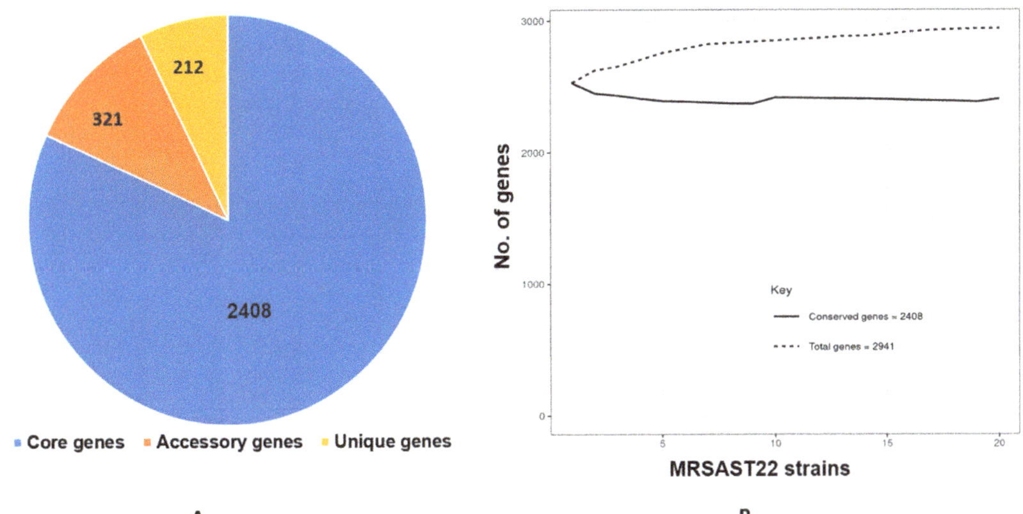

Figure 3. The pan genome of MRSA ST22 strains. (**A**) The pie chart shows number of genes in core, accessory, and unique genomes of twenty genomes of ST22. (**B**) The pan genome vs. core genome plot of ST22 strains.

The functional annotation of core genes shows that 960 (44.7%) Cluster of Orthologous Genes (COGs) are involved in metabolism and transport; 440 (20.5%) in information, storage, and processing; and 416 (19.4%) in cellular processing and signaling; 317 (14.8%) are poorly characterized, and 14 (0.7%) are found to be associated with mobilome (Figure 4A). The functional annotation of the genes unique to MRSA strain Lr2 ST22 revealed that 18 (32.1%) COGs are involved in metabolism and transport, 18 (32.1%) in information, storage, and processing, 9 (16.1%) are associated with mobilome, 6 (10.7%) are poorly characterized (either involve in general function and/or unknown function), and 5 (8.9%) are involved in cellular processing and signaling (Figure 4B).

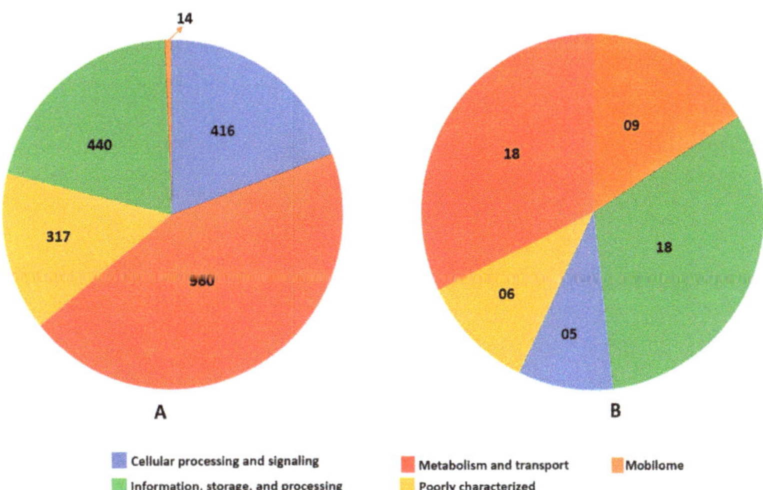

Figure 4. Functional annotation of core genome and genes unique to strain Lr2. (**A**) Distribution of COGs in core genome of MRSA strains ST22. (**B**) Distribution of COGs unique to strain Lr2.

The largest core-genome category of MRSA strains ST22 consists of genes with functions associated with metabolism and transport, which are further categorized as C (110 genes), involved in energy conversion and production; E (223 genes), involved in transport and metabolism of amino acid; F (75 genes), involved in transport and metabolism of nucleic acids; G (179 genes), involved in transport and metabolism of carbohydrates; H (121 genes), involved in transport and metabolism of coenzyme; I (80 genes), involved in lipid transport and metabolism; P (140 genes), involved in inorganic ions transport and metabolism; and Q (32 genes), involved in biosynthesis, transport, and catabolism of secondary metabolites (Figure 5). The second-largest COGs category with functions related to information, storage, and processing are individually categorized as A (2 genes), involved in modification and processing of RNAs; B (1 gene), involved in structure and dynamics of chromatins; J (200 genes), involved in ribosomal structure, biogenesis, and translation; K (139 genes), involved in transcription; and L (98 genes), involved in repair, recombination, and replication (Figure 5). The COGs category of core genome containing the lowest number of genes belonging to mobilome, prophages, and transposons are categorized as X, which accounts for about 0.7% of the total core genome (Figure 5).

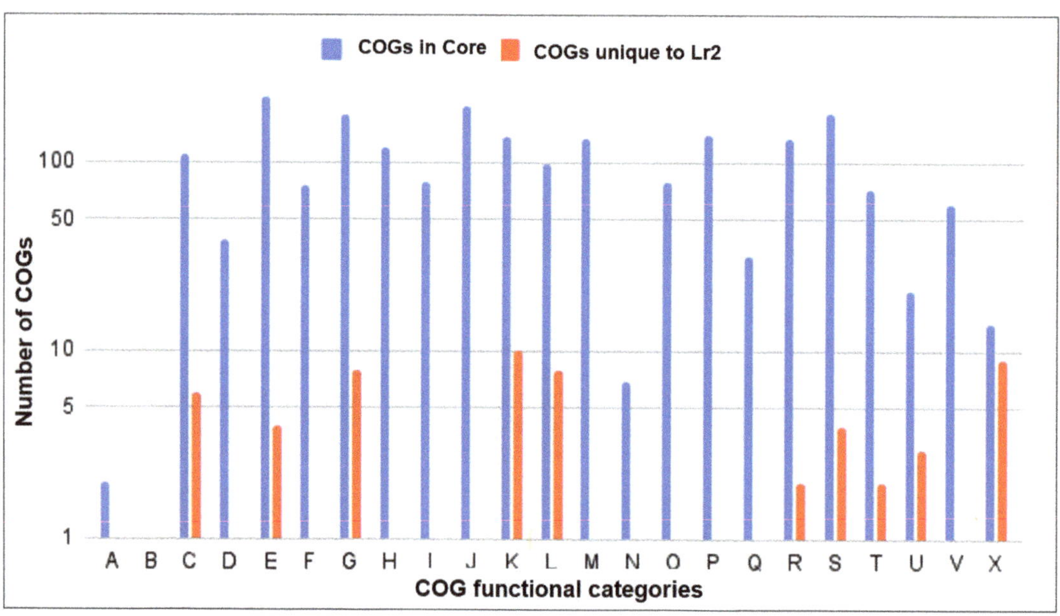

Figure 5. Comparative functional categories of COG in core genome of ST22 strains and COGs unique to strain Lr2.

As expected, the largest category of COGs unique to Lr2 involved in metabolism and transport and are individually categorized as six genes in the C category (involved in energy conversion and production), four in the E category (involved in transport and metabolism of amino acid), and eight in the G category (involved in transport and metabolism of carbohydrates). No protein of F (involved in transport and metabolism of nucleic acids), H (involved in transport and metabolism of coenzyme), I (involved in lipid transport and metabolism), P (involved in inorganic ions transport and metabolism), and Q (involved in biosynthesis, transport, and catabolism of secondary metabolites) category was found (Figure 5). Interestingly, the third-largest category of COGs unique to Lr2 belongs to the X category, which encodes genes associated with prophages and transposons (Figure 5). Cellular processing and signaling is the lowest COGs category with five genes, which are individually classified as three genes in the U category (involved in intracellular trafficking, secretions, and vesicular transport) and two in the T category (involved in signal transduction mechanisms) (Figure 5).

In the pan genome of MRSA ST22 strains, 5 COGs were found to be present in all 19 strains while absent in the Lr2 strain. Functional annotation showed that those COGs are involved in transposase (COG3666) and putative lipase (COG5153), which is essential for the disintegration of autophagic bodies inside the vacuole.

The core-genome phylogenetic analysis grouped the Lr2 strain with a South African ST22 strain 71A S11 strain and an Indian ST22 strain VB31683 (Figure 6). The phylogenetic relationships determined from the core genome are in agreement with the phylogenetic tree from whole-genome-based SNPs (Figure ?) The strains share more than 80% of genes; however, the accessory genome of MRSA strain Lr2 contains many genes that are absent in most of the strains (Figure 6).

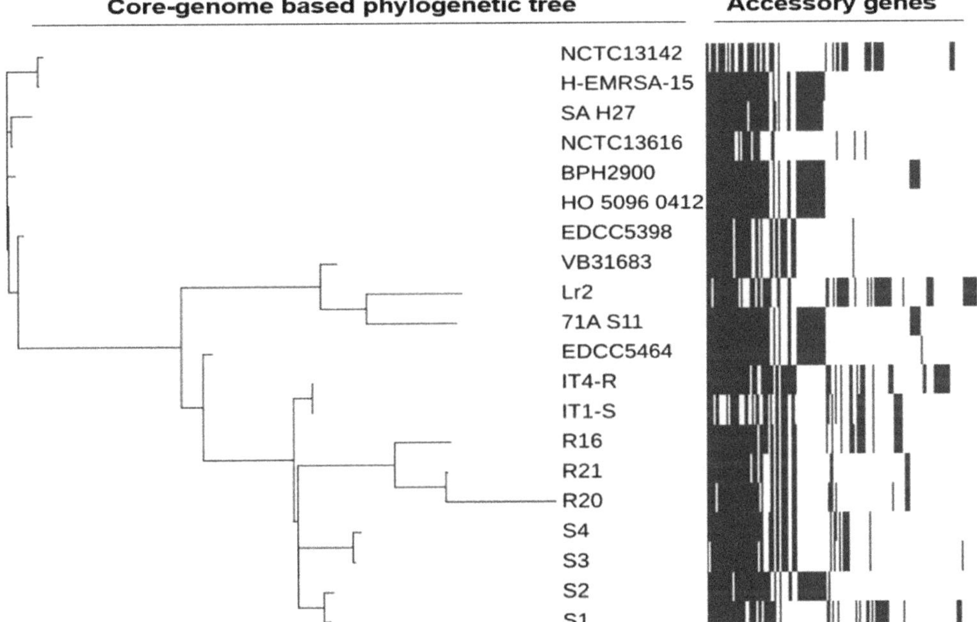

Figure 6. Core-genome SNPs-based tree and heatmap of presence and absence of accessory genes in 20 strains of MRSA ST22.

3. Discussion

The MRSA SCC*mec* type IV ST22 is a clinically important clone (EMRSA-15), and several outbreaks of ST22 CA-MRSA and/or CA-MSSA have been documented worldwide [24 26]. The ST22 clone is considered a genotype of HA-EMRSA-15; however, the presence of ST22 clones in the general population and its correlation with community settings have been reported recently [27]. The dissemination of MRSA from hospitals to the community and vice versa and the emergence of strains resistant to β-lactams is a major cause of concern worldwide. Consequently, tracking of emerging clones of MRSA at the genomic level is required to prevent further spread and to guide the development of rapid diagnostic tools and therapy. Although a small number of studies from Pakistan molecularly characterized MRSA, including SCC*mec*, PFGE, and MLST typing, as well as some studies that investigated PVL genes [19,28–30], none of them performed a genome-based analysis. This shows that there are limited epidemiological and genomic data on MRSA reported from Pakistan, leaving a substantial knowledge gap in our understanding of this important human pathogen that what are the dominant clones of MRSA circulating in Pakistan and what makes them antibiotic-resistant and virulent. Therefore, we performed an in-depth genome-based analysis of a CA-MRSA ST22 strain from Pakistan and its comparative genomic analysis with other MRSA ST22 available in the NCBI database.

Genome analysis indicated that Lr2 belongs to MLST type ST22, spa-type t2986, and SCC*mec* type IVa(2B). According to previous studies, *SCCmec* types IV is associated with CA-MRSA, whereas HA-MRSA mostly exhibits *SCCmec* types I, II, or III [22,31]. The strain was also found to be positive for PVL (*lukF-PV* and *lukS-PV*), which is commonly considered a CA-MRSA marker [22,32]. Another study reported a high rate of MRSA PVL-positive isolates from Pakistan, which were classified as community-associated [28]. The PVL-positive strains are responsible for abscesses formation, tissue necrosis, and increased inflammatory responses [33]. In addition to PVL, TSST-producing MRSA strains tend to cause more complex infections. The presence of both *tsst* and PVL (*lukF-PV* and *lukS-PV*) genes poses a concern for increased virulence of Lr2. A recent study also reported both PVL (*lukF-PV* and *lukS-PV*) genes and the *tsst* gene in MRSA isolates from this region [19]; however, the presence of PVL genes in combination with *tsst* gene appears to be extremely rare. The *tsst* and PVL (*lukF-PV* and *lukS-PV*) genes were found on PHAGE_Staphy_PT1028 and PHAGE_Staphy_phiPVL_CN125, respectively. The genes encoding staphylococcal enterotoxins (i.e., *selm*, *seg*, *yent2*, *sei*, *selo*, and *seln*) were found on PHAGE_Staphy_96, which are involved in staphylococcal food poisoning and belong to the enterotoxin gene cluster (*egc*) [34]. In addition, the staphylococcal complement inhibitor (SCIN) encoding gene *scn* as well as chemotaxis inhibiting protein of staphylococcus (CHIPS) encoding gene *chp* were also found on prophages of the sequenced strain. These proteins have a significant role in the host-pathogen interaction and help the pathogen to evade the host's immune response [35]. Therefore, the prophages predicted in this strain can act as a reservoir of virulence factors and could contribute to strain evolution towards high virulency and pose a serious threat [36]. Aside from the prophage-associated virulence factors, genes involved in the *S. aureus* type VII secretion system, such as *essA*, *essB*, *essC*, *esxA*, *esaA*, *esaB*, and *esaG*, were also present in PVL-positive CA-MRSA strain Lr2 (Figure 1), which promote bacterial persistence [37]. The *esxA* gene is involved in the colonization and dissemination of *S. aureus* as well as triggering T-cell immune response [38]. The genes encoding adhesion factors, including fibrinogen-binding protein (*efb*), collagen-binding protein (*cna*), Ser-Asp rich fibrinogen-binding proteins C (*sdrC*), D (*sdrD*), and E (*sdrE*), elastin-binding protein (*ebp*), clumping factors A (*clfA*), and fibronectin-binding protein (*fnbA*) and (*fnbB*), were also found in the studied genome. These surface components have several functions, including adhesion to host cells, evasion of immunological responses, and biofilm formation [39].

In silico screening of antibiotic-resistant determinants shows that *mecA* gene, *aac(6')-Ie-aph(2'')-Ia*, and a *gyrA* gene are present on the Lr2 chromosome. However, *erm(C)*, which confers resistance to lincosamide, macrolide (erythromycin), and streptogramin was found on plasmid pDLK1 (Figure 1). It is suggested that *erm(C)* is essential to this isolate for environmental adaptation [40]. It was noticed that *blaZ* gene is absent in the studied genome, and since the *blaZ* gene (β-lactamase) is carried by a transposon Tn*552* located on a large plasmid [41], it was probably eliminated due to the curing of that plasmid.

The SNPs phylogenetic relationship inferred from whole genome and core genome revealed that strain Lr2 exhibits the nearest identities to South African CA-MRSA ST22 strain 71A_S11 (CP010940) and makes a separate clade with Indian strain CA-MRSA ST22 VB31683 (CP035671) (Figure 2). The heatmap of accessory genes comparison revealed variation in accessory genes among different strains showing gain or loss of genes as well as mobile genetic elements (MGEs) such as plasmids, prophages, and *SCCmec* element, etc. (Figure 6). This pattern of variation in genes was also observed in previous studies [22,42]. These variations can be attributed to their distinct genetic makeup, as we noticed diverse distribution of antibiotic resistance determinants and virulence factors in these strains. We observed that *blaZ*, *mecA*, and *gyrA* antibiotic resistance genes are present in most ST22 strains. However, *blaZ* was absent in strains Lr2, R20, IT1-S, and IT4-R, while *aac(6')-Ie-aph(2'')-Ia* and *erm(C)* are unique to Lr2 and Indian strain VB31683. A similar pattern of variation in virulence factors was also observed as the PVL genes (*lukF-PV* and *lukS-PV*) were only found in Lr2, VB31683, and 71A_S11 strains. A virulence factor clumping factor B (*clfB*), was only present in Lr2 (Figure 2). These additional genes associated with antibiotic resistance and virulence are expected to make Lr2 more resistant to antibiotics and virulent.

The pan-genome analysis of MRSA ST22 strains revealed that more than 80% of genes are part of the core genome, which shows high conservancy in these strains. The core genome annotation using the Cluster of Orthologous Gene (COG) database revealed the two largest core genome categories with functions associated with metabolism and transport as well as related to information, storage, and processing. Previous studies also reported high conservancy in the *S. aureus* core genome and that the core genes are mostly associated with metabolism, replication, and information storage and processing [22,43,44]. However, genes belonging to mobilome, prophages, and transposons showed the lowest proportion. Interestingly, mobilome is the third-largest category of COGs unique to strain Lr2 (Figure 4). This abundance of mobile genetic elements in the Lr2 strain likely contributes to its increased virulence and antibiotic resistance.

4. Materials and Methods

4.1. Isolation and Antibiotic Susceptibility Testing

S. aureus isolates were collected from hospitals in Peshawar, Rawalpindi/Islamabad, and Lahore, and 4 isolates were randomly selected for whole-genome sequencing. Of the 4 sequenced strains, P10 and R46 were found to belong to ST113 [20], Lr12 belong to a new sequence type (ST5352) [21], and Lr2 to ST22. The isolate Lr2 was collected from blood culture in a hospital in Lahore, Pakistan, in March 2019. The isolate was preliminarily identified by biochemical tests (catalase, oxidase, and coagulase) [45]. The antimicrobial susceptibility was performed by disc diffusion method as per CLSI guidelines against the following antibiotics: ampicillin (10 μg), oxacillin (1 μg), methicillin (10 μg), vancomycin (5 μg), gentamicin (10 μg), erythromycin (15 μg), streptomycin (25 μg), clindamycin (2 μg), ciprofloxacin (5 μg), chloramphenicol (30 μg), linezolid (30 μg), tetracycline (30 μg), rifampicin (5 μg), and fusidic acid (10 μg) [46]. Furthermore, resistance to methicillin was confirmed by PCR amplification of the *mecA* gene [20].

4.2. Genome Sequencing, Assembly, and Annotation

The genomic DNA (gDNA) was extracted from a fresh culture of Lr2 by Invitrogen® DNA extraction kit Cat no. K1820-01 (Thermo Fisher Scientific, Carlsbad, CA, USA). The gDNA integrity was checked by 0.7% agarose gel and quantified by Qubit 2.0 fluorometer Cat no. Q32866 (Manufactured by Tecan Austria GmbH, Grodig, Austria for Life Technologies). The gDNA libraries were prepared by Nextera XT Library Prep Kit (Illumina, San Diego, CA, USA), which was used as per instructions with a slight modification. Sequencing was performed in Illumina (HiSeq) system using a 250 bp paired-end protocol, and Trimmomatic 0.30 was used to trim adapters from raw reads [47]. The resultant reads were de novo assembled using SPAdes version 3.7 [48], and the generated contigs of Lr2 were annotated by NCBI's Prokaryotic Genome Annotation Pipeline (PGAP) [49]. Finally, the assembled contigs were reordered with *S. aureus* reference genome NCTC8325 using Mauve [50], and the genome was visualized using CGView server (http://cgview.ca/, accessed on 3 December 2021) [51].

4.3. Genome-Based Characterization

In silico epidemiological characteristics of the assembled genome were carried out using SCCmecFinder-1.2 for the identification of *SCCmec* type [52], MLST 1.8 [53] for Multilocus Sequence Typing, and *spa*Typer 1.0 [54] for *spa* typing available at Center for Genomic Epidemiology (CGE) webserver (https://cge.cbs.dtu.dk/services/, accessed on 22 August 2021).

4.4. Prediction of Resistome, Virulome, and Mobilome

The chromosomal mutations and acquired genes conferring antibiotic resistance were identified by ResFinder 4.1 at CGE (https://cge.cbs.dtu.dk/services/ResFinder/, accessed on 22 August 2021) [55]. Virulence factors in the genome were identified and annotated using the Virulence Factor Database (VFDB) at http://www.mgc.ac.cn/VFs/, accessed on 22 August 2021 [56]. The assembled genome was searched for plasmid replicons (*rep*) in PlasmidFinder 2.1 using default parameters [57]. The identified plasmid replicons (*rep*) were searched in PLSDB (Plasmid Database) for the identification of full-length plasmids, and the full-length plasmids were then BLASTed with the Lr2 genome [58]. The prophage sequences in the assembled genome were identified and annotated by PHASTER online tool [59]. The identified plasmids and prophages were also investigated for genes associated with antimicrobial resistance and virulence.

4.5. Whole-Genome Single Nucleotide Polymorphism (SNP)-Based Phylogenetic Analysis

The whole-genome SNPs phylogenetic analysis of MRSA Lr2 ST22 strain was analyzed against a set of publicly available ST22 complete genomes (n = 19, according to PATRIC, https://www.patricbrc.org/, accessed on 29 March 2021). The SNPs were called against the reference genome NCTC_8325 (GenBank accession no. CP000253.1), and the maximum-likelihood tree was established using FastTree 2 tool [60] in CSI Phylogeny (https://cge.cbs. dtu.dk/services/CSIPhylogeny/, accessed on 15 April 2021) [61]. The following default settings were applied: SNP positions minimum depth 10, SNP positions relative depth 10, a minimum distance between SNPs (prune) 10, minimum quality of SNP 30, and minimum Z-score of SNP 1.96. The phylogenetic tree was visualized, and a heatmap of the presence and absence of antibiotic resistance genes and virulence factors was generated using Interactive Tree of Life (iTOL) [62].

4.6. Pan-Genome and Cluster of Orthologous Genes (COGs) Analysis

The MRSA ST22 genomes were annotated by Prokka at default parameters for estimation of pan genome [63]. The pangenome analysis and core-genome SNPs-based phylogenetic analysis were performed using the pangenome estimation module (PGM) of an in-house pipeline PanRV, which makes use of Roary [64,65]. The functional annotation of core genome of ST22 strains and COGs unique to Lr2 ST22 strain was performed using the functional annotation analysis module (FAM) of the in-house PanRV pipeline with 0.001 E-value, bit score 100, and percentage identity 70 [66,67].

5. Conclusions

This study provides important epidemiological and genomic data on a PVL-positive CA-MRSA ST22 strain Lr2 from Pakistan. The comparative analysis of the resistome shows that, unlike other ST22 strains, the strain Lr2 carries a plasmid-associated antibiotic resistance gene *erm(C)* and an aminoglycoside resistance gene *aac(6′)-Ie-aph(2″)-Ia*. The strain also harbors several prophages with genes encoding important *S. aureus* virulence factors such as PVL and TSST. These antibiotic resistance- and virulence-associated genes possibly make this ST22 clone highly antibiotic-resistant and virulent, and these genes could be transferred to other MRSA clones through horizontal gene transfer. However, further studies are needed on a large collection of isolates to determine the molecular epidemiology, evolution, and dynamics of transmission of this clone in Pakistan.

Author Contributions: Conceptualization, N.U. and A.A. (Amjad Ali); formal analysis, N.U., S.N., F.A. and Z.I.; investigation, N.U. and T.R.; methodology, N.U., A.A. (Amjad Ali), A.A. (Abdulrahman Alshammari), M.A., T.B., M.R. and A.R.; project administration, A.A. (Amjad Ali); software, N.U., S.N., F.A. and Z.I.; supervision, A.A. (Amjad Ali); funding acquisition, A.A. (Amjad Ali) and A.A. (Abdulrahman Alshammari); writing—original draft, N.U.; writing—review and editing, T.B., A.A. (Abdulrahman Alshammari), M.R. and A.A. (Amjad Ali). All authors have read and agreed to the published version of the manuscript.

Funding: This study was financially supported by the Higher Education Commission (HEC), Pakistan, under the NRPU project (4774) and King Saud University, Riyadh, Saudi Arabia, under the Researchers Supporting Project (RSP2022R491).

Institutional Review Board Statement: The study was approved by Atta-ur-Rahman School of Applied Biosciences (ASAB), NUST (Ref. # IRB-133, Dated: 6 December 2018).

Informed Consent Statement: Not applicable.

Data Availability Statement: The PVL-positive CA-MRSA strain Lr2 ST22 whole-genome sequence reads has been submitted in the NCBI SRA (SRR15497842) and sequence data in NCBI GenBank (JAIGOF000000000) under the BioProject PRJNA520768.

Conflicts of Interest: The authors declare that they have no conflict of interest.

References

1. D'Souza, N.; Rodrigues, C.; Mehta, A. Molecular Characterization of Methicillin-Resistant *Staphylococcus aureus* with Emergence of Epidemic Clones of Sequence Type (ST) 22 and ST 772 in Mumbai, India. *J. Clin. Microbiol.* **2010**, *48*, 1806–1811. [CrossRef] [PubMed]
2. Taylor, T.A.; Unakal, C.G. *Staphylococcus Aureus*; StatPearls Publishing: Treasure Island, FL, USA, 2017.
3. Mitevska, E.; Wong, B.; Surewaard, B.G.J.; Jenne, C.N. The Prevalence, Risk, and Management of Methicillin-Resistant *Staphylococcus aureus* Infection in Diverse Populations across Canada: A Systematic Review. *Pathogens* **2021**, *10*, 393. [CrossRef] [PubMed]
4. Yeo, W.-S.; Jeong, B.; Ullah, N.; Shah, M.A.; Ali, A.; Kim, K.K.; Bae, T. Ftsh Sensitizes Methicillin-Resistant *Staphylococcus aureus* to β-Lactam Antibiotics by Degrading YpfP, a Lipoteichoic Acid Synthesis Enzyme. *Antibiotics* **2021**, *10*, 1198. [CrossRef]
5. Goering, R.V.; Shawar, R.M.; Scangarella, N.E.; O'Hara, F.P.; Amrine-Madsen, H.; West, J.M.; Dalessandro, M.; Becker, J.A.; Walsh, S.L.; Miller, L.A. Molecular epidemiology of methicillin-resistant and methicillin-susceptible *Staphylococcus aureus* isolates from global clinical trials. *J. Clin. Microbiol.* **2008**, *46*, 2842–2847. [CrossRef]
6. Leme, R.C.P.; Bispo, P.J.M.; Salles, M.J. Community-genotype methicillin-resistant *Staphylococcus aureus* skin and soft tissue infections in Latin America: A systematic review. *Braz. J. Infect. Dis.* **2021**, *25*, 101539. [CrossRef] [PubMed]
7. Cuny, C.; Wieler, L.H.; Witte, W. Livestock-associated MRSA: The impact on humans. *Antibiotics* **2015**, *4*, 521–543. [CrossRef] [PubMed]
8. Hayakawa, K.; Yamaguchi, T.; Ono, D.; Suzuki, H.; Kamiyama, J.; Taguchi, S.; Kiyota, K. Two Cases of Intrafamilial Transmission of Community-Acquired Methicillin-Resistant *Staphylococcus aureus* Producing Both PVL and TSST-1 Causing Fatal Necrotizing Pneumonia and Sepsis. *Infect. Drug Resist.* **2020**, *13*, 2921. [CrossRef]
9. Boakes, E.; Kearns, A.; Ganner, M.; Perry, C.; Hill, R.; Ellington, M.J. Distinct bacteriophages encoding Panton-Valentine leukocidin (PVL) among international methicillin-resistant *Staphylococcus aureus* clones harboring PVL. *J. Clin. Microbiol.* **2011**, *49*, 684–692. [CrossRef]
10. Vandenesch, F.; Naimi, T.; Enright, M.C.; Lina, G.; Nimmo, G.R.; Heffernan, H.; Liassine, N.; Bes, M.; Greenland, T.; Reverdy, M.-E. Community-acquired methicillin-resistant *Staphylococcus aureus* carrying Panton-Valentine leukocidin genes: Worldwide emergence. *Emerg. Infect. Dis.* **2003**, *9*, 978. [CrossRef]
11. Pokhrel, R.H.; Aung, M.S.; Thapa, B.; Chaudhary, R.; Mishra, S.; Kawaguchiya, M.; Urushibara, N.; Kobayashi, N. Detection of ST772 Panton-Valentine leukocidin-positive methicillin-resistant *Staphylococcus aureus* (Bengal Bay clone) and ST22 *S. aureus* isolates with a genetic variant of elastin binding protein in Nepal. *New Microbes New Infect.* **2016**, *11*, 20–27. [CrossRef]
12. DeLeo, F.R.; Otto, M.; Kreiswirth, B.N.; Chambers, H.F.J.T.L. Community-associated meticillin-resistant *Staphylococcus aureus*. *Lancet* **2010**, *375*, 1557–1568. [CrossRef]
13. Tenover, F.C.; McDougal, L.K.; Goering, R.V.; Killgore, G.; Projan, S.J.; Patel, J.B.; Dunman, P.M. Characterization of a Strain of Community-Associated Methicillin-Resistant *Staphylococcus aureus* Widely Disseminated in the UnitedStates. *J. Clin. Microbiol.* **2006**, *44*, 108–118. [CrossRef] [PubMed]
14. Takano, T.; Higuchi, W.; Otsuka, T.; Baranovich, T.; Enany, S.; Saito, K.; Isobe, H.; Dohmae, S.; Ozaki, K.; Takano, M.; et al. Novel characteristics of community-acquired methicillin-resistant *Staphylococcus aureus* strains belonging to multilocus sequence type 59 in Taiwan. *Antimicrob. Agents Chemother.* **2008**, *52*, 837–845. [CrossRef] [PubMed]
15. Yamamoto, T.; Dohmae, S.; Saito, K.; Otsuka, T.; Takano, T.; Chiba, M.; Fujikawa, K.; Tanaka, M. Molecular characteristics and in vitro susceptibility to antimicrobial agents, including the des-fluoro (6) quinolone DX-619, of Panton-Valentine leukocidin-positive methicillin-resistant *Staphylococcus aureus* isolates from the community and hospitals. *Antimicrob. Agents Chemother.* **2006**, *50*, 4077–4086. [CrossRef]
16. Gostev, V.; Ivanova, K.; Kruglov, A.; Kalinogorskaya, O.; Ryabchenko, I.; Zyryanov, S.; Kalisnikova, E.; Likholetova, D.; Lobzin, Y.; Sidorenko, S. Comparative genome analysis of global and Russian strains of community-acquired methicillin-resistant *Staphylococcus aureus* ST22, a 'Gaza clone'. *Int. J. Antimicrob. Agents* **2021**, *57*, 106264. [CrossRef]

17. Reuter, S.; Török, M.E.; Holden, M.T.; Reynolds, R.; Raven, K.E.; Blane, B.; Donker, T.; Bentley, S.D.; Aanensen, D.M.; Grundmann, H. Building a genomic framework for prospective MRSA surveillance in the United Kingdom and the Republic of Ireland. *Genome Res.* **2016**, *26*, 263–270. [CrossRef]

18. Holden, M.T.; Hsu, L.-Y.; Kurt, K.; Weinert, L.A.; Mather, A.E.; Harris, S.R.; Strommenger, B.; Layer, F.; Witte, W.; de Lencastre, H. A genomic portrait of the emergence, evolution, and global spread of a methicillin-resistant *Staphylococcus aureus* pandemic. *Genome Res.* **2013**, *23*, 653–664. [CrossRef]

19. Khan, S.; Marasa, B.S.; Sung, K.; Nawaz, M. Genotypic Characterization of Clinical Isolates of *Staphylococcus aureus* from Pakistan. *Pathogens* **2021**, *10*, 918. [CrossRef]

20. Ullah, N.; Dar, H.A.; Naz, K.; Andleeb, S.; Rahman, A.; Saeed, M.T.; Hanan, F.; Bae, T.; Ali, A.J.A. Genomic Investigation of Methicillin-Resistant *Staphylococcus aureus* ST113 Strains Isolated from Tertiary Care Hospitals in Pakistan. *Antibiotics* **2021**, *10*, 1121. [CrossRef]

21. Ullah, N.; Raza, T.; Dar, H.A.; Shehroz, M.; Zaheer, T.; Naz, K.; Ali, A. Whole-genome sequencing of a new sequence type (ST5352) strain of community-acquired methicillin-resistant *Staphylococcus aureus* from a hospital in Pakistan. *J. Glob. Antimicrob. Resist.* **2019**, *19*, 161–163. [CrossRef]

22. Naorem, R.S.; Blom, J.; Fekete, C.J.P. Genome-wide comparison of four MRSA clinical isolates from Germany and Hungary. *PeerJ* **2021**, *9*, e10185. [CrossRef] [PubMed]

23. Punina, N.; Makridakis, N.; Remnev, M.; Topunov, A. Whole-genome sequencing targets drug-resistant bacterial infections. *Hum. Genom.* **2015**, *9*, 19. [CrossRef] [PubMed]

24. Chang, Q.; Abuelaish, I.; Biber, A.; Jaber, H.; Callendrello, A.; Andam, C.P.; Regev-Yochay, G.; Hanage, W.P.; Eurosurveillance, P.S.G.J. Genomic epidemiology of meticillin-resistant *Staphylococcus aureus* ST22 widespread in communities of the Gaza Strip, 2009. *Eurosurveillance* **2018**, *23*, 1700592. [CrossRef] [PubMed]

25. Manoharan, A.; Zhang, L.; Poojary, A.; Bhandarkar, L.; Koppikar, G.; Robinson, D.A. An outbreak of post-partum breast abscesses in Mumbai, India caused by ST22-MRSA-IV: Genetic characteristics and epidemiological implications. *Epidemiol. Infect.* **2012**, *140*, 1809–1812. [CrossRef] [PubMed]

26. Tinelli, M.; Monaco, M.; Vimercati, M.; Ceraminiello, A.; Pantosti, A. Methicillin-susceptible *Staphylococcus aureus* in skin and soft tissue infections, Northern Italy. *Emerg. Infect. Dis.* **2009**, *15*, 250. [CrossRef]

27. Toleman, M.S.; Watkins, E.R.; Williams, T.; Blane, B.; Sadler, B.; Harrison, E.M.; Coll, F.; Parkhill, J.; Nazareth, B.; Brown, N.M.; et al. Investigation of a Cluster of Sequence Type 22 Methicillin-Resistant *Staphylococcus aureus* Transmission in a Community Setting. *Clin. Infect. Dis.* **2017**, *65*, 2069–2077. [CrossRef]

28. Madzgalla, S.; Syed, M.; Khan, M.; Rehman, S.; Müller, E.; Reissig, A.; Ehricht, R.; Monecke, S. Molecular characterization of *Staphylococcus aureus* isolates causing skin and soft tissue infections in patients from Malakand, Pakistan. *Eur. J. Clin. Microbiol. Infect. Dis.* **2016**, *35*, 1541–1547. [CrossRef]

29. Shabir, S.; Hardy, K.J.; Abbasi, W.S.; McMurray, C.L.; Malik, S.A.; Wattal, C.; Hawkey, P.M. Epidemiological typing of meticillin-resistant *Staphylococcus aureus* isolates from Pakistan and India. *J. Med. Microbiol.* **2010**, *59*, 330–337. [CrossRef]

30. Zafar, A.; Stone, M.; Ibrahim, S.; Parveen, Z.; Hasan, Z.; Khan, E.; Hasan, R.; Wain, J.; Bamford, K. Prevalent genotypes of meticillin-resistant Staphylococcus aureus: Report from Pakistan. *J. Med. Microbiol.* **2011**, *60*, 56–62. [CrossRef]

31. Chua, K.Y.; Howden, B.P.; Jiang, J.-H.; Stinear, T.; Peleg, A.Y. Genetics Evolution. Population genetics and the evolution of virulence in Staphylococcus aureus. *Infect. Genet. Evol.* **2014**, *21*, 554–562. [CrossRef]

32. Shukla, S.K.; Pantrang, M.; Stahl, B.; Briska, A.M.; Stemper, M.E.; Wagner, T.K.; Zentz, E.B.; Callister, S.M.; Lovrich, S.D.; Henkhaus, J.K.; et al. Comparative Whole-Genome Mapping to Determine *Staphylococcus aureus* Genome Size, Virulence Motifs, and Clonality. *J. Clin. Microbiol.* **2012**, *50*, 3526–3533. [CrossRef] [PubMed]

33. Yuan, W.; Liu, J.; Zhan, Y.; Wang, L.; Jiang, Y.; Zhang, Y.; Sun, N.; Hou, N. Molecular typing revealed the emergence of pvl-positive sequence type 22 methicillin-susceptible *Staphylococcus aureus* in Urumqi, Northwestern China. *Infect. Drug Resist.* **2019**, *12*, 1719. [CrossRef] [PubMed]

34. Chen, T.-R.; Chiou, C.-S.; Tsen, H.-Y. Use of novel PCR primers specific to the genes of staphylococcal enterotoxin G, H, I for the survey of *Staphylococcus aureus* strains isolated from food-poisoning cases and food samples in Taiwan. *Int. J. Food Microbiol.* **2004**, *92*, 189–197. [CrossRef] [PubMed]

35. Van Wamel, W.J.; Rooijakkers, S.H.; Ruyken, M.; van Kessel, K.P.; van Strijp, J.A.G. The innate immune modulators staphylococcal complement inhibitor and chemotaxis inhibitory protein of *Staphylococcus aureus* are located on β-hemolysin-converting bacteriophages. *J. Bacteriol.* **2006**, *188*, 1310–1315. [CrossRef]

36. Barrera-Rivas, C.I.; Valle-Hurtado, N.A.; González-Lugo, G.M.; Baizabal-Aguirre, V.M.; Bravo-Patiño, A.; Cajero-Juárez, M.; Valdez-Alarcón, J.J. Bacteriophage Therapy: An alternative for the treatment of *Staphylococcus aureus* infections in animals and animal models. *Front. Staphylococcus Aureus* **2017**, *10*, 179–201.

37. Tchoupa, A.K.; Watkins, K.E.; Jones, R.A.; Kuroki, A.; Alam, M.T.; Perrier, S.; Chen, Y.; Unnikrishnan, M. The type VII secretion system protects *Staphylococcus aureus* against antimicrobial host fatty acids. *Sci. Rep.* **2020**, *10*, 14838.

38. Burts, M.L.; Williams, W.A.; DeBord, K.; Missiakas, D.M. EsxA and EsxB are secreted by an ESAT-6-like system that is required for the pathogenesis of *Staphylococcus aureus* infections. *Proc. Natl. Acad. Sci. USA* **2005**, *102*, 1169–1174. [CrossRef]

39. Azmi, K.; Qrei, W.; Abdeen, Z. Screening of genes encoding adhesion factors and biofilm production in methicillin resistant strains of *Staphylococcus aureus* isolated from Palestinian patients. *BMC Genom.* **2019**, *20*, 1–12. [CrossRef]

40. Lim, S.; Lee, D.-H.; Kwak, W.; Shin, H.; Ku, H.-J.; Lee, J.-E.; Lee, G.E.; Kim, H.; Choi, S.-H.; Ryu, S.; et al. Comparative genomic analysis of *Staphylococcus aureus* FORC_001 and *S. aureus* MRSA252 reveals the characteristics of antibiotic resistance and virulence factors for human infection. *J. Microbiol. Biotechnol.* **2015**, *25*, 98–108. [CrossRef]

41. Pugazhendhi, A.; Michael, D.; Prakash, D.; Krishnamaurthy, P.P.; Shanmuganathan, R.; Al-Dhabi, N.A.; Duraipandiyan, V.; Arasu, M.V.; Kaliannan, T. Antibiogram and plasmid profiling of beta-lactamase producing multi drug resistant *Staphylococcus aureus* isolated from poultry litter. *J. King Saud Univ. -Sci.* **2020**, *32*, 2723–2727. [CrossRef]

42. Vogel, V.; Falquet, L.; Calderon-Copete, S.P.; Basset, P.; Blanc, D.S. Short term evolution of a highly transmissible methicillin-resistant *Staphylococcus aureus* clone (ST228) in a tertiary care hospital. *PLoS ONE* **2012**, *7*, e38969. [CrossRef] [PubMed]

43. Bosi, E.; Monk, J.M.; Aziz, R.K.; Fondi, M.; Nizet, V.; Palsson, B. Comparative genome-scale modelling of *Staphylococcus aureus* strains identifies strain-specific metabolic capabilities linked to pathogenicity. *Proc. Natl. Acad. Sci. USA* **2016**, *113*, E3801–E3809. [CrossRef] [PubMed]

44. Liao, F.; Mo, Z.; Gu, W.; Xu, W.; Fu, X.; Zhang, Y. A comparative genomic analysis between methicillin-resistant *Staphylococcus aureus* strains of hospital acquired and community infections in Yunnan province of China. *BMC Infect. Dis.* **2020**, *20*, 137. [CrossRef] [PubMed]

45. Karmakar, A.; Dua, P.; Ghosh, C. Biochemical and molecular analysis of *Staphylococcus aureus* clinical isolates from hospitalized patients. *Can. J. Infect. Dis. Med. Microbiol.* **2016**, *2016*, 9041636. [CrossRef]

46. Patel, J.B.; Cockerill, F.; Bradford, P.A. Performance standards for antimicrobial susceptibility testing: Twenty-fifth informational supplement. *Clin. Lab. Stand. Inst.* **2015**, *35*, 29–50.

47. Bolger, A.M.; Lohse, M.; Usadel, B. Trimmomatic: A flexible trimmer for Illumina sequence data. *Bioinformatics* **2014**, *30*, 2114–2120. [CrossRef]

48. Bankevich, A.; Nurk, S.; Antipov, D.; Gurevich, A.A.; Dvorkin, M.; Kulikov, A.S.; Lesin, V.M.; Nikolenko, S.I.; Pham, S.; Prjibelski, A.D. SPAdes: A new genome assembly algorithm and its applications to single-cell sequencing. *J. Comput. Biol.* **2012**, *19*, 455–477. [CrossRef]

49. Tatusova, T.; DiCuccio, M.; Badretdin, A.; Chetvernin, V.; Nawrocki, E.P.; Zaslavsky, L.; Lomsadze, A.; Pruitt, K.; Borodovsky, M.; Ostell, J. NCBI prokaryotic genome annotation pipeline. *Nucleic Acids Res.* **2016**, *44*, 6614–6624. [CrossRef]

50. Darling, A.C.E.; Mau, B.; Blattner, F.R.; Perna, N.T. Mauve: Multiple alignment of conserved genomic sequence with rearrangements. *Genome Res.* **2004**, *14*, 1394–1403. [CrossRef]

51. Grant, J.R.; Arantes, A.S.; Stothard, P. Comparing thousands of circular genomes using the CGView Comparison Tool. *BMC Genom.* **2012**, *13*, 202. [CrossRef]

52. Kaya, H.; Hasman, H.; Larsen, J.; Stegger, M.; Johannesen, T.B.; Allesøe, R.L.; Lemvigh, C.K.; Aarestrup, F.M.; Lund, O.; Larsen, A.R.; et al. SCCmecFinder, a Web-Based Tool for Typing of Staphylococcal Cassette Chromosome mec in *Staphylococcus aureus* Using Whole-Genome Sequence Data. *mSphere* **2018**, *3*, e00612-17. [CrossRef] [PubMed]

53. Larsen, M.V.; Cosentino, S.; Rasmussen, S.; Friis, C.; Hasman, H.; Marvig, R.L.; Jelsbak, L.; Sicheritz-Pontén, T.; Ussery, D.W.; Aarestrup, F.M.; et al. Multilocus Sequence Typing of Total-Genome-Sequenced Bacteria. *J. Clin. Microbiol.* **2012**, *50*, 1355–1361. [CrossRef] [PubMed]

54. Bartels, M.D.; Petersen, A.; Worning, P.; Nielsen, J.B.; Larner-Svensson, H.; Johansen, H.K.; Andersen, L.P.; Jarløv, J.O.; Boye, K.; Larsen, A.R.; et al. Comparing Whole-Genome Sequencing with Sanger Sequencing for spa Typing of Methicillin-Resistant Staphylococcus aureus. *J. Clin. Microbiol.* **2014**, *52*, 4305–4308. [CrossRef] [PubMed]

55. Bortolaia, V.; Kaas, R.S.; Ruppe, E.; Roberts, M.C.; Schwarz, S.; Cattoir, V.; Philippon, A.; Allesoe, R.L.; Rebelo, A.R.; Florensa, A.F. ResFinder 4.0 for predictions of phenotypes from genotypes. *J. Antimicrob. Chemother.* **2020**, *75*, 3491–3500. [CrossRef]

56. Chen, L.H.; Yang, J.; Yu, J.; Ya, Z.J.; Sun, L.L.; Shen, Y.; Jin, Q. VFDB: A reference database for bacterial virulence factors. *Nucleic Acids Res.* **2005**, *33*, D325–D328. [CrossRef]

57. Carattoli, A.; Zankari, E.; García-Fernández, A.; Larsen, M.V.; Lund, O.; Villa, L.; Aarestrup, F.M.; Hasman, H. In Silico Detection and Typing of Plasmids using PlasmidFinder and Plasmid Multilocus Sequence Typing. *Antimicrob. Agents Chemother.* **2014**, *58*, 3895–3903. [CrossRef]

58. Galata, V.; Fehlmann, T.; Backes, C.; Keller, A. PLSDB: A resource of complete bacterial plasmids. *Nucleic Acids Res.* **2018**, *47*, D195–D202. [CrossRef]

59. Arndt, D.; Grant, J.R.; Marcu, A.; Sajed, T.; Pon, A.; Liang, Y.; Wishart, D.S. PHASTER: A better, faster version of the PHAST phage search tool. *Nucleic Acids Res.* **2016**, *44*, W16–W21. [CrossRef]

60. Price, M.N.; Dehal, P.S.; Arkin, A.P. FastTree 2-Approximately Maximum-Likelihood Trees for Large Alignments. *PLoS ONE* **2010**, *5*, e9490. [CrossRef]

61. Kaas, R.S.; Leekitcharoenphon, P.; Aarestrup, F.M.; Lund, O. Solving the Problem of Comparing Whole Bacterial Genomes across Different Sequencing Platforms. *PLoS ONE* **2014**, *9*, e104984. [CrossRef]

62. Letunic, I.; Bork, P. Interactive Tree of Life (iTOL) v4: Recent updates and new developments. *Nucleic Acids Res.* **2019**, *47*, W256–W259. [CrossRef] [PubMed]

63. Seemann, T. Prokka: Rapid prokaryotic genome annotation. *Bioinformatics* **2014**, *30*, 2068–2069. [CrossRef] [PubMed]

64. Naz, K.; Naz, A.; Ashraf, S.T.; Rizwan, M.; Ahmad, J.; Baumbach, J.; Ali, A. PanRV: Pangenome-reverse vaccinology approach for identifications of potential vaccine candidates in microbial pangenome. *BMC Bioinform.* **2019**, *20*, 123. [CrossRef] [PubMed]

65. Page, A.J.; Cummins, C.A.; Hunt, M.; Wong, V.K.; Reuter, S.; Holden, M.T.; Fookes, M.; Falush, D.; Keane, J.A.; Parkhill, J. Roary: Rapid large-scale prokaryote pan genome analysis. *Bioinformatics* **2015**, *31*, 3691–3693. [CrossRef] [PubMed]
66. Galperin, M.Y.; Makarova, K.S.; Wolf, Y.I.; Koonin, E.V. Expanded microbial genome coverage and improved protein family annotation in the COG database. *Nucleic Acids Res.* **2015**, *43*, D261–D269. [CrossRef]
67. Tatusov, R.L.; Galperin, M.Y.; Natale, D.A.; Koonin, E.V. The COG database: A tool for genome-scale analysis of protein functions and evolution. *Nucleic Acids Res.* **2000**, *28*, 33–36. [CrossRef]

Article

Alternative Pathways to Ciprofloxacin Resistance in *Neisseria gonorrhoeae*: An In Vitro Study of the WHO-P and WHO-F Reference Strains

Natalia González [1,*], Saïd Abdellati [2], Irith De Baetselier [2], Jolein Gyonne Elise Laumen [1,3], Christophe Van Dijck [1,3], Tessa de Block [2], Chris Kenyon [1,4,†] and Sheeba Santhini Manoharan-Basil [1,*,†]

1. STI Unit, Department of Clinical Sciences, Institute of Tropical Medicine, 2000 Antwerp, Belgium; jlaumen@itg.be (J.G.E.L.); cvandijck@itg.be (C.V.D.); ckeyon@itg.be (C.K.)
2. Clinical Reference Laboratory, Department of Clinical Sciences, Institute of Tropical Medicine, 2000 Antwerp, Belgium; sabdellati@itg.be (S.A.); idebaetselier@itg.be (I.D.B.); tdeblock@itg.be (T.d.B.)
3. Laboratory of Medical Microbiology, University of Antwerp, 2610 Wilrijk, Belgium
4. Division of Infectious Diseases and HIV Medicine, University of Cape Town, Anzio Road, Observatory, Cape Town 7700, South Africa
* Correspondence: ngonzalez@ext.itg.be (N.G.); sbasil@itg.be (S.S.M.-B.)
† These authors contributed equally to this work.

Abstract: Emerging resistance to ceftriaxone and azithromycin has led to renewed interest in using ciprofloxacin to treat *Neisseria gonorrhoeae*. This could lead to the rapid emergence and spread of ciprofloxacin resistance. Previous studies investigating the emergence of fluoroquinolone resistance have been limited to a single strain of *N. gonorrhoeae*. It is unknown if different genetic backgrounds affect the evolution of fluoroquinolone resistance in *N. gonorrhoeae*, as has been shown in other bacterial species. This study evaluated the molecular pathways leading to ciprofloxacin resistance in two reference strains of *N. gonorrhoeae*—WHO-F and WHO-P. Three clones of each of the two strains of *N. gonorrhoeae* were evolved in the presence of ciprofloxacin, and isolates from different time points were whole-genome sequenced. We found evidence of strain-specific differences in the emergence of ciprofloxacin resistance. Two out of three clones from WHO-P followed the canonical pathway to resistance proceeding via substitutions in GyrA-S91F, GyrA-D95N and ParC. None of the three WHO-F clones followed this pathway. In addition, mutations in *gyrB*, *uvrA* and *rne* frequently occurred in WHO-F clones, whereas mutations in *yhgF*, *porB* and *potA* occurred in WHO-P.

Keywords: *N. gonorrhoeae*; fluoroquinolone; AMR; resistance; ciprofloxacin

Citation: González, N.; Abdellati, S.; De Baetselier, I.; Laumen, J.G.E.; Van Dijck, C.; de Block, T.; Kenyon, C.; Manoharan-Basil, S.S. Alternative Pathways to Ciprofloxacin Resistance in *Neisseria gonorrhoeae*: An In Vitro Study of the WHO-P and WHO-F Reference Strains. *Antibiotics* **2022**, *11*, 499. https://doi.org/10.3390/antibiotics11040499

Academic Editor: Teresa V. Nogueira

Received: 22 March 2022
Accepted: 6 April 2022
Published: 8 April 2022

Publisher's Note: MDPI stays neutral with regard to jurisdictional claims in published maps and institutional affiliations.

1. Introduction

The World Health Organization (WHO) has categorized *Neisseria gonorrhoeae* as being at high risk for the emergence of further antimicrobial resistance (AMR) [1]. In particular, the emergence of combined gonococcal resistance to ceftriaxone and azithromycin has renewed interest in reintroducing fluoroquinolones (FQs) as therapeutic agents [1,2]. The resistance mechanisms developed by *N. gonorrhoeae* to ciprofloxacin (CIP) are due to point mutations in the quinolone resistance determining region (QRDR) of the *gyrA* and *parC* genes [3–6]. One study found that CIP could be successfully used to treat gonococcal isolates that were negative for the S91F mutation in DNA gyrase subunit A [2]. An understandable concern of this approach is that this will result in a renewed increase in fluoroquinolone resistance in *N. gonorrhoeae*.

The initial introduction of fluoroquinolones as a treatment for gonorrhoea was followed by an explosive increase in the prevalence of resistance to FQs [1,7]. The prevalence of resistance has declined in recent years in a number of locales, and it would be prudent only to reintroduce FQ to treat gonorrhoea in a way that minimizes the selection pressure for resistance [8]. An important unknown in this regard is if all gonococcal strains have the same propensity to develop FQ resistance. Studies in *Mycobacterium tuberculosis*, *Staphylococcus aureus* and *Clostridium difficile* have found large variations in this propensity [9–11]. For example, in the case of *M. tuberculosis*, in vitro experiments have confirmed that differences in the genetic background can result in two orders of magnitude differences in the frequency of the emergence of FQ resistance [9].

In a recent phylogenetic analysis of a global collection containing 17,871 isolates of *N. gonorrhoeae*, we found that the prevalence of FQ resistance varied considerably according to multilocus sequence type (MLST) [12]. About twenty-five MLSTs were represented by 100 or more isolates. If the GyrA-S91F substitution is used to define ciprofloxacin resistance, then of these 25 MLSTs, 13 MLSTs were predominantly resistant (median 97.4% [IQR 95.6–98.4%] resistance). The remaining 12 MLSTs were predominantly susceptible to ciprofloxacin (median 2.7% [IQR 0.2–10.1%] resistance). The same was true for high-level resistance (HLR; MIC to ciprofloxacin $\geq$16 mg/L). Of the 2384 ciprofloxacin HLR isolates detected, half of the isolates (50.4%) belonged to ST1901, followed by ST7363 (11.49%) and ST1579 (2.5%).

The molecular pathway to ciprofloxacin resistance in *N. gonorrhoeae* has been previously been assessed in vitro [3]. However, this study was limited to a single strain (FA-19), was not performed in duplicate, and the molecular sequencing was limited to the *gyrA/B* and *parC/E* genes [3]. In preliminary work, we noted that despite identical ciprofloxacin MICs (0.004 mg/L), the WHO-P strain of *N. gonorrhoeae* appeared to acquire resistance to ciprofloxacin more rapidly than WHO-F. Therefore, in this work, we aimed to assess if there is a difference between WHO-F and WHO-P in (1) order of acquisition of mutations in the target genes, (2) time to ciprofloxacin resistance, and (3) the molecular pathway to ciprofloxacin resistance.

2. Methods

2.1. WHO-Reference Strains, Genetic Characteristics and Comparative Genomics

The strains used in this study were *N. gonorrhoeae* WHO reference strains (WHO-F and WHO-P), both susceptible to ciprofloxacin with a minimum inhibitory concentration (MIC) of 0.004 mg/L. The genetic background of the two strains are as follows: (i) WHO-F belongs to *porB1a* serogroup with a WT *mtrR* promoter region, *rpoB* and *rpsJ* genes (ii) WHO-P belongs to *porB1b* serogroup, with A to C substitution in *mtrR* promoter region, and substitutions in RpoB- H552N and RpsJ-V57M [13]. Furthermore, the clades of WHO-F and WHO-P were further investigated to delineate the molecular pathways to FQ resistance using the previously defined whole genome MLST (wgMLST) and core genome MLST (cgMLST) schemes according to Manoharan-Basil et al. 2022. Briefly, WGS data that comprised of 17, 871 *N. gonorrhoeae* isolates were analyzed using chewBBACA version 2.8.5 [14]. A training file using the complete genome of *N. gonorrhoeae* FA1090 was created using Prodigal [15] and was followed by creating a study specific schema using two complete *N. gonorrhoeae* genomes (FA1090 and MS11). A FASTA file for each coding sequence (CDS) was generated, followed by the creation of wgMLST loci. The cgMLST loci were then extracted from the wgMLST loci and visualized using a grape tree [16].

2.2. Plating Experiment and MIC Determination

All of the strains were grown on a gonococcal (GC) agar plate (Gonococcal Medium Base, BD Difco™) supplemented with 1% IsoVitaleX (BD BBL™) and incubated at 36 °C in an atmosphere of 5% CO_2. Three independent clonal lineages of WHO-F (henceforth referred to as WHO-F_1, WHO-F_2 and WHO-F_3) and WHO-P (henceforth referred to as WHO-P_1, WHO-P_2 and WHO-P_3) strains were evolved from single colonies grown on a GC agar plate. A single colony from each strain was inoculated on a GC agar plate containing 0.004 mg/L ciprofloxacin and incubated at 36 °C with 5% CO_2. After visible growth was attained, typically after 36–72 h, one colony was inoculated onto a GC agar plate in which the drug concentration had been increased two-fold relative to the previous concentration. The selection was continued until a ciprofloxacin (CIP) MIC concentration of 32 mg/L was

Antibiotics **2022**, 11, 499

attained, or no visible growth was seen. The cultures from each time point were stored in skimmed milk supplemented with 20% of glycerol and stored at −80 °C. The number of passages for each strain is presented in Table 1.

Table 1. Evolution of ciprofloxacin resistance over time. The isolates that were subjected to WGS are colored in grey. Asterisk denotes bacterial cell death. X: data not recorded. Asterisk denotes bacterial cell death.

WHO Reference Strains	CIP MIC (mg/L)																					Total No. of Passages
1-23 WHO-F1	0.004	X	X	X	X	0.032	0.032	0.047	X	0.047	X	0.125	*									5
WHO-F2	0.004	0.064	X	X	0.064	0.064	X	0.094	X	0.094	X	X	0.38	X	0.75	4	X	12	*			9
WHO-F3	0.004	0.006	X	X	0.047	0.032	X	0.064	X	X	0.125	*										5
1-23 WHO-P1	0.004	0.008	X	X	0.012	0.125	*															3
WHO-P2	0.004	0.125	X	X	0.125	0.25	X	1.5	1	X	1.5	X	X	X	X	4	*					7
WHO-P3	0.004	0.125	X	X	0.125	X	0.5	X	X	X	2	4	4	X	X	16	X	32	X	X	32	9
1-23 Days	1	2	3	4	5	6	7	8	9	10	11	12	13	14	15	16	17	18	19	20	21	

MICs were determined using E-test strips (BioMerieux, Marcy-l'Étoile, France) on GC agar according to the manufacturer's instructions. The MIC for CIP $\geq$ 16 mg/L, >0.06 mg/L, $0.03 \leq 0.06$ and <0.03 mg/L were classified as being high-level resistant (HLR), resistant (R), intermediate (I) and susceptible (S) to CIP, respectively [12,17]. Geometric mean (GM) MIC are provided where appropriate. The doubling time was estimated to be ~60 min [18].

2.3. Whole-Genome Sequencing (WGS) and SNP Analysis

In this study, 11 clones from WHO-F [WHO-F$_1$ (n = 3); WHO-F$_2$ (n = 4); WHO-F$_3$ (n = 4)] and 12 clones from WHO-P [WHO-P$_1$ (n = 4); WHO-P$_2$ (n = 3); WHO-P$_3$ (n = 5)] strains were subjected to WGS (Table 1). Genomic DNA was extracted using the MasterPure Complete DNA and RNA Purification Kit (Epicenter, Madison, WI, USA) and eluted in nuclease-free water. DNA was outsourced for WGS (GENEWIZ, Germany and Eurofins, Leipzig, Germany) and was sequenced on an Illumina instrument using 150 bp paired-end sequencing chemistry (Illumina, San Diego, CA, USA). The quality of the raw reads was assessed using FASTQC [19]. The raw reads were trimmed for quality (Phred $\geq$ 20) and length ($\geq$32 bases) using trimmomatic (v0.39) [20]). The processed reads from the baseline isolates were assembled using Shovill (v1.0.4) [21], which uses SPAdes for the *denovo* assembly (v3.14.0). The parameters used for *denovo* assembly are as follows: —trim —depth 150 —opts —isolate. The quality of the *denovo* contigs was evaluated using Quast (v5.0.2) [22]. Finally, the draft genome was annotated using Prokka (v1.14.6) [23]. The quality controlled reads were mapped to baseline draft genome using BWA MEM and single nucleotide polymorphisms (SNPs) were determined using freebayes implemented in Snippy (v4.6.0) with default parameters (10× minimum read coverage and 90% read concordance at the variant locus) [24–26]. The raw reads generated are deposited at https://www.ncbi.nlm.nih.gov/sra/PRJNA815938 (accessed on 14 March 2022).

Genetic characterization of the relevant genes associated with ciprofloxacin resistance in WHO-F, WHO-P, and global *Neisseria* spp. collection.

The putative SNPs identified in the relevant genes associated with ciprofloxacin resistance were further examined in the global collection comprising the genomes of *N. gonorrhoeae* (n = 17,871), including the WHO-F and WHO-P, and commensal *Neisseria* spp. (n = 1136) whose provenance and metadata are described elsewhere [12].

2.4. Statistical Analysis

Stata 16.0 (StataCorp, College Station, TX, USA) was used for all analyses. A *p*-value of <0.05 was considered statistically significant. Mann-Whitney *U*-test was conducted to test for significant changes in the MIC.

3. Results

3.1. Association of WHO-F and WHO-P and Ciprofloxacin MICs

WHO-F belongs to ST10934, whereas WHO-P belongs to ST8127 (Figure 1). ST10934 (n = 9, including WHO-F) originated from ST7359 (n = 653) and ST8127 (n = 6, including WHO-P) originated from ST1580 (n = 560). Isolates belonging to ST7359 were all *gyrA* and *parC* wild type (WT). Out of 560 isolates belonging to ST1580, 23 isolates had mutations conferring resistance to ciprofloxacin, 257 were WT, and 280 isolates had no available data. Out of these 22 isolates, different combinations of mutations in the target genes were identified and are as follows: twelve isolates had substitutions in GyrA-S91F, GyrA-D95A and ParC-S87R (GM MIC—10.5 mg/L), four isolates had the GyrA-S91F and ParC-D86N substitutions (GM MIC—0.38 mg/L), two isolates had the GyrA-S91F, GyrA-D95G, ParC-E91Q substitutions (GM MIC—0.07 mg/L), two isolates had the GyrA-S91F, GyrA-D95G and ParC-S87R (GM MIC—4 mg/L), one isolate had the substitutions in GyrA-S91F, and GyrA-D95G (MIC—8 mg/L) and one isolate had only the ParC-S87R substitutions (MIC— 0.02 mg/L).

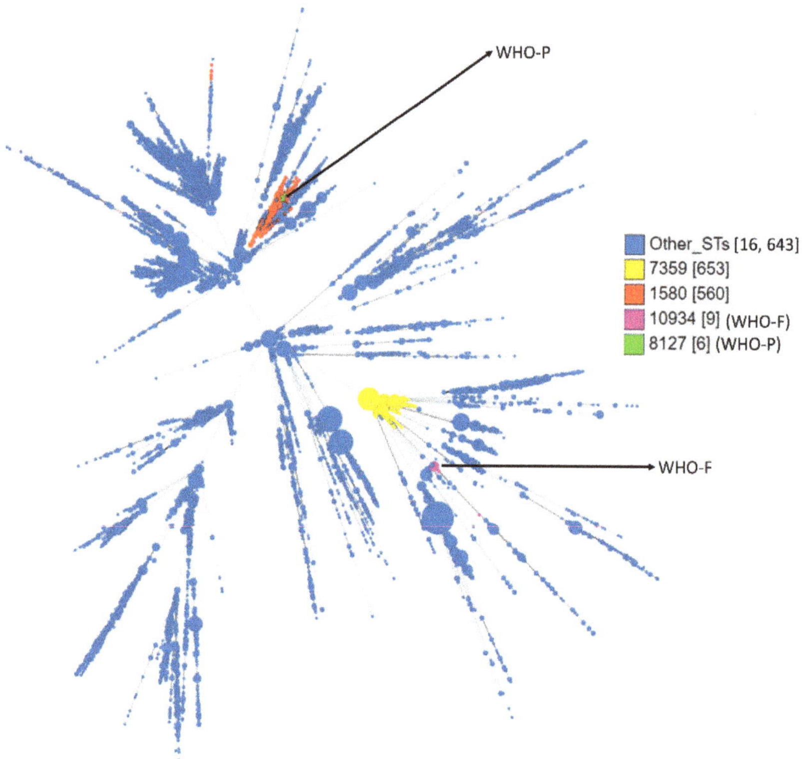

Figure 1. Minimum spanning tree comparing core-genome allelic profiles with MLST resulting in isolates with similar allelic profiles forming clusters. Isolates are displayed as circles. The size of each circle indicates the number of isolates of this particular type. Numbers in brackets refer to the number of isolates. WHO-F (ST10934) is denoted in magenta colour and WHO-P (ST8127) in green color.

3.2. In-Vitro Selection of Ciprofloxacin Resistance in WHO-F and WHO-P

In all six experiments, clones of the WHO-F and WHO-P strains acquired ciprofloxacin resistance (MIC > 0.06 mg/L). Only one clone, WHO-P$_3$, acquired high-level ciprofloxacin resistance and attained a MIC of 32 mg/L by day 18 (~432 generations), representing an increase in MIC compared to baseline of about 8000-fold (Table 1).

3.3. Mutations in Fluoroquinolone Target Genes (gyrA, gyrB, parC and parE)—GyrA S91F Pathway Is Associated with Higher Ciprofloxacin MICs

Mutations were detected in 9 out of the 11 clones of the WHO-F strain [WHO-F$_1$ (2/3); WHO-F$_2$ (3/4); WHO-F3 (4/4)] and in 9 out of the 12 clones of the WHO-P strains [WHO-P$_1$ (3/4); WHO-P$_2$ (2/3); WHO-P$_3$ (4/5)].

The genetic mechanisms underlying the antibiotic resistance in the evolved CIP resistant isolates are as follows:

I. WHO-F
WHO-F$_1$ and WHO-F$_3$ developed resistance after five passages, whereas the isolates of WHO-F$_2$ attained resistance after four passages (Table 1).

(a) WHO-F$_1$ & WHO-F$_3$: WHO-F$_1$ and WHO-F$_3$ acquired the GyrA-D95N substitution at days 5 (~120 generations) and 6 (~144 generations), respectively (MIC−0.032 mg/L). The highest MICs attained by these clones was 0.125 mg/L.

(b) WHO-F$_2$: WHO-F$_2$ acquired the S91Y substitution in the quinolone resistance-determining regions (QRDR) in GyrA at day 2 (~48 generations, MIC−0.064 mg/L) which was followed by additional substitutions in GyrA-D95N (MIC−0.38 mg/L; Day-13 and MIC−12 mg/L; Day-18) and ParC- E91K (MIC−12 mg/L; Day 18).

Additional substitutions outside the QRDR region were also observed and are as follows: GyrA-D80Y in WHO-F$_1$ (MIC−0.125 mg/L; Day 12) and ParC-R537S in WHO-F$_3$ (MIC-0.125 mg/L; Day 11).

II. WHO-P
WHO-P$_1$ took three passages to reach a MIC of 0.125 mg/L while isolates from WHO-P$_2$ and WHO-P$_3$ required one passage of ciprofloxacin exposure to attain the same MIC (Table 1).

(a) WHO-P$_1$: An insertion in GyrB (S467_G468ins) emerged by day 2 in WHO-P$_1$ (MIC−0.008 mg/L) and persisted till day 3 (MIC−0.012 mg/L). On day 6, WHO-P$_1$ acquired the D95N substitution in GyrA, and its MIC increased to 0.125 mg/L, which was the highest MIC attained by WHO-P$_1$.

(b) WHO-P$_2$ & WHO-P$_3$: The WHO-P$_2$ and WHO-P$_3$ clones acquired the GyrA-S91F substitution by day 2 (MIC−0.125 mg/L), followed by substitutions in ParC (ParC-D86N substitution in WHO-P$_2$ by day 16 [MIC 4 mg/L] and ParC-R537L substitution outside the QRDR at day 11 [MIC−0.125 mg/L]). WHO-P$_2$ and WHO-P$_3$ attained higher MICs (4 mg/L and 32 mg/L, respectively) than WHO-P$_1$ (0.125 mg/L).

Among all WHO-F and WHO-P clones, those that acquired the S91F/Y substitution in GyrA (WHO-F$_2$, WHO-P$_2$ & WHO-P$_3$) attained higher ciprofloxacin MICs (12, 4 & 32 mg/L, respectively) than the clones that did not acquire this mutation (WHO-F$_1$, WHO-F$_3$ & WHO-P$_1$, all MIC 0.125 mg/L; $p = 0.037$). The S91F/Y clones also survived for longer (16, 18 & 21 days) than those that did not acquire S91F/Y (7, 12 & 13 days; Table 1; $p = 0.049$).

3.4. Mutations in gyrB, uvrA and rne Frequently Occurred in WHO-F Strains, Whereas Mutations in yhgF, porB and potA Occurred in WHO-P during the Selection for Ciprofloxacin Resistance

WGS analysis revealed that in addition to acquiring the known resistance associated mutations (RAMs- GyrA-S91Y/F, GyrA-D95N, ParC-E91K and ParC-D86N) additional substitutions were differentially detected in WHO-F and WHO-P. In WHO-F, substitutions in *gyrB* [WHO-F$_1$ (n = 1), WHO-F$_3$ (n = 3)], *rne* [WHO-F$_1$ (n = 2)] and *uvrA* [WHO-F$_1$ (n = 1), WHO-F$_3$ (n = 2)] were identified (Figure 2 A, C, E), whereas in WHO-P substitutions were detected in *yhgF* [WHO-P$_2$ (n = 2), WHO-P$_3$ (n = 2)], *porB* [WHO-P$_1$ (n = 1), WHO-P$_3$ (n = 1)] and *potA* [WHO-P$_3$ (n = 1)] (Figure 2B,D,F).

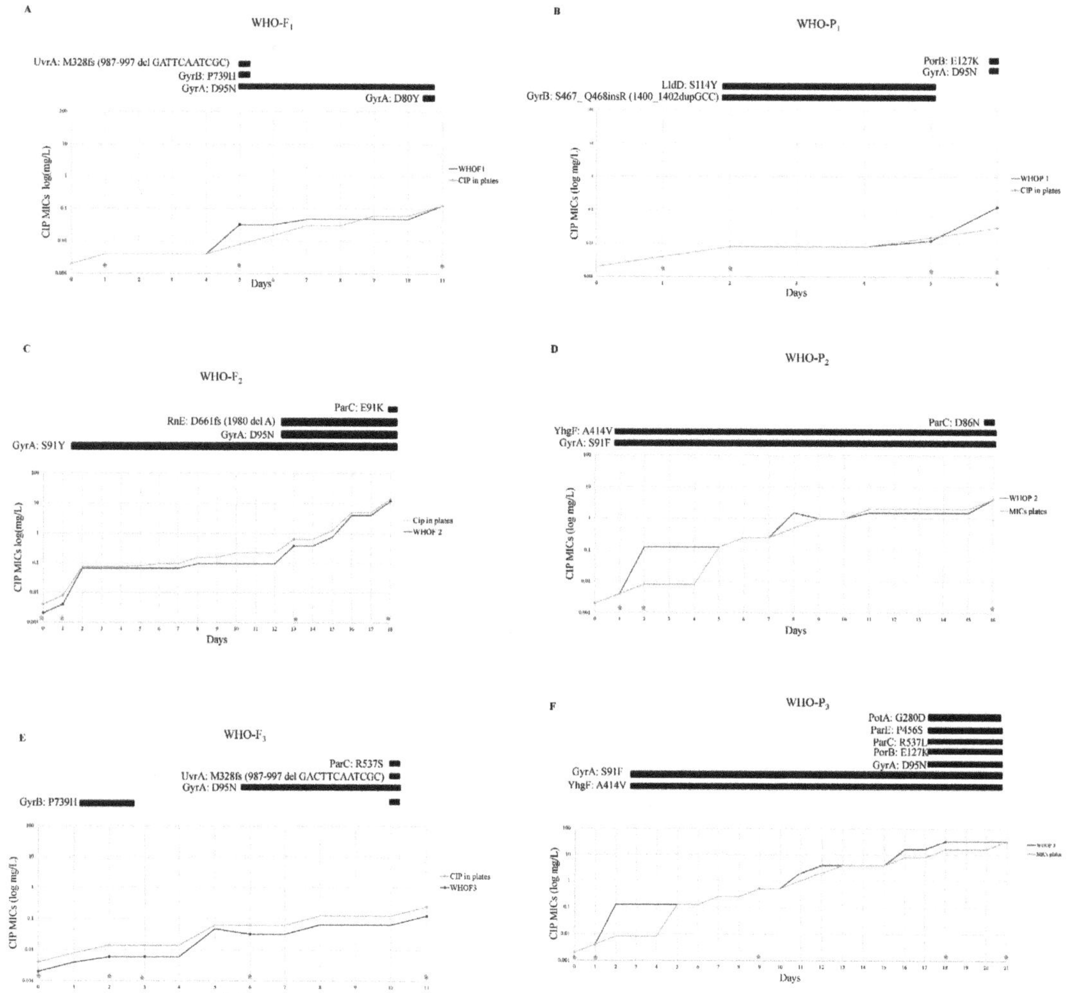

Figure 2. Ciprofloxacin resistance evolution of three *N. gonorrhoeae* clones of WHO-F (**A,C,E**) and three clones of WHO-P (**B,D,F**) over time. The MIC's of ciprofloxacin were tested using E-test (bioMerieux, Marcy-l'Étoile, France) once there was visible growth. Once there was visible growth, the ciprofloxacin MIC's were tested using E-Test (bioMerieux, Marcy-l'Étoile, France) and the next plate was inoculated. The initial MIC of all clones was 0.004 mg/L. The red cross indicates the sampling time that was subjected to WGS. The black squares represent the duration of the mutation noted next to them for as long as the days displayed in the axis below them.

3.4.1. Mutations in *gyrB*, *uvrA* and *rne* in WHO-F

Two clones (WHO-F$_1$ and -F$_3$) acquired substitutions in UvrA. In both cases, this involved a frameshift (fs) caused by a deletion (del) (987–997del GAC TTC AAT CGC; Figure 2A,E), UvrA-M329fsdel, leading to a truncated protein. These substitutions occurred at the same time as P739H substitutions in GyrB and D95N substitutions in GyrA that were associated with intermediate resistance (MICs—0.032 to 0.125 mg/L). Neither WHO-F$_1$ nor WHO-F$_3$ acquired the S91F GyrA substitution. The P739H substitution in GyrB was first observed in WHO-F$_3$ (n = 3) on day one and in WHO-F$_1$ (n = 1) on day five.

In two of the four-time points at which the P739H substitution in GyrB was detected in WHO-F$_3$, no other substitutions were found (Figure 2E). These time points were the first two time points for WHO-F$_3$. The ciprofloxacin MICs for these time points were not elevated compared to baseline (0.004 mg/L on baseline and day 1 and 0.006 mg/L on day 2).

In contrast, WHO-F$_2$ did not acquire these substitutions in GyrB or UvrA but did acquire a frameshift mutation in Rne (M329fs) at two-time points that were contemporaneous with the S91Y substitution in GyrA (Figure 2C).

3.4.2. Mutations in *lldD, porB, potA* and *yhgF* in WHO-P

In two WHO-P$_1$ clones, an insertion in GyrB (S467_G468ins) was observed at the same time a missense mutation in LldD-S114Y that emerged at day 2 (MIC–0.008 mg/L) and persisted until the MIC had increased three-fold (MIC–0.012 mg/L; Figure 2B). E127K substitutions were found to occur in PorB in WHO-P$_1$ and WHO-P$_3$ clones coincident with D95N substitutions in GyrA (MIC-0.125 to 32 mg/L; Figure 2B,F). The YhgF-A414V substitution was observed in two and four isolates belonging to WHO-P$_2$ and WHO-P$_3$ lineages, respectively (Figure 2D,F). This YhgF substitution was always accompanied by the GyrA-S91F substitution (MIC baseline-0.004 mg/L, mutants–0.125 mg/L, 0.5 mg/L, 4 mg/L and 32 mg/L).

The PotA-G280D substitution was found in the last two clones belonging to the WHO-P$_3$ lineage and was accompanied by a number of known RAMs such as S91F and D95N in GyrA as well ParC-R537L and ParE- P456S substitutions (MIC—32 mg/L; Figure 2F).

3.5. Other Mutations

A number of other mutations at single time points were identified and are as follows: (i) a duplication (dup) in transcriptional regulatory protein, ZraR-D5fs (12_13dupCG) that was detected in sample WHOP$_2$ (MIC-4 mg/L) after 16 days of experiment; (ii) a deletion in pilin glycosyltransferase, PglA-R253fs (756delG) was identified after a day of exposure in WHO-F$_1$ (MIC–0.004 mg/L); and (iii) a deletion in ribosomal protein L11 methyltransferase, PrmA-E93del (276–277 del GC) in WHOF$_1$ at day 12 (MIC–0.125 mg/L). The relevance of these mutations to the genesis of ciprofloxacin resistance requires further experimentation.

3.6. Genetic Characterization of Relevant Genes (potA, rne, uvrA and yhgf) in the Genomes of WHO-F and WHO-P

The baselines genomes of WHO-F and WHO-P were aligned, and the following amino acid changes were observed in the relevant genes considered in this manuscript. (i) PotA: Amino acid at positions 34 and 75 in WHO-P were N (Asn) and D (Asp), whereas, in WHO-F, they were D (Asp) and N (Asn), respectively. (ii) RnE: In WHO-P amino acids at positions 399 and 437 were A (Ala) and (R) Arg (R), whereas, in WHO-F, they were V (Val) and S (Ser) (iIi) UvrA: In WHO-P the amino acids at positions 163, 619, 931, and 935 were (G) Gly, D (Asp), (E) Glu and (V) Val, whereas in WHO-F they were A (Ala), (G) Gly, Q (Gln) and (I) Ile, respectively (iv) YhgF: Amino acids at positions 96 and 116 in WHO-P were A (Ala) and R (Arg), and in WHO-F they were T (Thr) and H (His), respectively (Table S1).

3.7. gyrB, parC, porB and yhgF Mutations in the GLOBAL Collection

The GyrB-P739H substitution was not observed in the 17,871 *N. gonorrhoeae* isolates. This substitution was, however, identified in two commensal *Neisseria* spp.—*N. brasiliensis* (PATRIC ID-2666100.4) and *Neisseria* sp. (PubMLST ID-94179). The ParC-R537S/L substitutions were not observed in any of the *Neisseria* spp. (n = 19,007). ParC-R53C and ParC-R537H substitutions were observed in three (GM MIC—0.75 mg/L) *N. gonorrhoeae* isolates. The PorB-E127K substitution in loop 3 was not present in any of the *Neisseria* spp. However, the PorB-E127D substitution was identified in three *N. gonorrhoeae* (MIC– not available) and 102 commensal *Neisseria* spp. (MIC- not available) Furthermore, PorB-E127G (n = 1, MIC–32 mg/L) substitution was found in one *N. gonorrhoeae* isolate (Pathogenwatch ID—ECDC-FR13-066).

4. Discussion

The impacts of different genetic backgrounds in *N. gonorrhoeae* on the evolution of FQ resistance are poorly understood. In this study, we evaluated the pathway to the FQ resistance of two gonococcal WHO reference strains with the same ciprofloxacin MIC (0.004 mg/L). Our global phylogenetic analysis revealed that none of the 653 isolates from ST7359 from which WHO-F emerged had acquired the FQ RAMs. In contrast, 21 out of 560 isolates from the WHO-P lineage associated ST1580 had acquired FQ RAMS.

Our experimental findings are compatible with strain- and pathway-specific differences in the emergence of ciprofloxacin resistance.

4.1. Pathway Specific Differences

Two clones of WHO-P followed the GyrA S91F pathway and one clone of WHO-F followed the similar GyrA S91Y pathway. These three clones attained higher ciprofloxacin MICs than three other clones from WHO-P and -F that acquired D95N- and not S91F-substitutions.

Furthermore, two of the three clones that followed the S91F/Y pathway went on to acquire known RAMs in ParC—D86N in WHO-P_2 and E91K in WHO-F_2—compared to none of the three clones that did not acquire S91F/Y substitutions. These findings are compatible with previous studies that have found that the development of FQ resistance in *N. gonorrhoeae* commences with the GyrA S91F substitution and that subsequent increases in MIC are related to D95N substitutions in GyrA and various mutations in ParC [3,27]. We found two other substitutions in ParC that were outside the QRDR but may be associated with increased MICs—R537L in WHO-P_3 and R537S in WHO-F_3. These substitutions were not found in our global dataset, and the relevance of these substitutions thus requires further investigation.

4.2. Strain-Specific Differences

Whilst one clone of WHO-F followed the S91Y GyrA pathway, this substitution was found far less commonly than the S91F substitution in an analysis of 17, 871 isolates of *N. gonorrhoeae* from around the world (n = 27 versus n = 17,871, respectively). This, plus the finding noted above that the GyrA S91F substitution has not been detected in the WHO-F-like sequence types, suggests that the probability of the S91F substitution emerging in WHO-F may be lower than in other strains such as WHO-P. In addition, we found strain-specific differences in the mutations outside the topoisomerase genes. In WHO-F, mutations were detected in *gyrB*, *uvrA*, and *rne*, whereas in WHO-P, mutations were detected in *yhgF*, *porB*, and *potA* (Figure 3).

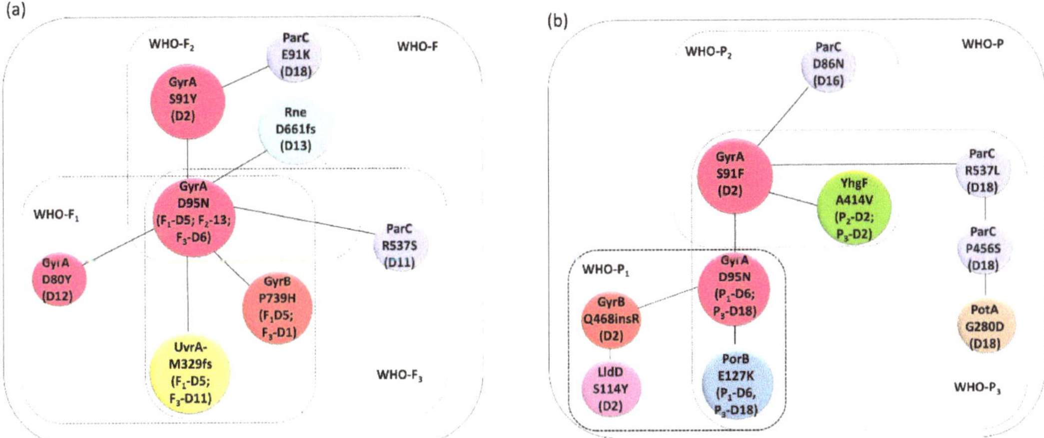

Figure 3. Association network of genes putatively associated with ciprofloxacin resistance in (**a**) WHO-F and (**b**) WHO-P strains. The clones and the days are denoted in parenthesis.

4.3. gyrB, uvrA and rne Mutations Detected in WHO-F

In WHO-F_1 and -F_3, the D95N substitution in GyrA was accompanied by a P739H substitution in GyrB and a M329 frameshift deletion in UvrA. This substitution in GyrB is outside the QRDR and not observed in the 17,871 isolates, and it may therefore reflect a transient mutation, appearing only temporarily and being lost at later stage due to fitness costs or other factors.

The nucleotide excision repair (NER) gene, *uvrA*, is part of the SOS regulon in many bacteria that catalyzes the recognition and processing of DNA damage [28–31]. Ciprofloxacin has been shown to increase the transcription of *uvrA* via the SOS pathway which could play a role in facilitating the subsequent acquisition of antimicrobial resistance [29]. Further experimentation is necessary to better characterize the role of the frameshift mutation in *uvrA* in the genesis of ciprofloxacin resistance in *N. gonorrhoeae*. Notably, in the baseline isolates the amino acids at positions 163, 619, 931 and 935 varied between WHO-F (A163, G619, Q931 and I935) and WHO-P (G163, D619, E31 and V935).

By way of contrast, the pathway to ciprofloxacin resistance in WHO-F_2 included a frameshift mutation in *rne* (Rne-D661fs), ribonuclease E, resulting in a premature stop codon (Figure 2C). In *E. coli* RNase deletion or inactivation precludes the normal initiation of the SOS response [32].

4.4. yhgF, porB and potA Mutations Detected in WHO-P

The A414V substitution in YhgF was detected at the first time point when ciprofloxacin MICs increased in both WHO-P_2 and -P_3 and persisted until the end of both experiments. YhgF has been shown to play a role in *E. coli's* ability to survive ionizing radiation [33]. A genetic interaction screen also established that YhgF has interactions with a number of proteins involved in translation and ribosome biogenesis, such as S1 [34]. For example, Δ*yhgF* mutants exhibit increased levels of stop codon readthrough [34]. In our experiments, there was 100% concordance between the detection of this A414V substitution in YhgF and the S91F substitution in GyrA. This contrasts with the complete absence of the A414V substitution in YhgF in the global collection of *N. gonorrhoeae*, a high proportion of whom have the S91F substitution in GyrA. One way to explain the apparent temporary emergence of the A414V substitution is that it acts as a stepping stone to FQ resistance. Gomez et al. have produced compelling experimental evidence that mutations in ribosomal proteins act as stepping stones to FQ resistance in *Mycobacterium smegmatis* [35]. They found that ciprofloxacin exposure first selected for mutations in ribosomal proteins, which facilitated the subsequent acquisition of resistance-associated mutations in *gyrA*. The ribosomal mutations conferred a fitness cost and thus were lost at some point after acquiring the *gyrA* mutations. Further experiments are required in *N. gonorrhoeae* to test this hypothesis.

The PorB-E127K substitution was observed at the last time point of WHO-P$_1$ and the last two time points of WHO-P$_3$. In both cases, it emerged at the same timepoint as the D95N substitution in GyrA. Once again, it was not found in any of the global collections. The *porB* gene encodes an outer membrane porin that has two mutually exclusive alleles—*PorB1a* in WHO-F and *PorB1b* in WHO-P [36]. Isolates with PorB1a tend to be more susceptible to penicillin and tetracycline [37]. Loop 3 of *PorB1b* folds into the barrel of the porin, constricting the pore. Amino acid substitutions in this loop, such as G120K and A121D, result in reduced susceptibility to cephalosporins, penicillin and tetracyclines [37]. Although speculative, it is possible that the E127K substitution, which is also in loop 3, has a similar effect for FQ. The fact that this substitution was not detected in the global collections would once again be compatible with this substitution having a fitness cost.

Polyamines are involved in several cellular processes such as energy metabolism, oxidative stress tolerance, biofilm formation, and iron transport [38–40]. PotA is a cytoplasmic protein with an ATP-binding motif that couples ATP hydrolysis to translocation of polyamines, part of the *potABCD* operon, encoding the spermidine-preferential uptake system PotABCD [41]. Mutation in *potA* (PotA-Q208L) alters the ATPase activity of the transporter, generating a 4-fold increase in gentamicin in *E. coli* [42]. In our experiments, we found that substitution in PotA-G280D was detected at two-time points when WHO-P$_3$ had acquired HLR (MIC-32 mg/L).

There are a number of limitations to this study. We tested the pathways to FQ resistance in a limited number of sequence types. We did not conduct knockout/complementation studies to assess the biological effect of the various mutations detected. Neither were fitness costs or cross-resistance to other antimicrobials assessed. These limitations notwithstanding, we found strain and pathway-specific variations in the genesis of FQ resistance in *N. gonorrhoeae*. WHO-F type strains may be less prone to develop ciprofloxacin resistance. This finding suggests that the proposed reintroduction of ciprofloxacin for the treatment of gonorrhea may result in less FQ resistance in regions where WHO-F type strains are more prevalent [12]. Hence, surveillance of *N. gonorrhoeae* genotypes may play an ancillary role in the safe reintroduction of FQ for the treatment of gonorrhea.

Supplementary Materials: The following supporting information can be downloaded at: https://www.mdpi.com/article/10.3390/antibiotics11040499/s1, Table S1: Difference of aminoacids between WHO-F and WHO-P strains in relevant genes.

Author Contributions: Conceptualization, N.G., S.A., J.G.E.L., S.S.M.-B. and C.K.; methodology, N.G. and S.A.; formal analysis, N.G., S.S.M.-B. and C.K.; investigation, N.G., S.A., S.S.M.-B. and C.K.; data curation, N.G., S.A., S.S.M.-B. and C.K.; writing—original draft preparation, N.G., S.S.M.-B. and C.K.; writing—review and editing, I.D.B., J.G.E.L., T.d.B., C.V.D. and S.A. All authors have read and agreed to the published version of the manuscript.

Funding: The study was funded by SOFI 2021 grant—"PReventing the Emergence of untreatable STIs via radical Prevention" (PRESTIP).

Data Availability Statement: WGS sequences: https://www.ncbi.nlm.nih.gov/sra/PRJNA815938 (accessed on 14 March 2022).

Conflicts of Interest: The authors declare no conflict of interest.

References

1. Unemo, M.; Bradshaw, C.S.; Hocking, J.S.; de Vries, H.J.C.; Francis, S.C.; Mabey, D.; Marrazzo, J.M.; Sonder, G.J.B.; Schwebke, J.R.; Hoornenborg, E.; et al. Sexually transmitted infections: Challenges ahead. *Lancet Infect. Dis.* **2017**, *17*, e235–e279. [CrossRef]
2. Allan-Blitz, L.-T.; Humphries, R.M.; Hemarajata, P.; Bhatti, A.; Pandori, M.W.; Siedner, M.J.; Klausner, J.D. Implementation of a Rapid Genotypic Assay to Promote Targeted Ciprofloxacin Therapy of *Neisseria gonorrhoeae* in a Large Health System. *Clin. Infect. Dis. Off. Publ. Infect. Dis. Soc. Am.* **2017**, *64*, 1268–1270. [CrossRef] [PubMed]

3. Belland, R.J.; Morrison, S.G.; Ison, C.; Huang, W.M. *Neisseria gonorrhoeae* acquires mutations in analogous regions of gyrA and parC in fluoroquinolone-resistant isolates. *Mol. Microbiol.* **1994**, *14*, 371–380. [CrossRef] [PubMed]

4. Piddock, L.J. Mechanisms of fluoroquinolone resistance: An update 1994–1998. *Drugs* **1999**, *58* (Suppl. 2), 11–18. [CrossRef]

5. Alcalá, B.; Arreaza, L.; Salcedo, C.; Antolín, I.; Borrell, N.; Cacho, J.; Cuevas, C.D.L.; Otero, L.; Sauca, G.; Vázquez, F.; et al. Molecular characterization of ciprofloxacin resistance of gonococcal strains in Spain. *Sex. Transm. Dis.* **2003**, *30*, 395–398. [CrossRef]

6. Bodoev, I.N.; Il'ina, E.N. Molecular mechanisms of formation of drug resistance in *Neisseria gonorrhoeae*: History and prospects. *Mol. Genet. Microbiol. Virol.* **2015**, *30*, 132–140. [CrossRef]

7. Kenyon, C.; Buyze, J.; Wi, T. Antimicrobial Consumption and Susceptibility of *Neisseria gonorrhoeae*: A Global Ecological Analysis. *Front. Med.* **2018**, *5*, 329. [CrossRef]

8. Kenyon, C.; Laumen, J.; Van Dijck, C.; De Baetselier, I.; Abdelatti, S.; Manoharan-Basil, S.S.; Unemo, M. Gonorrhea treatment combined with population-level general cephalosporin and quinolone consumption may select for *Neisseria gonorrhoeae* antimicrobial resistance at the levels of NG-MAST genogroup: An ecological study in Europe. *J. Glob. Antimicrob. Resist.* **2020**, *23*, 377–384. [CrossRef]

9. Castro, R.A.D.; Ross, A.; Kamwela, L.; Reinhard, M.; Loiseau, C.; Feldmann, J.; Borrell, S.; Trauner, A.; Gagneux, S. The Genetic Background Modulates the Evolution of Fluoroquinolone-Resistance in Mycobacterium tuberculosis. *Mol. Biol. Evol.* **2020**, *37*, 195–207. [CrossRef]

10. He, M.; Miyajima, F.; Roberts, P.; Ellison, L.; Pickard, D.J.; Martin, M.J.; Connor, T.; Harris, S.; Fairley, D.; Bamford, K.B.; et al. Emergence and global spread of epidemic healthcare-associated Clostridium difficile. *Nat. Genet.* **2013**, *45*, 109–113. [CrossRef]

11. Holden, M.T.G.; Hsu, L.Y.; Kurt, K.; Weinert, L.A.; Mather, A.E.; Harris, S.R.; Strommenger, B.; Layer, F.; Witte, W.; de Lencastre, H.; et al. A genomic portrait of the emergence, evolution, and global spread of a methicillin-resistant Staphylococcus aureus pandemic. *Genome Res.* **2013**, *23*, 653–664. [CrossRef]

12. Manoharan-Basil, S.S.; González, N.; Laumen, J.G.E.; Kenyon, C. Horizontal gene transfer of fluoroquinolone resistance-conferring genes from commensal Neisseria to *Neisseria gonorrhoeae*: A global phylogenetic analysis of 20,047 isolates. *Front. Microbiol.* **2022**, *13*, 793612. [CrossRef]

13. Unemo, M.; Golparian, D.; Sánchez-Busó, L.; Grad, Y.; Jacobsson, S.; Ohnishi, M.; Lahra, M.M.; Limnios, A.; Sikora, A.E.; Wi, T.; et al. The novel 2016 WHO *Neisseria gonorrhoeae* reference strains for global quality assurance of laboratory investigations: Phenotypic, genetic and reference genome characterization. *J. Antimicrob. Chemother.* **2016**, *71*, 3096–3108. [CrossRef]

14. Silva, M.; Machado, M.P.; Silva, D.N.; Rossi, M.; Moran-Gilad, J.; Santos, S.; Ramirez, M.; Carriço, J.A. chewBBACA: A complete suite for gene-by-gene schema creation and strain identification. *Microb. Genom.* **2018**, *4*, e000166. [CrossRef]

15. Hyatt, D.; Chen, G.L.; LoCascio, P.F.; Land, M.L.; Larimer, F.W.; Hauser, L.J. Prodigal: Prokaryotic gene recognition and translation initiation site identification. *BMC Bioinform.* **2010**, *11*, 119. [CrossRef]

16. Zhou, Z.; Alikhan, N.F.; Sergeant, M.J.; Luhmann, N.; Vaz, C.; Francisco, A.P.; Carriço, J.A.; Achtman, M. Grapetree: Visualization of core genomic relationships among 100,000 bacterial pathogens. *Genome Res.* **2018**, *28*, 1395–1404. [CrossRef]

17. Sánchez-Busó, L.; Yeats, C.A.; Taylor, B.; Goater, R.J.; Underwood, A.; Abudahab, K.; Argimón, S.; Ma, K.C.; Mortimer, T.D.; Golparian, D.; et al. A community-driven resource for genomic epidemiology and antimicrobial resistance prediction of *Neisseria gonorrhoeae* at Pathogenwatch. *Genome Med.* **2021**, *13*, 61. [CrossRef]

18. Dillard, J.P. Genetic Manipulation of *Neisseria gonorrhoeae*. *Curr. Protoc. Microbiol.* **2021**, *23*, Unit4A.2. [CrossRef]

19. Andrews, S. FastQC: A Quality Control Tool for High Throughput Sequence Data [Online]. 2015. Available online: http://www.bioinformatics.babraham.ac.uk/projects/fastqc/ (accessed on 14 March 2022).

20. Bolger, A.M.; Lohse, M.; Usadel, B. Trimmomatic: A flexible trimmer for Illumina sequence data. *Bioinformatics* **2014**, *30*, 2114–2120. [CrossRef]

21. Seemann, T. Shovill. 2019. Available online: https://github.com/tseemann/shovill (accessed on 14 March 2022).

22. Gurevich, A.; Saveliev, V.; Vyahhi, N.; Tesler, G. QUAST: Quality assessment tool for genome assemblies. *Bioinformatics* **2013**, *29*, 1072–1075. [CrossRef]

23. Seemann, T. Prokka: Rapid prokaryotic genome annotation. *Bioinformatics* **2014**, *30*, 2068–2069. [CrossRef] [PubMed]

24. Seemann, T. Snippy-Rapid Haploid Variant Calling and Core Genome Alignment. 2015. Available online: https://github.com/tseemann/snippy (accessed on 14 March 2022).

25. Li, H.; Durbin, R. Fast and accurate short read alignment with Burrows-Wheeler transform. *Bioinformatics* **2009**, *25*, 1754–1760. [CrossRef] [PubMed]

26. Lee, B.M.; Harold, L.K.; Almeida, D.V.; Afriat-Jurnou, L.; Aung, H.L.; Forde, B.M.; Hards, K.; Pidot, S.J.; Ahmed, F.H.; Mohamed, A.E.; et al. Predicting nitroimidazole antibiotic resistance mutations in Mycobacterium tuberculosis with protein engineering. *PLoS Pathog.* **2020**, *16*, e1008287. [CrossRef] [PubMed]

27. Lindbäck, E.; Rahman, M.; Jalal, S.; Wretlind, B. Mutations in gyrA, gyrB, parC, and parE in quinolone-resistant strains of *Neisseria gonorrhoeae*. *APMIS* **2002**, *110*, 651–657. [CrossRef] [PubMed]

28. Drlica, K.; Zhao, X. DNA gyrase, topoisomerase IV, and the 4-quinolones. *Microbiol. Mol. Biol. Rev.* **1997**, *61*, 377–392. [CrossRef]

29. Cirz, R.T.; Jones, M.B.; Gingles, N.A.; Minogue, T.D.; Jarrahi, B.; Peterson, S.N.; Romesberg, F.E. Complete and SOS-mediated response of Staphylococcus aureus to the antibiotic ciprofloxacin. *J. Bacteriol.* **2007**, *189*, 531–539. [CrossRef]

30. Au, N.; Kuester-Schoeck, E.; Mandava, V.; Bothwell, L.E.; Canny, S.P.; Chachu, K.; Colavito, S.A.; Fuller, S.N.; Groban, E.S.; Hensley, L.A.; et al. Genetic composition of the Bacillus subtilis SOS system. *J. Bacteriol.* **2005**, *187*, 7655–7666. [CrossRef]
31. Courcelle, J.; Khodursky, A.; Peter, B.; Brown, P.O.; Hanawalt, P.C. Comparative gene expression profiles following UV exposure in wild-type and SOS-deficient *Escherichia coli*. *Genetics* **2001**, *158*, 41–64. [CrossRef]
32. Manasherob, R.; Miller, C.; Kim, K.; Cohen, S.N. Ribonuclease E modulation of the bacterial SOS response. *PLoS ONE* **2012**, *7*, e38426. [CrossRef]
33. Byrne, R.T.; Chen, S.; Wood, E.A.; Cabot, E.L.; Cox, M.M. *Escherichia coli* Genes and Pathways Involved in Surviving Extreme Exposure to Ionizing Radiation. *J. Bacteriol.* **2014**, *196*, 3534–3545. [CrossRef]
34. Gagarinova, A.; Stewart, G.; Samanfar, B.; Phanse, S.; White, C.A.; Aoki, H.; Deineko, V.; Beloglazova, N.; Yakunin, A.F.; Golshani, A.; et al. Systematic Genetic Screens Reveal the Dynamic Global Functional Organization of the Bacterial Translation Machinery. *Cell Rep.* **2016**, *17*, 904–916. [CrossRef] [PubMed]
35. Gomez, J.E.; Kaufmann-Malaga, B.B.; Wivagg, C.N.; Kim, P.B.; Silvis, M.R.; Renedo, N.; Ioerger, T.R.; Ahmad, R.; Livny, J.; Fishbein, S.; et al. Ribosomal mutations promote the evolution of antibiotic resistance in a multidrug environment. *Elife* **2017**, *6*, e20420. [CrossRef]
36. Giles, J.A.; Falconio, J.; Yuenger, J.D.; Phanse, S.; White, C.A.; Aoki, H.; Deineko, V.; Beloglazova, N.; Yakunin, A.F.; Golshani, A.; et al. Quinolone resistance-determining region mutations and por type of *Neisseria gonorrhoeae* isolates: Resistance surveillance and typing by molecular methodologies. *J. Infect. Dis.* **2004**, *189*, 2085–2093. [CrossRef] [PubMed]
37. Olesky, M.; Hobbs, M.; Nicholas, R.A. Identification and analysis of amino acid mutations in porin IB that mediate intermediate-level resistance to penicillin and tetracycline in *Neisseria gonorrhoeae*. *Antimicrob. Agents Chemother.* **2002**, *46*, 2811–2820. [CrossRef]
38. Tkachenko, A.G.; Nesterova, L.Y. Polyamines as modulators of gene expression under oxidative stress in *Escherichia coli*. *Biochemistry* **2003**, *68*, 850–856. [CrossRef]
39. Tenaillon, O.; Rodríguez-Verdugo, A.; Gaut, R.L.; McDonald, P.; Bennett, A.F.; Long, A.D.; Gaut, B.S. The molecular diversity of adaptive convergence. *Science* **2012**, *335*, 457–461. [CrossRef]
40. Yoshida, M.; Kashiwagi, K.; Shigemasa, A.; Taniguchi, S.; Yamamoto, K.; Makinoshima, H.; Ishihama, A.; Igarashi, K. A Unifying Model for the Role of Polyamines in Bacterial Cell Growth, the Polyamine Modulon. *J. Biol. Chem.* **2004**, *279*, 46008–46013. [CrossRef] [PubMed]
41. Igarashi, K.; Kashiwagi, K. Polyamine transport in bacteria and yeast. *Biochem. J.* **1999**, *344 Pt 3*, 633–642. [CrossRef]
42. Ibacache-Quiroga, C.; Oliveros, J.C.; Couce, A.; Blázquez, J. Parallel Evolution of High-Level Aminoglycoside Resistance in *Escherichia coli* Under Low and High Mutation Supply Rates. *Front. Microbiol.* **2018**, *9*, 427. [CrossRef]

Article

First Description of Ceftazidime/Avibactam Resistance in a ST13 KPC-70-Producing *Klebsiella pneumoniae* Strain from Portugal

Gabriel Mendes [1], João F. Ramalho [1], Ana Bruschy-Fonseca [2], Luís Lito [2], Aida Duarte [3,4], José Melo-Cristino [2,5] and Cátia Caneiras [1,3,6,*]

1 Microbiology Research Laboratory on Environmental Health (EnviHealthMicro Lab), Institute of Environmental Health (ISAMB), Faculty of Medicine, Universidade de Lisboa (ULisboa), 1649-028 Lisboa, Portugal; gabriel-mendes@campus.ul.pt (G.M.); jfrancisko.ramalho@gmail.com (J.F.R.)
2 Microbiology Laboratory, Clinical Pathology Department, Centro Hospitalar Universitário Lisboa Norte, 1649-035 Lisboa, Portugal; ana.bruschy@chln.min-saude.pt (A.B.-F.); lmlito@chln.min-saude.pt (L.L.); melo_cristino@medicina.ulisboa.pt (J.M.-C.)
3 Faculty of Pharmacy, Universidade de Lisboa (ULisboa), 1649-033 Lisboa, Portugal; aduarte@ff.ulisboa.pt
4 Egas Moniz Interdisciplinary Research Center, Egas Moniz University Institute, 2829-511 Monte da Caparica, Portugal
5 Institute of Microbiology, Faculty of Medicine, Universidade de Lisboa (ULisboa), 1649-028 Lisboa, Portugal
6 Institute of Preventive Medicine and Public Health, Faculty of Medicine, Universidade de Lisboa (ULisboa), 1649-028 Lisboa, Portugal
* Correspondence: ccaneiras@medicina.ulisboa.pt

Citation: Mendes, G.; Ramalho, J.F.; Bruschy-Fonseca, A.; Lito, L.; Duarte, A.; Melo-Cristino, J.; Caneiras, C. First Description of Ceftazidime/Avibactam Resistance in a ST13 KPC-70-Producing *Klebsiella pneumoniae* Strain from Portugal. *Antibiotics* **2022**, *11*, 167. https://doi.org/10.3390/antibiotics11020167

Academic Editor: Teresa V. Nogueira

Received: 28 December 2021
Accepted: 25 January 2022
Published: 27 January 2022

Publisher's Note: MDPI stays neutral with regard to jurisdictional claims in published maps and institutional affiliations.

Abstract: The combination of ceftazidime/avibactam (CZA) is a novel β-lactam/β-lactamase inhibitor with activity against *Klebsiella pneumoniae* carbapenemase (KPC)-producing Enterobacterales. Emerging cases caused by CZA-resistant strains that produce variants of KPC genes have already been reported worldwide. However, to the best of our knowledge, no CZA-resistant strains were reported in Portugal. In September 2019, a *K. pneumoniae* CZA-resistant strain was collected from ascitic fluid at a surgery ward of a tertiary University Hospital Center in Lisboa, Portugal. The strain was resistant to ceftazidime/avibactam, as well as to ceftazidime, cefoxitin, gentamicin, amoxicillin/clavulanic acid, and ertapenem, being susceptible to imipenem and tigecycline. A hypermucoviscosity phenotype was confirmed by string test. Whole-genome sequencing (WGS) analysis revealed the presence of an ST13 KPC70-producing *K. pneumoniae*, a KPC-3 variant, differing in two amino-acid substitutions (D179Y and T263A). The D179Y mutation in the KPC Ω-loop region is the most common amino-acid substitution in KPC-2 and KPC-3, further leading to CZA resistance. The second mutation causes a KPC-70 variant in which threonine replaces alanine (T263A). The CZA-resistant strain showed the capsular locus KL3 and antigen locus O1v2. Other important virulence factors were identified: fimbrial adhesins type 1 and type 3, as well as the cluster of iron uptake systems aerobactin, enterobactin, salmochelin, and yersiniabactin included in integrative conjugative element 10 (ICE*Kp10*) with the genotoxin colibactin cluster. Herein, we report the molecular characterization of the first hypervirulent CZA-resistant ST13 KPC-70-producing *K. pneumoniae* strain in Portugal. The emergence of CZA-resistant strains might pose a serious threat to public health and suggests an urgent need for enhanced clinical awareness and epidemiologic surveillance.

Keywords: KPC-70; *Klebsiella pneumoniae*; ceftazidime/avibactam resistance; KPC-3; ST13; KPC-variant; hypermucoviscosity; hypervirulence; Portugal

1. Introduction

Klebsiella pneumoniae belongs to the order Enterobacterales and is one of the most common nosocomial pathogens worldwide, representing a serious threat to clinical and

Table 2. Molecular characterization of the ceftazidime/avibactam-resistant *K. pneumoniae*.

ID	MLST	*bla*_Carb	*bla*_Narrow Spectrum	Aminogly-cosides	tmp	sul	Quino-lones	fos	OmpK	*wzi*	K_locus	O_locus	Fimbriae	Iron Uptake	ICE*Kp*	Geno-toxin	Plasmid
FMUL94	ST13	KPC-70	TEM-1; SHV-1; OXA-9	*aac(6′)-Ib*; *aadA*; *aph(3″)-Ib*; *aph(6)-Id*	*dfrA14*	*sul2*	*gyr*A-87N	*fosA*	*OmpK* 35–70%	*wzi*40	KL3	O1v2	*fimA-fimK*; *mrkA-mrkJ*	*kfu*, *enterobactin* (*entA–fes*), *aerobactin* (*iutA*), *salmochelin* (*iroN–iroE*), *yersini-abactin* (*fyuA–ybtX*)	*ybt*17; *ICEKp10*	*colibactin* (*clb-c,2-i,l-q*)	ColRNAI; IncFIA (pBK30683); IncFII (pBK30683)

bla: β-lactamase; tmp: trimetoprim; sul: sulfanamides; fos: fosfomycin.

3. Discussion

Ceftazidime/avibactam is a recent therapeutic option for treating serious infections caused by carbapenem-resistant *K. pneumoniae* (CRK) [9]. However, increased reports of CZA resistance in CRK are worrisome [16,17,21]. In Portugal, no reports of CZA resistance were found to date. In this study, we report a *K. pneumoniae* KPC-70 CZA^R strain that produces a *bla*$_{KPC-70}$ gene, a *bla*$_{KPC-3}$ variant with two single-amino-acid variants (D179Y and T263A).

The first mutation in the KPC Ω-loop region (D179Y) is the most common amino-acid substitution of KPC in KPC-2 [21] and KPC-3 [22,23]. It has been previously shown that mutations occurring at the 164–179 site result in flexibility of the Ω-loop region, which decreases the binding of avibactam and ceftazidime hydrolysis activity, leading to CZA resistance [24]. The KPC-70 variant, recently described in Italy [12], presents a D179Y mutation in the KPC Ω-loop region and a second mutation in which threonine replaces alanine (T263A).

Multilocus sequence typing (MLST) revealed that the *K. pneumoniae* clinical strain belonged to sequence type 13 (ST13). A previous study detected seven KPC-3-producers strains belonging to ST13 in Portugal [25] despite the genomic population structure being apparently dominated by ST147 [25,26]. In the remainder of the European continent, most ST13 *K. pneumoniae* strains were identified as OXA-48-like producers [27–29]. In contrast, KPC-producing clones were recently reported among South American public hospitals [30–32]. Nevertheless, none of these studies found CZA-resistant strains. In fact, although there has been an evident increase in CZA-resistant *K. pneumoniae* reports in the last few years [12,15,18,21] none of them exhibited an ST13 allelic profile, which highlights the importance of genomic surveillance of this ST13 clone.

Concerning virulence factors that have been characterized for *K. pneumoniae*, there are different types according to infection source, as well as *K. pneumoniae* ST strain. However, there are four major classes of virulence factors: capsule, including the production of hypermucoviscosity, lipopolysaccharide (LPS), siderophores, and fimbriae [33]. The iron acquisition systems are essential in *K. pneumoniae* infections, generally to control host defenses by inducing the dissemination through chelation of host cellular iron and regulating the production of multiple bacterial virulence factors [34]. Four principal siderophore systems were found in our strain. Firstly, enterobactin (*ent*) is common to all *K. pneumoniae* and conserved in the chromosome as core gene. Secondly, yersiniabactin is the most common virulence factor associated with human *K. pneumoniae* infections and enhances bacterial survival in the host [33,34]. Usually, this siderophore is included in an integrative and conjugative element (ICE) ICE*Kp10*, which in our study was harboring yersiniabactin (*ybt17*) and colibactin (*clb3*), which have a genotoxin effect on host cells by crosslinking DNA and disrupting host immune response [35]. Most ICEs are extremely prevalent in hypervirulent *K. pneumoniae* lineages. In particular, ICE*Kp10* is the most common type found in ST23 and ST258 [36]. The remaining siderophore gene clusters found encode for the biosynthesis of aerobactin and salmochelin. They are associated with invasive disease and are common amongst hypervirulent *K. pneumoniae* clones that cause severe community-associated infections such as liver abscesses and pneumonia. These siderophores were considered as the primary virulence determinant among iron acquisition systems in hypervirulent *K. pneumoniae* species [37], representing key genes for high virulence scores [38]. The other iron uptake system, *kfu*, is associated with invasive infections, capsule formation, and hypermucoviscosity [33]. Our CZA-resistant strain also presented a hypermucoviscosity (HMV) phenotype and genes coding for biofilm production, which are two key factors for colonization and persistence in the host. Notably, infection with biofilm-producing KPC-*Kp* strains was an independent predictor of mortality [39]. Worryingly, our strain revealed the presence of the type 3 fimbriae *mrkA–mrkJ* cluster, including the type 3 fimbriae-related coding genes (*mrkA* and *mrkD*) and regulation gene (*mrkH*). The biofilm formation capacity of strains that carried the *mrkH* cassette was previously reported as significantly higher than other strains considering that the expression of *mrkA* in *K. pneumoniae* carrying the *mrkH* gene was significantly upregulated [40]. Indeed, the assessment of biofilm production may provide a key element in supporting the clinical management of high-risk patients with KPC-*Kp* infection [39] and a relevant emerging problem [39–41].

Additionally, the CZA-resistant strain reported herein produced a capsular type K3 (considered the usual cause of rhinoscleroma) [42], not previously associated with hvKp, since it is not amongst the most common capsule loci associated with hvKp—K1, K2, K5, and K57 [36]. Nevertheless, the string test confirmed the hypermucoviscosity phenotype of our strain; the PCR and WGS analyses were negative for *magA* gene (considered to be restricted to K1 strains) [36], as well as for *rmpA* and *rmpA2*. However, given that our strain was isolated from an ascitic fluid, the number of virulence factors found (including relevant siderophores as aerobactin and yersiniabactin), associated with resistance to last-line antibiotic ceftazidime/avibactam, it can be characterized as an hvKp strain. However, this is still a controversial concept. In the literature, hypermucoviscous and hypervirulent are often used synonymously, but studies found that the string test performed could be a predictor of hypervirulent strains, and that most hypermucoviscosity phenotypes likely contribute to the majority of hvKp strains [36]. The emergence of hvKp strains is a great cause for concern, as they successfully escape immune system mechanisms due to their high number of virulence factors. Because genes that encode for virulence or antibiotic resistance are often on mobile genetic elements such as plasmids, transposons, and ICE, it is not surprising that we are observing a convergence of virulence and antibiotic resistance. Despite this fact, the interplay between antimicrobial resistance and virulence still remains poorly understood worldwide, particularly in the convergence of hvKp and CZA-resistant strains.

Interestingly, it has been described that, during CZA administration, the MIC of meropenem (MEM) can decrease and some KPC-producing *K. pneumoniae* strains can result in CZA-resistant and MEM-susceptible strains [43,44]. Indeed, our CZA-resistant strain is MEM-susceptible considering the EUCAST category. However, this information should be interpreted with caution by the clinicians. Previous studies confirmed the difficulty of treating infections caused by KPC-*Kp*, which, under CZA-treatment, can rapidly evolve from CZA-susceptible to CZA-resistant via the emergence of KPC-3 variants. These KPC-3 variant-producing strains can be treated with MEM, but the colonization persistence of MEM-resistant KPC-3 producers can lead to treatment failure [19] with high mortality rates [18]. In fact, the role of carbapenems in treating such infections remains uncertain, because meropenem resistance is readily selected during passage experiments [43].

Ceftazidime/avibactam has definitely become an important first-line option due the limited therapeutic options available for KPC-*Kp* infections, which was recently recognized by Infectious Diseases Society of America (IDSA) guidelines that indicate CZA, meropenem/vaborbactam, and imipenem/cilastatin/relebactam as the preferred agents for KPC-*Kp* infections outside of the urinary tract [11]. Previous existing options, such as polymyxins, aminoglycosides, tigecycline, and carbapenems, were of limited use due to inferior efficacy, resistance rates, suboptimal doses, and high toxicity [45]. Therefore, it is imperative to encourage CZA use from an antimicrobial stewardship perspective in order to retain the activity against KPC-*Kp* strains. However, the data currently available highlight that an optimal therapeutic regimen for CZA or for patients infected by CZA-resistant strains remains lacking [18]. In Italy, a case of a 68 year old man with recurrent bacteremia caused by a KPC-*Kp* strain resistant to ceftazidime/avibactam and cefiderocol was reported. A KPC-3 variant enzyme (D179Y; KPC-31) was identified, which confers resistance to ceftazidime/avibactam and restores meropenem susceptibility. The patient that was successfully treated with meropenem/vaborbactam [46], which previously demonstrated potential to be an effective option as salvage therapy for ceftazidime/avibactam-resistant KPC-producing *K. pneumoniae* infections [47].

4. Materials and Methods

4.1. Bacterial Strain

The strain was recovered using standard clinical operating procedures. The identification was performed by microbiology laboratories using conventional methods or automated systems such as Vitek2® (BioMérieux, Marcy, l'Étoile, France) or MicroScan (Snap-on, Kenosha, WI, USA). The strain was frozen in brain heart infusion (BHI) broth (VWR Prolabo, Lisboa, Portugal) with 15% glycerol at -80 °C. For analysis, the strain was grown using BHI broth (18 h, 37 °C) and later seeded in Mueller–Hinton agar (VWR Prolabo, Lisboa, Portugal).

4.2. Antimicrobial Susceptibility Testing and Hypermucoviscosity Phenotype

Antimicrobial susceptibility testing was performed using the standardized Kirby–Bauer disc diffusion technique, in accordance with the European Committee on Antimicrobial Susceptibility Testing (EUCAST). The detailed methodology is available at http://www.eucast.org/ast_of_bacteria/disk_diffusion_methodology/ (accessed on 17 December 2021). Detailed instructions for Mueller–Hinton agar medium (VWR Prolabo®, Lisboa, Portugal), including preparation and storage, are also available in the same EUCAST guidelines document. Quality control was carried out in accordance with EUCAST (version 11.0, 2021). Susceptibility was tested by a panel of antibiotics: amoxicillin/clavulanic acid (20/10 µg), cefoxitin (30 µg), cefotaxime (5 µg), ceftazidime (10 µg), imipenem (10 µg), gentamicin (10 µg), ciprofloxacin (5 µg), tigecycline (15 µg), aztreonam (30 µg), ertapenem (10 µg), doripenem (10 µg), meropenem (10 µg), and ceftazidime/avibactam (10/4 µg) (Biorad, Portugal). The strains were categorized as susceptible to standard dosing regimen (S), susceptible to increased exposure (I), and resistant (R) by applying the breakpoints in the phenotypic test results. A complementary Etest® (BioMérieux, Marcy l' Étoile, France) for ceftazidime/avibactam was also performed. The inhibition zones and MIC values were interpreted according to the EUCAST breakpoints (version 11.0, 2021) (available at https://eucast.org/clinical_breakpoints/ (accessed on 17 December 2021)).

The hypermucoviscosity phenotype is a known virulent factor which was determined using a simple method, the string test. A positive string test is defined as the formation of viscous strings of >5 mm in length when a loop is used to stretch the colony on an agar plate. Multidrug-resistant (MDR) bacteria were defined as those that acquired nonsusceptibility to at least one agent in three or more antimicrobial categories, in accordance with the United States Center for Disease Control and Prevention (CDC) and the European Center for Disease Prevention and Control (ECDC) consensual definition [48].

4.3. Resistance and Virulence Determinants

PCR-based screening was performed to identify carbapenemase genes with primers designed for *bla*$_{OXA-48}$ (F: 5′–GGCTGTGTTTTTGGTGGCATC–3′; R: 5′–GCAGCCCTAAACC-ATCCGATG–3′), *bla*$_{KPC}$ [49], *bla*$_{VIM}$ [50], *bla*$_{NDM}$ [51], and *bla*$_{GES}$ [52]. The virulence factors were also assessed by PCR with specific primers for the fimbrial adhesins type 1 (*fimH*) [53], hemolysin (*khe*), iron uptake system (*kfu*) [54], siderophore (*enterobactin, entB*), and the hypermucoviscosity phenotype (*magA*) [55].

4.4. Molecular Methods

Polymerase chain reaction (PCR) was performed using the NZYTaq II 2× Green Master Mix (NZYTech, Lisboa, Portugal) following the manufacturer's instructions. The PCR products were resolved in 1% agarose gel in 10× concentrated Tris–borate–EDTA (TBE buffer) (NZYTech, Lisboa, Portugal). Positive and negative controls were included in all PCR assays. The positive controls were used with strains containing the respective genes and previously sequenced. The purification of PCR amplification products was performed using the ExoCleanUp FAST (VWR Prolabo®, Lisboa, Portugal) kit, and they were sequenced at STABVida Portugal. Searches for nucleotide sequences were performed with the BLAST program, which is available at the National Center for Biotechnology

Information website (http://www.ncbi.nim.nih.gov/ accessed on 17 December 2021). Multiple-sequence alignments were performed with the Clustal Omega program, which is available at https://cge.cbs.dtu.dk/services (accessed on 17 December 2021).

4.5. Multilocus Sequence Typing (MLST)

MLST was performed by analyses of internal fragments of seven housekeeping genes (*gapA, infB, mdh, pgi, phoE, rpoB,* and *tonB*), as previously described [56]; the sequencing was performed at STABVida Portugal and submitted to the MLST database for allele attribution. The *K. pneumoniae* database is available at the Pasteur MLST website (http://www.pasteur.fr/mlst/ (accessed on 17 December 2021)).

4.6. Whole-Genome Sequencing

The genomic DNA of the clinical strain was extracted from cultures grown overnight, in Mueller–Hinton agar, using the NZY Tissue gDNA Isolation kit (NZYTech, Lisboa, Portugal), as per the manufacturer's recommendations. The sequencing was performed at STABVida Portugal. Sequencing libraries were prepared using the KAPA HyperPrep Library Preparation Kit (Roche, Switzerland) following the manufacturer's recommended protocol and sequenced using an Illumina HiSeq Novaseq 6000 platform with paired-end reads (2 × 151 bp). The raw data quality control was performed using FASTQC v0.11.9, and the trimming and de novo assembly were performed using CLC Genomics Workbench 12.0.3 (QIAGEN, Aarhus, Denmark). All assemblies were carried out with automatic word size, a similarity fraction of 0.95, a length fraction of 0.95, and a minimum contig size of 500 bp.

4.7. Drug Resistance-Associated Genes, Virulence Genes, Capsular Types, and Plasmid Replicons

Antimicrobial resistance (AMR) K and O antigen loci were identified using Kleborate (https://github.com/katholt/Kleborate (accessed on 17 December 2021)), a *K. pneumoniae*-specific genomic typing tool, and virulence factors using the virulence factor database VFDB (http://www.mgc.ac.cn/VFs/ accessed on 17 December 2021). Plasmid analyses were identified using the PlasmidFinder database with the following cutoff values: minimum of 60% coverage and 95% identity (https://cge.cbs.dtu.dk/services/PlasmidFinder/ (accessed on 17 December 2021)).

4.8. Accession Number

The sequence was submitted in the NCBI database under the GenBank accession number MZ893465.

5. Conclusions

Herein, we reported a hypervirulent multidrug-resistant ST13 *K. pneumoniae* strain producing KPC-70, a KPC-3 variant conferring resistance to ceftazidime/avibactam combination. To the best of our knowledge, this is the first report of a CZA-resistant strain in Portugal and the second KPC-70 variant description worldwide. The application of genomic characterization and molecular surveillance promotes the understanding of *K. pneumoniae* population structure, AMR, pathogenicity, and transmission in clinical environments and should be widely encouraged. Further studies are needed to elucidate the molecular features involved in CZA resistance development and spread.

Author Contributions: Conceptualization, C.C.; data curation, G.M. and C.C.; formal analysis, G.M., J.F.R., A.D. and C.C.; investigation, G.M., J.F.R., A.D. and C.C.; methodology, C.C.; project administration, C.C.; resources, A.B.-F., L.L. and J.M.-C.; software, G.M. and J.F.R.; supervision, C.C.; writing—original draft, G.M.; writing—review and editing, J.F.R., A.B.-F., L.L., A.D., J.M.-C. and C.C. All authors have read and agreed to the published version of the manuscript.

Funding: This research was partially funded by the Fundação para a Ciência e a Tecnologia (FCT), grant number UIDB/04295/2020 and UIDP/04295/2020. Gabriel Mendes (G.M.) is supported by the Fundação para a Ciência e Tecnologia (FCT), Portugal, through a PhD Research Studentship Contract (Contrato de Bolsa de Investigação para Doutoramento 2020.07736.BD). Cátia Caneiras (C.C.) acknowledges the funding provided by the "Research Award in Healthcare Associated Infections" granted by the Escola Superior de Saúde Norte da Cruz Vermelha Portuguesa (2019) and by the "BInov award", an innovation award granted by Southern Regional and Autonomous Regions Section of the Portuguese Pharmaceutical Society (2021).

Institutional Review Board Statement: The strains were obtained as part of routine diagnostic testing and were analyzed anonymously. The study proposal was analyzed and dismissed from evaluation by the Ethics Committee of the Lisboa Academic Medical Centre of the Faculty of Medicine, University of Lisboa, Portugal (Nr. 248/21).

Informed Consent Statement: Not applicable.

Data Availability Statement: Not applicable.

Conflicts of Interest: J.M.C. received research grants administered through his university and honoraria for serving on the speaker's bureaus of Pfizer and MSD, not related to the present study. The other authors declare that the research was conducted in the absence of any commercial or financial relationships that could be construed as potential conflicts of interest.

References

1. Wyres, K.L.; Lam, M.M.C.; Holt, K.E. Population genomics of *Klebsiella pneumoniae*. *Nat. Rev. Microbiol.* **2020**, *18*, 344–359. [CrossRef] [PubMed]
2. ECDC. Country Summaries EARS-Net 2019. *EARSS Report*. 2019. Available online: https://www.ecdc.europa.eu/en/publications-data/surveillance-antimicrobial-resistance-europe-2019 (accessed on 13 December 2021).
3. Caneiras, C.; Lito, L.; Mayoralas-Alises, S.; Diaz-Lobato, S.; Melo-Cristino, J.; Duarte, A. Virulence and resistance determinants of *Klebsiella pneumoniae* isolated from a Portuguese tertiary university hospital centre over a 31-year period. *Enfermedades Infecciosas y Microbiología Clínica* **2019**, *37*, 387–393. [CrossRef] [PubMed]
4. Cassini, A.; Hogberg, L.D.; Plachouras, D.; Quattrocchi, A.; Hoxha, A.; Simonsen, G.S.; Colomb-Cotinat, M.; Kretzschmar, M.E.; Devleesschauwer, B.; Cecchini, M.; et al. Attributable deaths and disability-adjusted life-years caused by infections with antibiotic-resistant bacteria in the EU and the European Economic Area in 2015: A population-level modelling analysis. *Lancet Infect. Dis.* **2019**, *19*, 56–66. [CrossRef]
5. Brolund, A.; Lagerqvist, N.; Byfors, S.; Struelens, M.J.; Monnet, D.L.; Albiger, B.; Kohlenberg, A.; European Antimicrobial Resistance Genes Surveillance Network (EURGen-Net) Capacity Survey Group. Worsening epidemiological situation of carbapenemase-producing Enterobacteriaceae in Europe, assessment by national experts from 37 countries, July 2018. *Eurosurveillance* **2019**, *24*, 1900123. [CrossRef]
6. Shirley, M. Ceftazidime-Avibactam: A Review in the Treatment of Serious Gram-Negative Bacterial Infections. *Drugs* **2018**, *78*, 675–692. [CrossRef] [PubMed]
7. EML Secretariat on behalf of the EML Antibiotic Working Group. Application for Inclusion of Ceftazidime—Avibactam (J01DD52) as a Reserve Antibiotic on the WHO Model List of Essential Medicines (EML) and Model List of Essential Medicines for Children (EMLc). Available online: https://www.who.int/selection_medicines/committees/expert/22/applications/s6.2_new-antibiotics-AWaRe.pdf?ua=1 (accessed on 26 December 2021).
8. King, M.; Heil, E.; Kuriakose, S.; Bias, T.; Huang, V.; El-Beyrouty, C.; McCoy, D.; Hiles, J.; Richards, L.; Gardner, J.; et al. Multicenter Study of Outcomes with Ceftazidime-Avibactam in Patients with Carbapenem-Resistant *Enterobacteriaceae* Infections. *Antimicrob. Agents Chemother.* **2017**, *61*, e00449-17. [CrossRef] [PubMed]
9. Shields, R.K.; Nguyen, M.H.; Chen, L.; Press, E.G.; Potoski, B.A.; Marini, R.V.; Doi, Y.; Kreiswirth, B.N.; Clancy, C.J. Ceftazidime-Avibactam is Superior to Other Treatment Regimens against Carbapenem-Resistant *Klebsiella pneumoniae* Bacteremia. *Antimicrob. Agents Chemother.* **2017**, *61*, e00883-17. [CrossRef]
10. Tumbarello, M.; Trecarichi, E.M.; Corona, A.; De Rosa, F.G.; Bassetti, M.; Mussini, C.; Menichetti, F.; Viscoli, C.; Campoli, C.; Venditti, M.; et al. Efficacy of Ceftazidime-Avibactam Salvage Therapy in Patients with Infections Caused by *Klebsiella pneumoniae* Carbapenemase-producing *K. pneumoniae*. *Clin. Infect. Dis.* **2019**, *68*, 355–364. [CrossRef]
11. Tamma, P.D.; Aitken, S.L.; Bonomo, R.A.; Mathers, A.J.; van Duin, D.; Clancy, C.J. Infectious Diseases Society of America Guidance on the Treatment of Extended-Spectrum beta-lactamase Producing Enterobacterales (ESBL-E), Carbapenem-Resistant Enterobacterales (CRE), and *Pseudomonas aeruginosa* with Difficult-to-Treat Resistance (DTR-P. aeruginosa). *Clin. Infect. Dis.* **2021**, *72*, 1109–1116. [CrossRef]
12. Carattoli, A.; Arcari, G.; Bibbolino, G.; Sacco, F.; Tomolillo, D.; Di Lella, F.M.; Trancassini, M.; Faino, L.; Venditti, M.; Antonelli, G.; et al. Evolutionary Trajectories toward Ceftazidime-Avibactam Resistance in *Klebsiella pneumoniae* Clinical Isolates. *Antimicrob. Agents Chemother.* **2021**, *65*, e0057421. [CrossRef]

13. Gaibani, P.; Re, M.C.; Campoli, C.; Viale, P.L.; Ambretti, S. Bloodstream infection caused by KPC-producing *Klebsiella pneumoniae* resistant to ceftazidime/avibactam: Epidemiology and genomic characterization. *Clin. Microbiol. Infect.* **2020**, *26*, 516.e1–516.e4. [CrossRef] [PubMed]
14. Gottig, S.; Frank, D.; Mungo, E.; Nolte, A.; Hogardt, M.; Besier, S.; Wichelhaus, T.A. Emergence of ceftazidime/avibactam resistance in KPC-3-producing *Klebsiella pneumoniae* in vivo. *J. Antimicrob. Chemother.* **2019**, *74*, 3211–3216. [CrossRef] [PubMed]
15. Mueller, L.; Masseron, A.; Prod'Hom, G.; Galperine, T.; Greub, G.; Poirel, L.; Nordmann, P. Phenotypic, biochemical and genetic analysis of KPC-41, a KPC-3 variant conferring resistance to ceftazidime-avibactam and exhibiting reduced carbapenemase activity. *Antimicrob. Agents Chemother.* **2019**, *63*, e01111-19. [CrossRef] [PubMed]
16. Shields, R.K.; Chen, L.; Cheng, S.; Chavda, K.D.; Press, E.G.; Snyder, A.; Pandey, R.; Doi, Y.; Kreiswirth, B.N.; Nguyen, M.H.; et al. Emergence of Ceftazidime-Avibactam Resistance Due to Plasmid-Borne *bla*KPC-3 Mutations during Treatment of Carbapenem-Resistant *Klebsiella pneumoniae* Infections. *Antimicrob. Agents Chemother.* **2017**, *61*, e02097-16. [CrossRef] [PubMed]
17. Zhang, P.; Shi, Q.; Hu, H.; Hong, B.; Wu, X.; Du, X.; Akova, M.; Yu, Y. Emergence of ceftazidime/avibactam resistance in carbapenem-resistant Klebsiella pneumoniae in China. *Clin. Microbiol. Infect.* **2020**, *26*, 124.e1–124.e4. [CrossRef] [PubMed]
18. Di Bella, S.; Giacobbe, D.R.; Maraolo, A.E.; Viaggi, V.; Luzzati, R.; Bassetti, M.; Luzzaro, F.; Principe, L. Resistance to ceftazidime/avibactam in infections and colonisations by KPC-producing Enterobacterales: A systematic review of observational clinical studies. *J. Glob. Antimicrob. Resist.* **2021**, *25*, 268–281. [CrossRef]
19. Arcari, G.O.A.; Sacco, F.; Di Lella, F.M.; Raponi, G.; Tomolillo, D.; Curtolo, A.; Venditti, M.; Carattoli, A. Interplay between *Klebsiella pneumoniae* producing KPC-31 and KPC-3 under treatment with high dosage meropenem: A case report. *Eur. J. Clin. Microbiol. Infect. Dis.* **2021**. [CrossRef]
20. Cuzon, G.; Naas, T.; Nordmann, P. Functional characterization of Tn*4401*, a Tn3-based transposon involved in *bla*KPC gene mobilization. *Antimicrob. Agents Chemother.* **2011**, *55*, 5370–5373. [CrossRef]
21. Giddins, M.J.; Macesic, N.; Annavajhala, M.K.; Stump, S.; Khan, S.; McConville, T.H.; Mehta, M.; Gomez-Simmonds, A.; Uhlemann, A.C. Successive Emergence of Ceftazidime-Avibactam Resistance through Distinct Genomic Adaptations in *bla*KPC-2-Harboring *Klebsiella pneumoniae* Sequence Type 307 Isolates. *Antimicrob. Agents Chemother.* **2018**, *62*, e02101-17. [CrossRef]
22. Gaibani, P.; Campoli, C.; Lewis, R.E.; Volpe, S.L.; Scaltriti, E.; Giannella, M.; Pongolini, S.; Berlingeri, A.; Cristini, F.; Bartoletti, M.; et al. In vivo evolution of resistant subpopulations of KPC-producing *Klebsiella pneumoniae* during ceftazidime/avibactam treatment. *J. Antimicrob. Chemother.* **2018**, *73*, 1525–1529. [CrossRef]
23. Shields, R.K.; Nguyen, M.H.; Press, E.G.; Chen, L.; Kreiswirth, B.N.; Clancy, C.J. In Vitro Selection of Meropenem Resistance among Ceftazidime-Avibactam-Resistant, Meropenem-Susceptible *Klebsiella pneumoniae* Isolates with Variant KPC-3 Carbapenemases. *Antimicrob. Agents Chemother.* **2017**, *61*, e00079-17. [CrossRef]
24. Wang, Y.; Wang, J.; Wang, R.; Cai, Y. Resistance to ceftazidime-avibactam and underlying mechanisms. *J. Glob. Antimicrob. Resist.* **2020**, *22*, 18–27. [CrossRef] [PubMed]
25. Aires-de-Sousa, M.; Ortiz de la Rosa, J.M.; Goncalves, M.L.; Pereira, A.L.; Nordmann, P.; Poirel, L. Epidemiology of Carbapenemase-Producing *Klebsiella pneumoniae* in a Hospital, Portugal. *Emerg Infect. Dis* **2019**, *25*, 1632–1638. [CrossRef] [PubMed]
26. Perdigao, J.; Caneiras, C.; Elias, R.; Modesto, A.; Spadar, A.; Phelan, J.; Campino, S.; Clark, T.G.; Costa, E.; Saavedra, M.J.; et al. Genomic Epidemiology of Carbapenemase Producing *Klebsiella pneumoniae* Strains at a Northern Portuguese Hospital Enables the Detection of a Misidentified Klebsiella variicola KPC-3 Producing Strain. *Microorganisms* **2020**, *8*, 1986. [CrossRef] [PubMed]
27. Fursova, N.K.; Astashkin, E.I.; Gabrielyan, N.I.; Novikova, T.S.; Fedyukina, G.N.; Kubanova, M.K.; Esenova, N.M.; Sharapchenko, S.O.; Volozhantsev, N.V. Emergence of Five Genetic Lines ST395(NDM-1), ST13(OXA-48), ST3346(OXA-48), ST39(CTX-M-14), and Novel ST3551(OXA-48) of Multidrug-Resistant Clinical *Klebsiella pneumoniae* in Russia. *Microb. Drug Resist.* **2020**, *26*, 924–933. [CrossRef]
28. Osterblad, M.; Kirveskari, J.; Hakanen, A.J.; Tissari, P.; Vaara, M.; Jalava, J. Carbapenemase-producing *Enterobacteriaceae* in Finland: The first years (2008–2011). *J. Antimicrob. Chemother.* **2012**, *67*, 2860–2864. [CrossRef]
29. Wrenn, C.; O'Brien, D.; Keating, D.; Roche, C.; Rose, L.; Ronayne, A.; Fenelon, L.; Fitzgerald, S.; Crowley, B.; Schaffer, K. Investigation of the first outbreak of OXA-48-producing *Klebsiella pneumoniae* in Ireland. *J. Hosp. Infect.* **2014**, *87*, 41–46. [CrossRef]
30. Cejas, D.; Elena, A.; Guevara Nunez, D.; Sevillano Platero, P.; De Paulis, A.; Magarinos, F.; Alfonso, C.; Berger, M.A.; Fernandez-Canigia, L.; Gutkind, G.; et al. Changing epidemiology of KPC-producing *Klebsiella pneumoniae* in Argentina: Emergence of hypermucoviscous ST25 and high-risk clone ST307. *J. Glob. Antimicrob. Resist.* **2019**, *18*, 238–242. [CrossRef]
31. Rodrigues, A.C.S.; Chang, M.R.; Santos, I.C.O.; Carvalho-Assef, A.P.D. Molecular Epidemiology of blaKPC-Encoding *Klebsiella pneumoniae* Isolated from Public Hospitals in Midwest of Brazil. *Microb. Drug Resist.* **2021**, *28*, 1–6. [CrossRef]
32. Vargas, J.M.; Moreno Mochi, M.P.; Lopez, C.G.; Alarcon, J.A.; Acosta, N.; Soria, K.; Nunez, J.M.; Villafane, S.; Ramacciotti, J.; Del Campo, R.; et al. Impact of an active surveillance program and infection control measures on the incidence of carbapenem-resistant Gram-negative bacilli in an intensive care unit. *Revista Argentina de Microbiologia* **2021**. [CrossRef]
33. Paczosa, M.K.; Mecsas, J. *Klebsiella pneumoniae*: Going on the Offense with a Strong Defense. *Microbiol. Mol. Biol. Rev.* **2016**, *80*, 629–661. [CrossRef] [PubMed]
34. Holden, V.I.; Bachman, M.A. Diverging roles of bacterial siderophores during infection. *Metallomics* **2015**, *7*, 986–995. [CrossRef] [PubMed]
35. Vizcaino, M.I.; Crawford, J.M. The colibactin warhead crosslinks DNA. *Nat. Chem.* **2015**, *7*, 411–417. [CrossRef] [PubMed]

36. Choby, J.E.; Howard-Anderson, J.; Weiss, D.S. Hypervirulent *Klebsiella pneumoniae*—Clinical and molecular perspectives. *J. Intern. Med.* **2020**, *287*, 283–300. [CrossRef]

37. Russo, T.A.; Marr, C.M. Hypervirulent Klebsiella pneumoniae. *Clin. Microbiol. Rev.* **2019**, *32*, e00001-19. [CrossRef] [PubMed]

38. Lam, M.M.C.; Wick, R.R.; Watts, S.C.; Cerdeira, L.T.; Wyres, K.L.; Holt, K.E. A genomic surveillance framework and genotyping tool for *Klebsiella pneumoniae* and its related species complex. *Nat. Commun.* **2021**, *12*, 4188. [CrossRef] [PubMed]

39. Di Domenico, E.G.; Cavallo, I.; Sivori, F.; Marchesi, F.; Prignano, G.; Pimpinelli, F.; Sperduti, I.; Pelagalli, L.; Di Salvo, F.; Celesti, I.; et al. Biofilm Production by Carbapenem-Resistant *Klebsiella pneumoniae* Significantly Increases the Risk of Death in Oncological Patients. *Front. Cell Infect. Microbiol.* **2020**, *10*, 561741. [CrossRef]

40. Fang, R.; Liu, H.; Zhang, X.; Dong, G.; Li, J.; Tian, X.; Wu, Z.; Zhou, J.; Cao, J.; Zhou, T. Difference in biofilm formation between carbapenem-resistant and carbapenem-sensitive *Klebsiella pneumoniae* based on analysis of mrkH distribution. *Microb. Pathog.* **2021**, *152*, 104743. [CrossRef]

41. Chung, P.Y. The emerging problems of *Klebsiella pneumoniae* infections: Carbapenem resistance and biofilm formation. *FEMS Microbiol. Lett.* **2016**, *363*, fnw219. [CrossRef]

42. Fevre, C.; Passet, V.; Deletoile, A.; Barbe, V.; Frangeul, L.; Almeida, A.S.; Sansonetti, P.; Tournebize, R.; Brisse, S. PCR-based identification of *Klebsiella pneumoniae* subsp. rhinoscleromatis, the agent of rhinoscleroma. *PLoS Negl. Trop. Dis.* **2011**, *5*, e1052. [CrossRef]

43. Shields, R.K.; Nguyen, M.H.; Press, E.G.; Chen, L.; Kreiswirth, B.N.; Clancy, C.J. Emergence of Ceftazidime-Avibactam Resistance and Restoration of Carbapenem Susceptibility in *Klebsiella pneumoniae* Carbapenemase-Producing K pneumoniae: A Case Report and Review of Literature. *Open Forum Infect. Dis.* **2017**, *4*, ofx101. [CrossRef] [PubMed]

44. Haidar, G.C.C.; Shields, R.K.; Hao, B.; Cheng, S.; Nguyen, M.H. Mutations in *bla*KPC-3 that confer ceftazidime-avibactam resistance encode novel KPC-3 variants that function as extended spectrum β-lactamases. *Antimicrob. Agents Chemother.* **2017**, *61*, e02534-16. [CrossRef] [PubMed]

45. Van Duin, D.; Lok, J.J.; Earley, M.; Cober, E.; Richter, S.S.; Perez, F.; Salata, R.A.; Kalayjian, R.C.; Watkins, R.R.; Doi, Y.; et al. Colistin Versus Ceftazidime-Avibactam in the Treatment of Infections Due to Carbapenem-Resistant *Enterobacteriaceae*. *Clin. Infect. Dis.* **2018**, *66*, 163–171. [CrossRef] [PubMed]

46. Tiseo, G.; Falcone, M.; Leonildi, A.; Giordano, C.; Barnini, S.; Arcari, G.; Carattoli, A.; Menichetti, F. Meropenem-Vaborbactam as Salvage Therapy for Ceftazidime-Avibactam-, Cefiderocol-Resistant ST-512 *Klebsiella pneumoniae*-Producing KPC-31, a D179Y Variant of KPC-3. *Open Forum Infect. Dis.* **2021**, *8*, ofab141. [CrossRef]

47. Athans, V.; Neuner, E.A.; Hassouna, H.; Richter, S.S.; Keller, G.; Castanheira, M.; Brizendine, K.D.; Mathers, A.J. Meropenem-Vaborbactam as Salvage Therapy for Ceftazidime-Avibactam-Resistant *Klebsiella pneumoniae* Bacteremia and Abscess in a Liver Transplant Recipient. *Antimicrob. Agents Chemother.* **2019**, *63*, e01551-18. [CrossRef]

48. Magiorakos, A.P.; Srinivasan, A.; Carey, R.B.; Carmeli, Y.; Falagas, M.E.; Giske, C.G.; Harbarth, S.; Hindler, J.F.; Kahlmeter, G.; Olsson-Liljequist, B.; et al. Multidrug-resistant, extensively drug-resistant and pandrug-resistant bacteria: An international expert proposal for interim standard definitions for acquired resistance. *Clin. Microbiol. Infect.* **2012**, *18*, 268–281. [CrossRef]

49. Yigit, H.; Queenan, A.M.; Anderson, G.J.; Domenech-Sanchez, A.; Biddle, J.W.; Steward, C.D.; Alberti, S.; Bush, K.; Tenover, F.C. Novel carbapenem-hydrolyzing beta-lactamase, KPC-1, from a carbapenem-resistant strain of *Klebsiella pneumoniae*. *Antimicrob. Agents Chemother.* **2001**, *45*, 1151–1161. [CrossRef]

50. Poirel, L.; Naas, T.; Nicolas, D.; Collet, L.; Bellais, S.; Cavallo, J.D.; Nordmann, P. Characterization of VIM-2, a carbapenem-hydrolyzing metallo-beta-lactamase and its plasmid- and integron-borne gene from a *Pseudomonas aeruginosa* clinical isolate in France. *Antimicrob. Agents Chemother.* **2000**, *44*, 891–897. [CrossRef]

51. Yong, D.; Toleman, M.A.; Giske, C.G.; Cho, H.S.; Sundman, K.; Lee, K.; Walsh, T.R. Characterization of a new metallo-beta-lactamase gene, *bla*(NDM-1), and a novel erythromycin esterase gene carried on a unique genetic structure in *Klebsiella pneumoniae* sequence type 14 from India. *Antimicrob. Agents Chemother.* **2009**, *53*, 5046–5054. [CrossRef]

52. Poirel, L.; Le Thomas, I.; Naas, T.; Karim, A.; Nordmann, P. Biochemical sequence analyses of GES-1, a novel class A extended-spectrum beta-lactamase, and the class 1 integron In52 from *Klebsiella pneumoniae*. *Antimicrob. Agents Chemother.* **2000**, *44*, 622–632. [CrossRef]

53. Struve, C.; Bojer, M.; Krogfelt, K.A. Identification of a conserved chromosomal region encoding *Klebsiella pneumoniae* type 1 and type 3 fimbriae and assessment of the role of fimbriae in pathogenicity. *Infect. Immun.* **2009**, *77*, 5016–5024. [CrossRef] [PubMed]

54. Ku, Y.H.; Chuang, Y.C.; Yu, W.L. Clinical spectrum and molecular characteristics of *Klebsiella pneumoniae* causing community-acquired extrahepatic abscess. *J. Microbiol. Immunol. Infect.* **2008**, *41*, 311–317. [PubMed]

55. Wu, K.M.; Li, L.H.; Yan, J.J.; Tsao, N.; Liao, T.L.; Tsai, H.C.; Fung, C.P.; Chen, H.J.; Liu, Y.M.; Wang, J.T.; et al. Genome sequencing and comparative analysis of *Klebsiella pneumoniae* NTUH-K2044, a strain causing liver abscess and meningitis. *J. Bacteriol.* **2009**, *191*, 4492–4501. [CrossRef] [PubMed]

56. Diancourt, L.; Passet, V.; Verhoef, J.; Grimont, P.A.; Brisse, S. Multilocus sequence typing of *Klebsiella pneumoniae* nosocomial isolates. *J. Clin. Microbiol.* **2005**, *43*, 4178–4182. [CrossRef] [PubMed]

Article

Colistin and Carbapenem-Resistant *Acinetobacter baumannii* Aci46 in Thailand: Genome Analysis and Antibiotic Resistance Profiling

Nalumon Thadtapong [1], Soraya Chaturongakul [2,3], Sunhapas Soodvilai [1] and Padungsri Dubbs [2,3,*]

[1] Department of Physiology, Faculty of Science, Mahidol University, Bangkok 10400, Thailand; nalumon.tha@mahidol.ac.th (N.T.); sunhapas.soo@mahidol.ac.th (S.S.)

[2] Department of Microbiology, Faculty of Science, Mahidol University, Bangkok 10400, Thailand; soraya.cha@mahidol.ac.th

[3] Center of Microbial Genomics (CENMIG), Faculty of Science, Mahidol University, Bangkok 10400, Thailand

* Correspondence: padungsri.vic@mahidol.ac.th

Abstract: Resistance to the last-line antibiotics against invasive Gram-negative bacterial infection is a rising concern in public health. Multidrug resistant (MDR) *Acinetobacter baumannii* Aci46 can resist colistin and carbapenems with a minimum inhibitory concentration of 512 µg/mL as determined by microdilution method and shows no zone of inhibition by disk diffusion method. These phenotypic characteristics prompted us to further investigate the genotypic characteristics of Aci46. Next generation sequencing was applied in this study to obtain whole genome data. We determined that Aci46 belongs to Pasture ST2 and is phylogenetically clustered with international clone (IC) II as the predominant strain in Thailand. Interestingly, Aci46 is identical to Oxford ST1962 that previously has never been isolated in Thailand. Two plasmids were identified (pAci46a and pAci46b), neither of which harbors any antibiotic resistance genes but pAci46a carries a conjugational system (type 4 secretion system or T4SS). Comparative genomics with other polymyxin and carbapenem-resistant *A. baumannii* strains (AC30 and R14) identified shared features such as CzcCBA, encoding a cobalt/zinc/cadmium efflux RND transporter, as well as a drug transporter with a possible role in colistin and/or carbapenem resistance in *A. baumannii*. Single nucleotide polymorphism (SNP) analyses against MDR ACICU strain showed three novel mutations i.e., Glu229Asp, Pro200Leu, and Ala138Thr, in the polymyxin resistance component, PmrB. Overall, this study focused on Aci46 whole genome data analysis, its correlation with antibiotic resistance phenotypes, and the presence of potential virulence associated factors.

Keywords: *Acinetobacter baumannii*; colistin; carbapenems; multidrug resistant; WGS

Citation: Thadtapong, N.; Chaturongakul, S.; Soodvilai, S.; Dubbs, P. Colistin and Carbapenem-Resistant *Acinetobacter baumannii* Aci46 in Thailand: Genome Analysis and Antibiotic Resistance Profiling. *Antibiotics* **2021**, *10*, 1054. https://doi.org/10.3390/antibiotics10091054

Academic Editor: Teresa V. Nogueira

Received: 29 July 2021
Accepted: 27 August 2021
Published: 30 August 2021

Publisher's Note: MDPI stays neutral with regard to jurisdictional claims in published maps and institutional affiliations.

1. Introduction

Acinetobacter baumannii is an opportunistic pathogenic bacterium that causes nosocomial infections in immunocompromised patients, especially patients treated in the intensive care unit (ICU) [1,2]. *A. baumannii* infections usually occur following: trauma, surgery, catheterization, or endotracheal intubation [3]. Moreover, this bacterium is well known for its multidrug resistant (MDR) characteristics, defined as resistance to at least one agent in three or more antibiotic categories [4], and as a nosocomial ESKAPE pathogen, a group including: *Enterococcus faecium, Staphylococcus aureus, Klebsiella pneumoniae, A. baumannii, Pseudomonas aeruginosa,* and *Enterobacter* species [5]. *A. baumannii* can resist almost all available antibiotics and it is possible for a strain to be pan drug resistant (PDR), which is defined as resistant to all agents in all antibiotic categories including last-resort antibiotics (carbapenems and polymyxins) [4].

Carbapenem-resistant *A. baumannii* or CRAB is considered by WHO (World Health Organization) as one of the leading threats to global human healthcare [6]. During the

COVID-19 pandemic, CRAB infections have increased among COVID-19 patients who are subjected to long-term stays in the ICU [7,8]. The incidence of *A. baumannii* infection in Thailand is widely distributed in all regions of the country [9]. Clinical isolates of *A. baumannii* accounted for 15–16% of hospital-acquired bacteremia and CRAB comprises 70–88% of total *A. baumannii* clinical isolates [10]. Presence of the carbapenemase encoding genes, *blaOXA-23* or *blaOXA-51*, in combination with insertion sequence elements is frequently found in CRAB [11].

Unfortunately, colistin-resistant *A. baumannii* strains that are either MDR or PDR have been reported worldwide [12,13]. Polymyxins, including polymyxin B and colistin, are alternative last-line drugs used against CRAB [14]. Cationic polymyxin molecules target the polyanionic portions of the outer membrane of the bacterial envelope, specifically lipid A in lipopolysaccharide (LPS) [15]. Polymyxins bind lipid A and disrupt the outer membrane, causing cytoplasm leakage [16]. In Thailand, colistin-resistant *A. baumannii* isolates are found in 35–44% of pneumonia patients [17,18]. Among the colistin-resistant *A. baumannii* strains, four known mechanisms have been identified: (i) modification of lipid A, (ii) loss of LPS, (iii) disruption of outer membrane asymmetry, and (iv) efflux pumps [19]. Modification of lipid A involves increased addition of phosphoethanolamine (PEtN) to lipid A resulting from a mutation in the PmrAB two-component system. Increased expression of *pmrC*, which encodes lipid A phosphoethanolamine transferase, enhances addition of PEtN to either the 4′-phosphate or 1′-phosphate of lipid A [20]. Inactivation or complete loss of LPS occurs as a result of mutations in LPS biosynthesis genes such as: *lpxA*, *lpxC*, and *lpxD* [21]. Mutations in *vacJ*, encoding an outer membrane lipoprotein, and *pldA*, encoding an outer membrane phospholipase, result in accumulation of phospholipid, disrupting LPS organization, membrane asymmetry, and colistin binding [22]. Efflux pumps have also been shown to play important roles in the osmotic stress response and colistin resistance, specifically the AdeRS two-component system, which regulates expression of the AdeABC [23] and EmrAB efflux pumps [24].

Our previous report characterized MDR *A. baumannii* Aci46 that was isolated from a pus sample from a Thai patient at Ramathibodi hospital [25]. This strain was reported as a CRAB, i.e., resistant to imipenem. In the current study, we further explored the Aci46 resistance profile against other antibiotics including the alternative last-line drug, colistin. Since the genotypic characteristics of the MDR Aci46 were still unknown, genome studies were applied to identify antibiotic resistance genes, sequence type, and international clonal (IC) group of Aci46. In addition, the Aci46 genome was compared with other MDR *A. baumannii* and drug sensitive *A. baumannii* to characterize the unique gene features of the MDR (or CRAB) Aci46.

2. Results and Discussion

2.1. Antibiotic Resistance Phenotypes of Aci46

A previous report has shown that Aci46 is susceptible to amikacin and resistant to cefoperazone-sulbactam, ceftazidime, ciprofloxacin, and imipenem [25]. To expand the antibiotic profile of Aci46, twenty drugs from eight classes (i.e., aminoglycosides, beta-lactams (and beta-lactam combined), carbapenems, quinolones, folate pathway blocks, phenicol, tetracycline, and colistin) were used in disk diffusion and microdilution assays. We found that Aci46 was resistant to all twenty drugs (Table 1, Figure S1 and Supplement Table S1) with a minimum inhibitory concentration (MIC) for colistin of 512 µg/mL (Figure 1). We have demonstrated that Aci46 is an XDR (extensively drug resistant) strain according to the definition described by Magiorakos et al. (non-susceptible to at least one agent in all but two antimicrobial categories specified for *Acinetobacter* spp.) [4].

Table 1. Antibiotic susceptibility profiling of *A. baumannii* Aci46 by disk diffusion method.

Class of Antibiotic	Drug	ID	Dose	Zone of Inhibition (mm) *	Interpretation
Aminoglycosides	Gentamicin	CN10	10 µg	0	R
	Kanamycin	K30	30 µg	0	R
	Streptomycin	S10	10 µg	0	R
Beta-lactams	Ampicillin	AMP10	10 µg	0	R
	Cephalothin	KF30	30 µg	0	R
	Cefoxitin	FOX30	30 µg	0	R
	Cefotaxime	CTX30	30 µg	0	R
	Ceftazidime	CAZ30	30 µg	0	R
	Ceftriaxone	CRO30	30 µg	0	R
Beta-lactam combined	Amoxycillin/clavulanic acid	AMC30	30 µg	0	R
Carbapenems	Imipenem	IPM10	10 µg	7–8	R
	Meropenem	MEM10	10 µg	0	R
Quinolones	Ciprofloxacin	CIP5	5 µg	0	R
	Nalidixic acid	NA30	30 µg	0	R
	Norfloxacin	NOR10	10 µg	0	R
Folate pathway blocks	Trimethoprim	W5	5 µg	0	R
	Trimethoprim-sulfamethoxazole	SXT25	25 µg	0	R
Phenical	Chloramphenicol	C30	30 µg	10–11	R
Tetracycline	Tetracycline	TE30	30 µg	0	R

* The ranges of inhibition zones were calculated from three individual replicates and "0" means no inhibition zone.

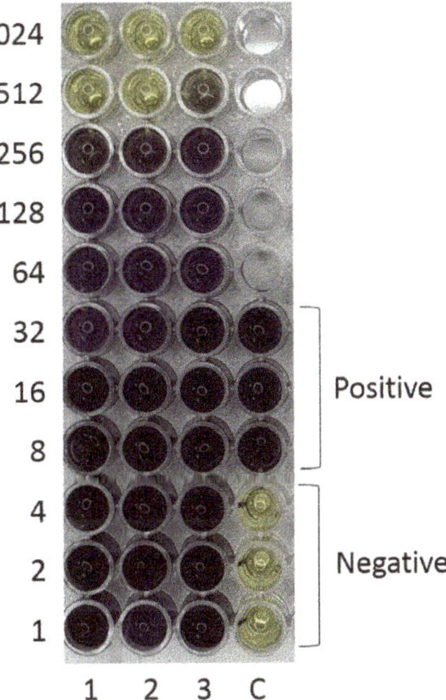

Figure 1. Minimum inhibitory concentration (MIC) determination by microdilution assay using MTT staining. Aci46 viable cells and positive control are shown in purple while dead cells or negative control are shown in yellow. The experiments were tested with three biological replicates.

2.2. Whole Genome Sequencing Data

To further investigate the genetic makeup of XDR Aci46, the chromosome and plasmids of Aci46 were subjected to next-generation sequencing. The summarized genome data is shown in Table 2. The Aci46 genome size is 3,887,827 bp with a GC content of 38.87%. The number of predicted protein coding sequences, rRNA genes, and tRNA genes were 3754, 3, and 63, respectively. We also identified two plasmids from the whole genome data, namely pAci46a and pAci46b. The size of pAci46a was 70,873 bp with a GC content of 33.39% while pAci46b was 8808 bp with a GC content of 34.31%. The number of predicted protein coding sequences for pAci46a and pAci46b were 102 and 11, respectively. No rRNA or tRNA genes were present in either case.

Table 2. Standard data of Aci46 genome features.

Feature	Chromosome	
Total number of bases (bp)	3,887,827	
G + C content (%)	38.87	
number of contigs	80	
genome coverage (x)	686.7	
number of coding sequences	3754	
rRNA	3	
tRNA	63	
N50	135,425	
L50	11	
Feature	**pAci46a**	**pAci46b**
Total number of bases (bp)	70,873	8808
G + C content (%)	33.39	34.31
number of contigs	1	1
genome coverage (x)	1189.0	7787.5
number of coding sequences	102	11
rRNA	0	0
tRNA	0	0

The microbial taxonomy of Aci46 was confirmed as *A. baumannii* at 100% identity based on variation of 54 genes encoding ribosomal protein subunits. Typing of Aci46 was classified by multi-locus sequence typing (MLST) using Oxford and Pasture schemes. The sequence type (ST) of Aci46 was ST1962 (*gltA*-1, *gyrB*-3, *gdhB*-189, *recA*-2, *cpn60*-2, *gpi*-140, *rpoD*-3) based on the Oxford scheme [26], while it belonged to ST2 (*cpn60*-2, *fusA*-2, *gltA*-2, *pyrG*-2, *recA*-2, *rplB*-2, *rpoB*-2) based on the Pasture scheme [27]. ST2 based on Pasture scheme is a predominant ST of CRAB found in Thailand and Southeast Asia [10,28]. However, ST1962 based on the Oxford scheme has never been reported in Thailand. ST1962 has been reported in the USA for only one strain (PubMLST database, to be published). From our previous report, we knew that Aci46 harbored class 1 integrase [25]. However, the class 1 integron can be transferred across two *A. baumannii* IC groups, IC I and IC II [29]. Phylogenetic analysis of the Aci46 genome compared with ten genomes of *A. baumannii* from three different IC's, with the *A. baylyi* ADP1 genome as an outgroup to root the tree, revealed that Aci46 is more closely related to *A. baumannii* MDR-ZJ06 and belongs to IC II (Figure 2).

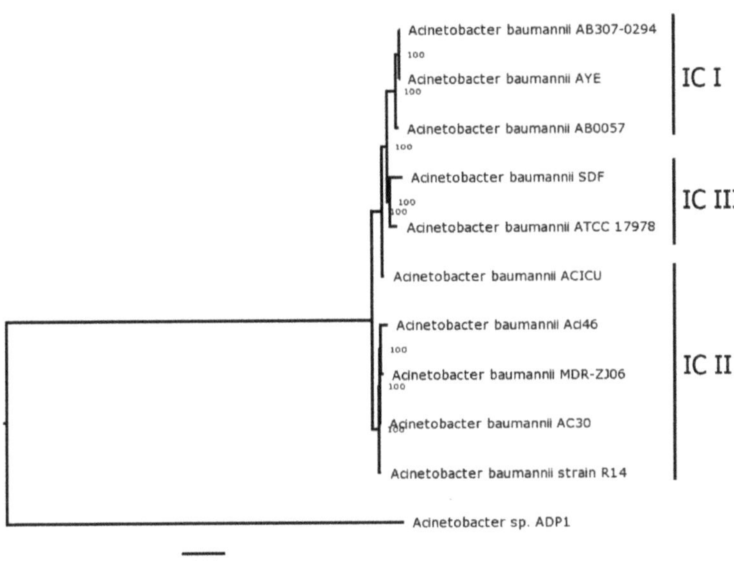

Figure 2. Phylogenetic analysis of *A. baumannii* Aci46 and ten sequences of *A. baumannii* genomes for identification of international clonal group (IC). The phylogenetic tree was constructed by RAxML version 8.2.11 using 100 single-copy genes with bootstrap values set to 100 replicates. Selected strains belonging to IC I, IC II, and IC III are labeled. The accession numbers of the strains are as follows: AYE (CU459141.1), AB0057 (CP001182.2), AB307-0294 (CP001172), AC30 (ALXD00000000), ACICU (CP031380.1), MDR-ZJ06 (CP001937.2), R14 (PUDB01000000), ATCC17978 (CP000521.1), SDF (CU468230.2), and *A. baylyi* ADP1 (CR543861).

2.3. Antibiotic Resistance Gene, Efflux Pump, and Virulence Gene Predictions

Based on the phenotypic characteristics of the antibiotic resistance profile, a search for the presence of antibiotic resistance genes and genes encoding efflux pumps associated with the XDR phenotype in Aci46 was undertaken. ResFinder, CARD, and NDARO databases were used to predict the antibiotic resistance genes and efflux pumps present in Aci46. We found that Aci46 harbored sixteen resistance genes against eight classes of drugs (i.e., aminoglycosides, beta-lactams/carbapenems, beta-lactams/cephalosporins, colistin, fluoroquinolones, macrolides, tetracycline, and sulfonamide) and twenty-two genes belonging to five classes of drug transporters (i.e., RND (resistance-nodulation-division) efflux systems, MFS (major facilitator superfamily) family transporter, ABC (ATP-binding cassette) transporter, MATE (multidrug and toxic compound extrusion) family transporter, and SMR (small multidrug resistance)) (Table 3). The genes, *blaOXA-23*, *blaOXA-66* or *blaOXA-51*-like, and *oprD* genes are present in Aci46 and they are known confer carbapenem-resistance [11]. These data correlate with presence of *blaOXA-23* and *blaOXA-51* in other CRAB isolates found in Thailand [30]. Moreover, class 1 and class 2 integrase genes and *blaOXA-23* are often found in XDR *A. baumannii* [31,32].

With regard to colistin resistance, we identified *lpxA* and *lpxC* in Aci46, these genes might play a role in loss of LPS and leading to colistin resistance [21]. In addition, we found four genes encoding efflux pumps (i.e., *adeR*, *adeS*, *emrA*, and *emrB*) and two genes with roles in lipid modification (i.e., *pmrA* and *pmrB*). AdeRS is a two-component system that regulates the expression of the AdeABC efflux pump, which is an RND efflux system [23]. The EmrAB efflux system belongs to the MFS family of transporters [24]. The PmrAB two-component system regulates PmrC expression. PmrC adds PEtN to lipid A [33]. In summary, the genotypic characteristics of Aci46 suggest that Aci46 resists colistin by way of lipid A modification, LPS loss, and AdeABC-mediated efflux.

Table 3. List of predicted antibiotic resistance and drug transporter genes.

Target	Location					Product
	Gene	Contig	Start	Stop	Length (bp)	
Antibiotics						
Aminoglycosides	*aph(3″)-Ib*	Aci46-0022	7765	8568	804	Aminoglycoside 3″-phosphotransferase
	aph(6)-Id	Aci46-0022	6929	7765	837	Aminoglycoside 6-phosphotransferase
	armA	Aci46-0031	687	1460	774	hypothetical protein
Beta-lactams/ carbapenems	*blaOXA-23*	Aci46-0058	1667	2488	882	Class D beta-lactamase OXA-23
	blaOXA-66	Aci46-0018	65,582	66,406	825	Class D beta-lactamase OXA-66 or OXA-51-like
	OprD family	Aci46-0035	24,815	26,131	1317	Outer membrane low permeability porin, OprD family
Beta-lactams/ cephalosporins	*blaADC-25*	Aci46-0017	86	1237	1152	Class C beta-lactamase ADC-25
Colistin	*lpxA*	Aci46-0001	58,724	59,512	789	Acyl-[acyl-carrier-protein]–UDP-N-acetylglucosamine O-acyltransferase
	lpxC	Aci46-0006	139,160	140,062	903	UDP-3-O-[3-hydroxymyristoyl] N-acetylglucosamine deacetylase
	pmrA	Aci46-0016	34,812	35,486	675	Polymyxin resistant component PmrA
	pmrB	Aci46-0016	35,512	36,846	1335	Polymyxin resistant component PmrB
Fluoroquinolones	*gyrA*	Aci46-0003	54,354	57,068	2715	DNA gyrase subunit A
	gyrB	Aci46-0028	32,254	34,722	2469	DNA gyrase subunit B
Macrolide	*mph(E)*	Aci46-0054	2308	3192	885	Mph(E) family macrolide 2′-phosphotransferase
	msr(E)	Aci46-0054	3248	4723	1476	ABC-F type ribosomal protection protein Msr(E)
Tetracycline	*tetA*	Aci46-0022	2357	3556	1200	Tetracycline resistance protein
	tetR	Aci46-0022	3638	4261	624	Tetracycline resistance regulatory protein TetR
Sulfonamide	*sul2*	Aci46-0064	97	912	816	Dihydropteroate synthase type-2
Drug transporters						
RND efflux system	*adeA*	Aci46-0019	60,468	61,658	1191	RND efflux system, membrane fusion protein
	adeB	Aci46-0019	61,655	64,765	3111	RND efflux system, inner membrane transporter
	adeF	Aci46-0011	61,731	62,951	1221	RND efflux system, membrane fusion protein
	adeG	Aci46-0011	62,958	66,137	3180	RND efflux system, inner membrane transporter
	adeH	Aci46-0011	6389	6835	447	Efflux transport system, outer membrane factor (OMF) lipoprotein
	adeI	Aci46-0016	53,597	54,847	1251	RND efflux system, membrane fusion protein
	adeJ	Aci46-0016	50,408	53,584	3177	RND efflux system, inner membrane transporter
	adeK	Aci46-0016	48,954	50,408	1455	Efflux transport system, outer membrane factor (OMF) lipoprotein
	adeN	Aci46-0001	43,293	43,946	654	Transcriptional regulator, AcrR family
	adeR	Aci46-0019	59,579	60,322	744	Two-component transcriptional response regulator, LuxR family
	adeS	Aci46-0019	58,474	59,547	1074	Osmosensitive K+ channel histidine kinase KdpD
	mexT	Aci46-0029	34,598	35,587	990	Transcriptional regulator, LysR family
	opmH	Aci46-0022	48,920	50,266	1347	Outer membrane channel TolC (OpmH)
MFS family transporter	*emrA*	Aci46-0024	20,758	21,909	1152	Multidrug efflux system EmrAB-OMF, membrane fusion component EmrA
	emrB	Aci46-0024	19,228	20,751	1524	Multidrug efflux system EmrAB-OMF, inner-membrane proton/drug antiporter EmrB
	mdfA	Aci46-0002	123,994	125,223	1230	Multidrug efflux pump MdfA/Cmr (of MFS type), broad spectrum
ABC transporter	*macA*	Aci46-0012	77,649	78,989	1341	Macrolide-specific efflux protein MacA
	macB	Aci46-0012	78,992	80,986	1995	Macrolide export ATP-binding/permease protein MacB
MATE family transporter	*abeM*	Aci46-0007	67,926	69,272	1347	Multidrug efflux transporter MdtK/NorM (MATE family)
SMR	*abeS*	Aci46-0011	56,864	57,193	330	small multidrug resistance family (SMR) protein

Several virulence factors of *A. baumannii* have been identified by genome-based analysis [34]. The outer membrane protein, OmpA, which functions as a porin, is a key factor in virulence where it plays particular roles in cell invasion, development of cytotoxicity, and apoptosis [35]. Capsular polysaccharides and LPS are also virulence factors and contribute to serum resistance, biofilm formation, and escape from the host immune response [36]. *A. baumannii* uses combined strategies, namely, bacterial fitness and pathogenicity, to cause disease in humans [35]. PAI (pathogenicity islands), such as prophages and secretion systems, have also been implicated in virulence and pathogenicity [37,38]. In Aci46, we have identified pathogenicity islands comprising four prophages, one T4SS (type four secretion system), one T6SS (type six secretion system), and one ICE (integrative and conjugation element) (Table 4 and Supplement Table S2). No antibiotic resistance genes were found on the plasmids and prophages. Genotypic characteristics underlying the antibiotic resistance profile of Aci46 are found on its chromosome, and not on plasmids or other mobile genetic elements. T4SS is located in plasmid pAci46a, similar to pAC30c in *A. baumannii* AC30 and pAC29b in *A. baumannii* AC29 [2]. Generally, T4SS plays a role in the transfer antibiotic resistance genes via horizontal gene transfer [38]. Based on comparative genome analysis, T4SS loci are found in clinical isolates associated with hospital outbreaks [39]. The function of T4SS is still unclear in *A. baumannii*, but it might be implicated in pathogenicity or host–pathogen interaction [38,40]. Thus, pAci46a might play a role in pathogenesis instead of drug resistance. Moreover, we found *attL* (gtaataacaaagcaatcccgcagggttgcgacaaatagcc-ctctaaatcgctctaattgcccctagattcaatttta) and *attR* (gtaataacaaagcaatcccgcagggttgcgacaaatagc-cctctaaatcgctctaattgcccctagattcaatttta) sites on pAci46a (or ICE region). It is possible that pAci46a could be a conjugative plasmid responsible for plasmid mobilization, similar to pAC30c and pAC29b [2]. T6SS injects toxic effectors into other bacteria in the same niche; therefore, it is an important factor for competitive killing and host colonization [34]. The plasmid, pAci46b, carries eleven genes encoding: one outer membrane receptor protein, one replication protein, and nine hypothetical proteins. The functions of pAci46b are still unclear.

Table 4. Predicted pathogenicity islands in WGS of *A. baumannii* Aci46.

Pathogenicity Island	Contig	Start	Stop	Length (bp)	GC Content (%)	Number of CDS
Prophage-1	Aci46-0033	9107	23,948	14,842	40.09	14
Prophage-2	Aci46-0033	27,330	37,053	9724	40.71	11
Prophage-3	Aci46-0036	785	30,936	30,152	38.52	46
Prophage-4	Aci46-0041	806	17,770	16,965	40.29	29
Type 4 secretion system, T4SS	pAci46a	20,944	48,690	27,747	34.60	32
Type 6 secretion system, T6SS	Aci46-0010	8673	27,581	19,412	36.92	15
Integrative and conjugation element, ICE	pAci46a	1	70,873	70,873	33.39	94

2.4. Comparative Pangenomic Analysis against Other A. baumannii Strains

In order to further explore the novel gene(s) that might be involved in the colistin and carbapenem-resistant phenotypes in Aci46, pangenome analysis was performed (Figure 3). Pangenome (all genes from all genomes) and core genes (present in all genomes) (Supplement Table S7) comprise 4605 and 2754 genes, respectively. We found that the number of strain-specific genes for Aci46, ACICU, ATCC17978, AC30, and R14 are 45, 193, 382, 91, and 145, respectively (Supplement Table S3). MDR (Aci46, ACICU, AC30, and R14) (Supplement Table S4), polymyxin and carbapenem resistance (renamed PCRAB) (Aci46, AC30, and R14), and colistin and carbapenem resistance (renamed CCRAB) (Aci46 and R14) groups contained 504, 34, and 8 accessory genes (i.e., genes present in specific genomes), respectively. We hypothesized that the accessory genes in PCRAB and/or CCRAB might be involved in resistance to polymyxins (polymyxin B and colistin). The thirty-four PCRAB specific genes encode: amidase, carbapenemase BlaOXA-23, Czc-CBA cobalt/zinc/cadmium efflux RND transporter, twenty-three hypothetical proteins, lysophospholipid, nickel-cobalt-cadmium resistance protein, oxidoreductase, transcriptional regulator, and DNA-methyltransferase subunit M (Supplement Table S5). The eight CCRAB specific genes encode: one primosomal protein I and seven hypothetical proteins (Supplement Table S6). One of the PCRAB specific genes, *blaOXA-23*, is frequently reported in CRAB [41,42]. Interestingly, three genes (*czcA*, *czcB*, and *czcC*) encode Czc-CBA cobalt/zinc/cadmium efflux RND transporters in PCRAB specific genes. CzcCBA cobalt/zinc/cadmium efflux RND transporters have functions in exporting cations (Co^{2+}, Zn^{2+}, and Cd^{2+}) from the cytoplasm and confer heavy metal resistance [43,44], and are reported to be associated with the XDR phenotype in *A. baumannii* [45]. This efflux system might function in colistin resistance in PCRAB, including Aci46, by exporting colistin and polymyxin (cationic molecules). In the case of CCRAB specific genes, their functions are unknown.

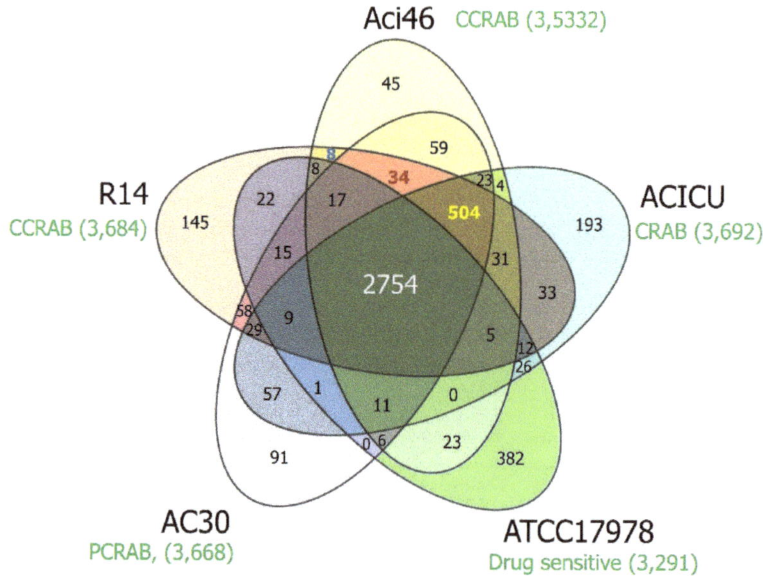

Figure 3. Venn diagram of comparative pangenome analysis among *A. baumannii* strains. The yellow, blue, green, white, and orange represent Aci46, ACICU, ATCC17978, AC30, and R14, respectively, with the number of total genes and the character of drug resistance in each genome (shown in green). The overlapped areas show genes encoding shared protein families among strains. The number of genes in core genes (Aci46, ACICU, ATCC17978, AC30, and R14), MDR (Aci46, ACICU, AC30, and R14), PCRAB (Aci46, AC30, and R14), and CCRAB (Aci46 and R14) the specific groups are labeled in white, yellow, red, and blue, respectively. CRAB: carbapenem-resistant *A. baumannii*, CCRAB: colistin and carbapenem-resistant *A. baumannii*, PCRAB: polymyxins and carbapenem-resistant *A. baumannii*.

2.5. Pairwise SNP Analysis

Although we identified genes that were specific to PCRAB that encoded CzcCBA efflux pumps, known MDR genes were among the core genes (Supplement Table S7). Thus, in order to identify resistance-associated mutations in MDR and CCRAB strains, non-synonymous SNPs between Aci46 and ATCC17978 and between Aci46 and ACICU were identified. SNPs within known MDR genes from the core genes are listed in Table 5. Twenty genes show SNPs in Aci46 vs. ATCC17978 i.e., seven antibiotic resistance genes (*blaADC-25*, *blaOXA-66*, *lpxA*, *lpxC*, *pmrB*, *gyrA*, and *gyrB*) and thirteen drug transporter genes (*adeA*, *adeB*, *adeF*, *adeG*, *adeH*, *adeJ*, *adeR*, *adeS*, *opmH*, *emrB*, *mdfA*, *macB*, and *abeS*). The deduced amino acid sequences of these genes among the MDR strains were similar. For example, the deduced amino acid sequences of *blaOXA-66* or *blaOXA-51*-like genes in Aci46, ACICU, AC30, and R14 showed conserved amino acids at Val36, Lys107, and Asn225 while ATCC17978 contained Glu36, Gln107, and Asp225 (Figure 4). In 2015, the Trp22Met mutation of *blaOXA-51* was linked with carbapenem resistance function in

A. baumannii [46]. This result suggested that amino acid sequences of antibiotic resistant and drug transporter proteins in MDR strains could be different from drug sensitive strains and might be linked to drug resistance level. For colistin resistance, we found three genes (*blaADC-25*, *pmrB*, and *gyrB*) that were mutated in Aci46 vs. ACICU. Of these only *pmrB* is related to colistin resistance. From comparisons of Aci46 vs. ACICU and Aci46 vs. ATCC17978, mutations in PmrB were detected in three positions: Ala138Thr, Pro200Leu, and Glu229Asp (Table 5). Known PmrB mutations that confer colistin resistance are Leu9-Gly12 deletion, Ala22Val, Ile232Thr, and Gln270Pro [47,48]. Hence, Ala138Thr, Pro200Leu, and Glu229Asp mutations in Aci46 PmrB might be novel mutations involved in colistin resistance.

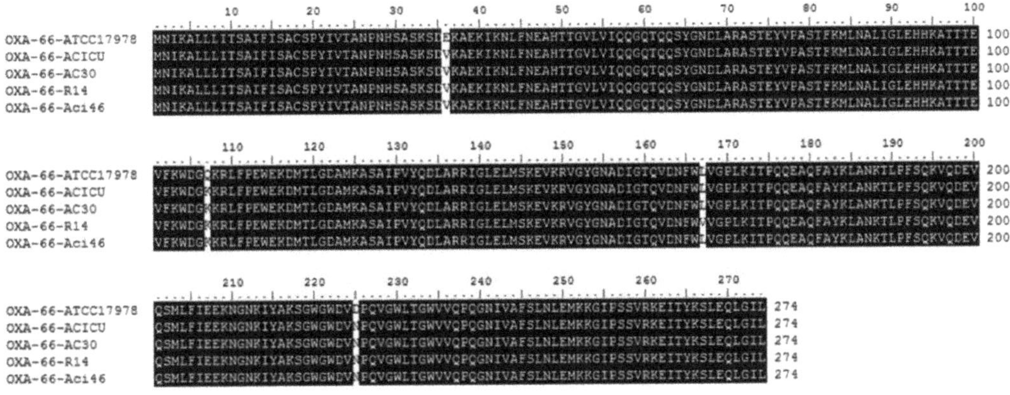

Figure 4. Comparisons of deduced amino acid sequences of *blaOXA-66* from five strains of *A. baumannii* by multiple sequence alignment.

Table 5. List of non-synonymous SNPs in antibiotic resistance genes and efflux pumps from core gene group.

Target	Gene	Aci46 vs. ATCC17978		Aci46 vs. ACICU	
		Nucleotide Change	Amino Acid Change	Nucleotide Change	Amino Acid Change
Antibiotic Resistance Genes					
Beta-lactams/ cephalosporins	*blaADC-25*	356TG > AA	Val119Glu	238C > A	Arg80Ser
		448C > A	Gln150Lys	448C > A	Gln150Lys
		499C > T	Pro167Ser	487C > A	Gln163Lys
		739G > A	Gly247Ser	499C > T	Pro167Ser
		843AGGGTT > GGGTCG	GlnGlyPhe281GlnGlyArg	547A > G	Arg183Gly
		1022A > C	Asn341Thr	739G > A	Gly247Ser
				843AGGGTT > GGGTCG	GlnGlyPhe281GlnGlyArg
				932G > A	Ser311Asn
				1020CAA > TAC	ThrAsn340ThrThr
				1135G > A	Asp379Asn
Beta-lactams/ carbapenems	*blaOXA-66*	107A > T	Glu36Val		
		315CGGGC > TGGTA	AspGlyGln105AspGlyLys		
		673G > A	Asp225Asn		
Colistin	*lpxA*	391T > C	Tyr131His		
	lpxC	358T > C	Cys120Arg		
		859A > G	Asn287Asp		
	pmrB	412G > A	Ala138Thr	412G > A	Ala138Thr
		599C > T	Pro200Leu	599C > T	Glu229Asp
		687A > C	Glu229Asp	687A > C	Pro200Leu
		1331C > T	Ala444Val		

Table 5. Cont.

Target	Gene	Aci46 vs. ATCC17978		Aci46 vs. ACICU	
		Nucleotide Change	Amino Acid Change	Nucleotide Change	Amino Acid Change
Fluoroquinolones	*gyrA*	173C > T	Ser58Leu		
	gyrB	2059A > G	Ile687Val	1738T > C	Tyr580His
Drug Transporters					
RND efflux system	*adeA*	70A > G	Lys24Glu		
	adeB	917G > T 1279AAT > TCG 1654G > A 1928C > A 1936A > T 2191C > T	Gly306Val Asn427Ser Ala552Thr Ala643Asp Thr646Ser Leu731Phe		
	adeF	1163A > G	Asn388Ser		
	adeG	1540G > T	Val514Leu		
	adeH	214A > G 683T > G	Thr72Ala Val228Gly		
	adeJ	2482A > G	Lys828Glu		
	adeR	358G > A 407C > T	Val120Ile Ala136Val		
	adeS	515T > C 557G > T 802A > C 908A > T 1042G > A	Leu172Pro Gly186Val Asn268His Tyr303Phe Val348Ile		
	opmH	386A > G 512A > G	Lys129Arg Asn171Ser		
MFS family transporter	*emrB*	715A > G	Ile239Val		
	mdfA	1157C > T	Ala386Val		
ABC transporter	*macB*	1462G > A	Val488Ile		
SMR	*abeS*	121A > G 165G > C 250G > T 268CTTA > TTGG 292ATC > GTG	Ile41Val Met55Ile Val84Leu LeuThr90LeuAla Ile98Val		

3. Materials and Methods

3.1. Bacterial Strains

A. baumannii Aci46 was isolated from a male Thai patient treated at Ramathibodi hospital, Thailand [25]. This strain was isolated from a pus sample, identified by routine biochemical test, and confirmed by *blaOXA51*-like gene detection [25,49]. Aci46 was cultured on MHA (Mueller Hinton Agar, BD Difco, Eysins, Switzerland) and incubated at 37 °C for overnight.

3.2. Antibiotic Susceptibility Testing by Disk Diffusion and Microdilution

Antibiotic susceptibility was determined by disk diffusion method for 19 drugs (gentamicin, kanamycin, streptomycin, cephalothin, cefoxitin, cefotaxime, ceftazidime, ceftriaxone, ampicillin-clavulanic acid, imipenem, meropenem, ciprofloxacin, nalidixic acid, norfloxacin, trimethoprim, trimethoprim-sulfamethoxazole, ampicillin, chloramphenicol, and tetracycline) (Oxoid, Thermo Fisher Scientific, Waltham, MA, USA) and microdilution method for colistin. Aci46 was streaked on MHA and incubated overnight. Colonies were picked and resuspended in normal saline solution at 1×10^8 cfu/mL (OD_{600} = 0.08–0.12 or 0.5 McFarland). The cell suspension was spread on MHA using a cotton swab. Antibiotic disks were placed on the agar surface. After incubation at 37 °C for 20–24 h., the zones of inhibition were measured and the results were interpreted following the CLSI (Clinical and Laboratory Standard Institute) guideline [50]. For the microdilution method, cell suspensions of Aci46 were diluted in CAMHB (Cation-Adjusted Mueller Hinton Broth, BD Difco, Eysins, Switzerland) to 1×10^6 cfu/mL. Two-fold serial dilutions of colistin were prepared (1–1024 µg/mL) and mixed with 5×10^5 cfu/mL of Aci46 in 200 µL total volume. After incubation at 37 °C for 20–24 h., the minimum inhibitory concentration was observed and cell viability was measured by MTT-based staining [51]. Ten µL of 5 mg/mL MTT (3-(4,5-dimethylthiazol-2-yl)-2,5-diphenyl-tetrazolium bromide (Invitrogen, Life Technologies, Carlsbad, CA, USA) in phosphate buffer saline was added in 100 µL of cell culture and incubated at 37 °C, 200 rpm, 1 h in the dark. One hundred µL of 10% SDS (sodium dodecyl sulfate, Merck, Darmstadt, Germany) and 50% DMSO (dimethyl sulfoxide, Sigma-Aldrich, St. Louis, MO, USA) was added and continually incubated at 37 °C, 200 rpm, 2 h in the dark. The absorbance of formazan dissolution was detected at 570 nm using a microplate reader (Azure Ao Absorbance Microplate Reader, Azure Biosystems, Dublin, CA, USA). Relative optical density at 570 nm was calculated by dividing OD_{570} of drug-containing wells with the OD_{570} of drug-free wells [52]. The cut-off for no detection was the relative OD_{570} of 0.1. Viable cells under MTT staining can also be observed by the naked eye i.e., color change from yellow to purple. Colistin resistance was determined by CLSI guideline (MIC $\geq$ 4 µg/mL; resistant) [50]. *Escherichia coli* ATCC25922 was selected to be a control strain for disk diffusion and microdilution methods.

3.3. Genomic DNA Extraction and Whole Genome Sequencing

Whole genomic DNA of Aci46 was extracted using a modified Marmur procedure [53]. Briefly, Aci46 cells were harvested from 3 mL of cell culture in CAMHB and resuspended in EDTA-saline (0.01 M EDTA and 0.15 M NaCl, pH 8.0). Thirty µL of 110 mg/mL lysozyme and 10 µL of 20 mg/mL RNase A were added and incubated at 37 °C for 2 h. After incubation, 80 µL of 20% SDS and 10 µL of 5 mg/mL proteinase K were added and incubated again at 65 °C, 30 min. Then, 5 M NaCl was added at 0.5 volume followed by phenol-chloroform extraction. The upper liquid phase was transferred to a new 1.5 mL microcentrifuge tube. A 0.25 volume of 5 M NaCl and a 0.1 volume of 3 M sodium acetate were added and mixed well. Ice-cold absolute ethanol was added at 2 volumes and inverted gently. The DNA pellet was hooked and transferred into a new 1.5 mL microcentrifuge tube, air dried, and resuspended in DNase-RNase-free water. Quality and quantity of DNA were measured by UV spectrophotometer (OD_{260}/OD_{280} and OD_{260}/OD_{230} ratio) (DeNovix DS-11 FX+ spectrophotometer, DeNovix, Wilmington, DE, USA), Qubit dsDNA BR assay kit (Invitrogen, Life Technologies, Carlsbad, CA, USA), and 1% agarose gel electrophoresis (Bio-rad, Hercules, CA, USA). One hundred ng of extracted DNA was used for library preparation using TruSeq Nano DNA Kit (Illumina, San Diego, CA, USA) followed by pair-end sequencing on Illumina HiSeq platform (Illumina, San Diego, CA, USA).

3.4. Genome Assembly, Annotation, and Pathogenicity Island Prediction

Raw sequence data of Aci46 were trimmed by Trim Galore version 0.6.3 [54] and the quality was checked using FastQC version 0.11.8 [55]. Trimmed reads were assembled using SPAdes version 3.12.0 [56], corrected assembly error by Pilon version 1.23 [57], and calculated genome coverage by SAMTools version 1.3 [58] in PATRIC (Pathosystems Resource Integration Center) version 3.6.9 [59]. The quality of de novo assembled contigs was assessed by QUAST version 5.0.2 [60] and visualized using Bandage version 0.8.1 [61]. Coding sequences and functional genes were annotated using RASTtk (Rapid Annotation using Subsystem Technology toolkit) [62]. Antibiotic resistance genes were predicted using ResFinder version 4.1 [63], CARD (Comprehensive Antibiotic Resistance Database) [64], and NDARO (National Database of Antibiotic Resistant Organisms) [65] databases. Pathogenicity islands (type 4 secretion system and type 6 secretion system) and prophages were predicted using VRprofile version 2.0 [66] and PHASTER (PHAge Search Tool Enhanced Release) [67], respectively. In plasmid analysis, trimmed reads were used for searching and assembling plasmid sequences using plasmidSPAdes version 3.12.0 [68] in PATRIC version 3.6.9 server [59]. Quality control, annotation, and pathogenicity island predictions of plasmids were assessed using the same tools as with genomic analysis.

3.5. MLST and Phylogenetic Analysis

Identification of *A. baumannii* Aci46 was confirmed by rMLST (ribosomal multilocus sequence typing) [69] in PubMLST server [70]. The ST (sequence typing) of Aci46 was identified by MLST (multilocus sequence typing) according to the Pasture scheme (based on seven housekeeping genes *cpn60, gdhB, gltA, gpi, gyrB, recA*, and *rpoD*) [27] and Oxford scheme (based on seven housekeeping genes *cpn60, fusA, gltA, pyrG, recA, rplB*, and *rpoB*) [26] in PubMLST server [70]. The IC (international clonal) group of Aci46 was determined by phylogenetic relationship analysis using RAxML version 8.2.11 [71] based on 100 single-copy genes selected by PATRIC PGFams program [72] in PATRIC version 3.6.9 server [59]. The phylogenetic tree was visualized by FigTree [73]. Genomes of *A. baumannii* from IC I, IC II, IC III groups were used to identify the IC of Aci46. IC I group included *A. baumannii* strain AYE (CU459141.1) [74], AB0057 (CP001182.2) [75], and AB307-0294 (CP001172) [75]. IC II group included *A. baumannii* strain AC30 (ALXD00000000) [2], ACICU (CP031380.1) [76], MDR-ZJ06 (CP001937.2) [77], and R14 (PUDB01000000) [78]. IC III included *A. baumannii* strain ATCC17978 (CP000521.1) [79] and SDF (CU468230.2) [74]. The outgroup strain was *A. baylyi* ADP1 (CR543861) [80].

3.6. Comparative Pangenome and Pairwise SNP Analysis

Genomic data from five strains comprised of colistin and carbapenem-resistant Aci46, polymyxin B and carbapenem-resistant AC30 (ALXD00000000) [2], colistin and carbapenem-resistant R14 (PUDB01000000) [78], wild-type or drug sensitive ATCC17978 (CP000521.1) [79], and colistin-sensitive and carbapenem-resistant ACICU (CP031380.1) [76] were compared using the Protein Family Sorter with PATRIC genus-specific families (PLfams) strategy in PATRIC server [59]. In pairwise SNP analysis, genome data of Aci46 was aligned with the ATCC17978 genome (CP000521.1) and ACICU (CP031380.1) using BWA-mem aligner [81] and SNP calling by FreeBayes [82]. The deduced amino acid sequences of *blaOXA-66* from five genomes were compared by Clustal Omega in EMBL-EBI [83].

4. Conclusions

In summary, this study reported the genome data of colistin and carbapenem-resistant *A. baumannii* Aci46, which was isolated from a patient in a Thai hospital. The MLST genotype of Aci46 is Pasture ST2 which is a predominant ST found in Thailand and Oxford ST1962 which has never been reported in Thailand. The predicted antibiotic resistance genes (for example, *blaOXA-23, blaOXA-66*, and *blaADC-25*) are on the chromosome, not plasmids. Based on pangenome analysis, we found that the CzcCBA cobalt/zinc/cadmium efflux RND transporter might be involved in conferring resistance to colistin and/or carbapenem. From SNP analysis, we identified three points of non-synonymous mutations in *pmrB* (412G > A, 599C > T, and 687A > C) that change amino acid sequences. These amino acid changes, specifically Glu229Asp, Pro200Leu, and Ala138Thr may confer colistin resistance in MDR *A. baumannii* strains.

Supplementary Materials: Supplementary data are available online at https://www.mdpi.com/article/10.3390/antibiotics10091054/s1, Supplement Figure S1: The relative optical density at 570 nm for determination of minimum inhibitory concentration (MIC) by microdilution assay using MTT staining, Supplement Table S1: Guideline for interpretation of disk diffusion results, Supplement Table S2: Details of pathogenicity islands, Supplement Table S3: List of strain-specific genes, Supplement Table S4: List of multidrug resistant-specific genes (Aci46-ACICU-AC30-R14), Supplement Table S5: List of polymyxins and carbapenem-resistant-specific genes (Aci46-AC30-R14), Supplement Table S6: List of colistin and carbapenem-resistant-specific genes (Aci46-R14), Supplement Table S7: List of core genes (present in all genomes).

Author Contributions: N.T., S.C., S.S. and P.D. conceptualized the study; N.T. designed the research, tested antibiotic resistant profiling, analyzed genomic sequences, and wrote the paper; P.D. collected bacterial samples and identified *A. baumannii* Aci46. All authors have read and agreed to the published version of the manuscript.

Funding: The research project was partially supported by Postdoctoral fellowship award from Mahidol University, grant number MU-PD_2020_9.

Institutional Review Board Statement: Ethical review and approval were waived for this study, because the isolate used in this study was obtained from a collection of isolates that has already been published.

Informed Consent Statement: Informed consent was obtained from all subjects involved in a previous study.

Data Availability Statement: The whole genome and plasmid sequences of *A. baumannii* Aci46 have been deposited at DDBJ/ENA/GenBank under the BioProject ID PRJNA739068.

Acknowledgments: The research project was partially supported by Postdoctoral fellowship award from Mahidol University, grant number MU-PD_2020_9. We thank James M. Dubbs for editing the manuscript.

Conflicts of Interest: The authors declare no conflict of interest.

References

1. Singh, H.; Thangaraj, P.; Chakrabarti, A. *Acinetobacter baumannii*: A Brief Account of Mechanisms of Multidrug Resistance and Current and Future Therapeutic Management. *J. Clin. Diagn. Res.* **2013**, *7*, 2602–2605. [CrossRef] [PubMed]
2. Lean, S.S.; Yeo, C.C.; Suhaili, Z.; Thong, K.L. Comparative Genomics of Two ST 195 Carbapenem-Resistant *Acinetobacter baumannii* with Different Susceptibility to Polymyxin Revealed Underlying Resistance Mechanism. *Front. Microbiol.* **2015**, *6*, 1445. [CrossRef] [PubMed]
3. Chopra, T.; Marchaim, D.; Johnson, P.C.; Awali, R.A.; Doshi, H.; Chalana, I.; Davis, N.; Zhao, J.J.; Pogue, J.M.; Parmar, S.; et al. Risk factors and outcomes for patients with bloodstream infection due to *Acinetobacter baumannii*-calcoaceticus complex. *Antimicrob. Agents Chemother.* **2014**, *58*, 4630–4635. [CrossRef] [PubMed]
4. Magiorakos, A.P.; Srinivasan, A.; Carey, R.B.; Carmeli, Y.; Falagas, M.E.; Giske, C.G.; Harbarth, S.; Hindler, J.F.; Kahlmeter, G.; Olsson-Liljequist, B.; et al. Multidrug-resistant, extensively drug-resistant and pandrug-resistant bacteria: An international expert proposal for interim standard definitions for acquired resistance. *Clin. Microbiol. Infect.* **2012**, *18*, 268–281. [CrossRef]
5. Mulani, M.S.; Kamble, E.E.; Kumkar, S.N.; Tawre, M.S.; Pardesi, K.R. Emerging Strategies to Combat ESKAPE Pathogens in the Era of Antimicrobial Resistance: A Review. *Front. Microbiol.* **2019**, *10*, 539. [CrossRef] [PubMed]
6. Tacconelli, E.; Carrara, E.; Savoldi, A.; Harbarth, S.; Mendelson, M.; Monnet, D.L.; Pulcini, C.; Kahlmeter, G.; Kluytmans, J.; Carmeli, Y.; et al. Discovery, research, and development of new antibiotics: The WHO priority list of antibiotic-resistant bacteria and tuberculosis. *Lancet. Infect. Dis.* **2018**, *18*, 318–327. [CrossRef]
7. Gottesman, T.; Fedorowsky, R.; Yerushalmi, R.; Lellouche, J.; Nutman, A. An outbreak of carbapenem-resistant *Acinetobacter baumannii* in a COVID-19 dedicated hospital. *Infect. Prev. Pract.* **2021**, *3*, 100113. [CrossRef] [PubMed]
8. Perez, S.; Innes, G.K.; Walters, M.S.; Mehr, J.; Arias, J.; Greeley, R.; Chew, D. Increase in Hospital-Acquired Carbapenem-Resistant *Acinetobacter baumannii* Infection and Colonization in an Acute Care Hospital During a Surge in COVID-19 Admissions—New Jersey, February-July 2020. *MMWR Morb. Mortal Wkly Rep.* **2020**, *69*, 1827–1831. [CrossRef] [PubMed]
9. Khuntayaporn, P.; Kanathum, P.; Houngsaitong, J.; Montakantikul, P.; Thirapanmethee, K.; Chomnawang, M.T. Predominance of international clone 2 multidrug-resistant *Acinetobacter baumannii* clinical isolates in Thailand: A nationwide study. *Ann. Clin. Microbiol. Antimicrob.* **2021**, *20*, 19. [CrossRef]
10. Loraine, J.; Heinz, E.; Soontarach, R.; Blackwell, G.A.; Stabler, R.A.; Voravuthikunchai, S.P.; Srimanote, P.; Kiratisin, P.; Thomson, N.R.; Taylor, P.W. Genomic and Phenotypic Analyses of *Acinetobacter baumannii* Isolates from Three Tertiary Care Hospitals in Thailand. *Front. Microbiol.* **2020**, *11*, 548. [CrossRef]
11. Wong, D.; Nielsen, T.B.; Bonomo, R.A.; Pantapalangkoor, P.; Luna, B.; Spellberg, B. Clinical and Pathophysiological Overview of *Acinetobacter* Infections: A Century of Challenges. *Clin. Microbiol. Rev.* **2017**, *30*, 409–447. [CrossRef] [PubMed]
12. Butler, D.A.; Biagi, M.; Tan, X.; Qasmieh, S.; Bulman, Z.P.; Wenzler, E. Multidrug Resistant *Acinetobacter baumannii*: Resistance by Any Other Name Would Still be Hard to Treat. *Curr. Infect. Dis. Rep.* **2019**, *21*, 46. [CrossRef]
13. Karyne, R.; Curty Lechuga, G.; Almeida Souza, A.L.; Rangel da Silva Carvalho, J.P.; Simoes Villas Boas, M.H.; De Simone, S.G. Pan-Drug Resistant *Acinetobacter baumannii*, but Not Other Strains, Are Resistant to the Bee Venom Peptide Mellitin. *Antibiotics* **2020**, *9*, 178. [CrossRef]
14. Huttner, B.; Jones, M.; Rubin, M.A.; Neuhauser, M.M.; Gundlapalli, A.; Samore, M. Drugs of last resort? The use of polymyxins and tigecycline at US Veterans Affairs medical centers, 2005–2010. *PLoS ONE* **2012**, *7*, e36649. [CrossRef] [PubMed]

15. Velkov, T.; Roberts, K.D.; Nation, R.L.; Thompson, P.E.; Li, J. Pharmacology of polymyxins: New insights into an 'old' class of antibiotics. *Future Microbiol.* **2013**, *8*, 711–724. [CrossRef]
16. Trimble, M.J.; Mlynarcik, P.; Kolar, M.; Hancock, R.E. Polymyxin: Alternative Mechanisms of Action and Resistance. *Cold Spring Harb. Perspect. Med.* **2016**, *6*, a025288. [CrossRef]
17. Thet, K.T.; Lunha, K.; Srisrattakarn, A.; Lulitanond, A.; Tavichakorntrakool, R.; Kuwatjanakul, W.; Charoensri, N.; Chanawong, A. Colistin heteroresistance in carbapenem-resistant *Acinetobacter baumannii* clinical isolates from a Thai university hospital. *World J. Microbiol. Biotechnol.* **2020**, *36*, 102. [CrossRef] [PubMed]
18. Lertsrisatit, Y.; Santimaleeworagun, W.; Thunyaharn, S.; Traipattanakul, J. *In vitro* activity of colistin mono- and combination therapy against colistin-resistant *Acinetobacter baumannii*, mechanism of resistance, and clinical outcomes of patients infected with colistin-resistant *A. baumannii* at a Thai university hospital. *Infect. Drug. Resist.* **2017**, *10*, 437–443. [CrossRef] [PubMed]
19. Da Silva, G.J.; Domingues, S. Interplay between Colistin Resistance, Virulence and Fitness in *Acinetobacter baumannii*. *Antibiotics* **2017**, *6*, 28. [CrossRef] [PubMed]
20. Beceiro, A.; Llobet, E.; Aranda, J.; Bengoechea, J.A.; Doumith, M.; Hornsey, M.; Dhanji, H.; Chart, H.; Bou, G.; Livermore, D.M.; et al. Phosphoethanolamine modification of lipid A in colistin-resistant variants of *Acinetobacter baumannii* mediated by the *pmrAB* two-component regulatory system. *Antimicrob. Agents Chemother.* **2011**, *55*, 3370–3379. [CrossRef]
21. Moffatt Jennifer, H.; Harper, M.; Harrison, P.; Hale John, D.F.; Vinogradov, E.; Seemann, T.; Henry, R.; Crane, B.; St. Michael, F.; Cox Andrew, D.; et al. Colistin Resistance in *Acinetobacter baumannii* Is Mediated by Complete Loss of Lipopolysaccharide Production. *Antimicrob. Agents Chemother.* **2010**, *54*, 4971–4977. [CrossRef] [PubMed]
22. Thi Khanh Nhu, N.; Riordan, D.W.; Do Hoang Nhu, T.; Thanh, D.P.; Thwaites, G.; Huong Lan, N.P.; Wren, B.W.; Baker, S.; Stabler, R.A. The induction and identification of novel Colistin resistance mutations in *Acinetobacter baumannii* and their implications. *Sci. Rep.* **2016**, *6*, 28291. [CrossRef] [PubMed]
23. Yilmaz, S.; Hasdemir, U.; Aksu, B.; Altinkanat Gelmez, G.; Soyletir, G. Alterations in AdeS and AdeR regulatory proteins in 1-(1-naphthylmethyl)-piperazine responsive colistin resistance of *Acinetobacter baumannii*. *J. Chemother.* **2020**, *32*, 286–293. [CrossRef]
24. Lin, M.F.; Lin, Y.Y.; Lan, C.Y. Contribution of EmrAB efflux pumps to colistin resistance in *Acinetobacter baumannii*. *J. Microbiol.* **2017**, *55*, 130–136. [CrossRef] [PubMed]
25. Kansakar, P.; Dorji, D.; Chongtrakool, P.; Mingmongkolchai, S.; Mokmake, B.; Dubbs, P. Local dissemination of multidrug-resistant *Acinetobacter baumannii* clones in a Thai hospital. *Microb. Drug. Resist.* **2011**, *17*, 109–119. [CrossRef] [PubMed]
26. Bartual, S.G.; Seifert, H.; Hippler, C.; Luzon, M.A.; Wisplinghoff, H.; Rodriguez-Valera, F. Development of a multilocus sequence typing scheme for characterization of clinical isolates of *Acinetobacter baumannii*. *J. Clin. Microbiol.* **2005**, *43*, 4382–4390. [CrossRef] [PubMed]
27. Diancourt, L.; Passet, V.; Nemec, A.; Dijkshoorn, L.; Brisse, S. The population structure of *Acinetobacter baumannii*: Expanding multiresistant clones from an ancestral susceptible genetic pool. *PLoS ONE* **2010**, *5*, e10034. [CrossRef]
28. Wareth, G.; Linde, J.; Nguyen, N.H.; Nguyen, T.N.M.; Sprague, L.D.; Pletz, M.W.; Neubauer, H. WGS-Based Analysis of Carbapenem-Resistant *Acinetobacter baumannii* in Vietnam and Molecular Characterization of Antimicrobial Determinants and MLST in Southeast Asia. *Antibiotics* **2021**, *10*, 563. [CrossRef]
29. Liu, C.C.; Tang, C.Y.; Chang, K.C.; Kuo, H.Y.; Liou, M.L. A comparative study of class 1 integrons in *Acinetobacter baumannii*. *Gene* **2014**, *544*, 75–82. [CrossRef]
30. Niumsup, P.R.; Boonkerd, N.; Tansawai, U.; Tiloklurs, M. Carbapenem-resistant *Acinetobacter baumannii* producing OXA-23 in Thailand. *Jpn. J. Infect. Dis.* **2009**, *62*, 152–154.
31. Kyriakidis, I.; Vasileiou, E.; Pana, Z.D.; Tragiannidis, A. *Acinetobacter baumannii* Antibiotic Resistance Mechanisms. *Pathogens* **2021**, *10*, 373. [CrossRef]
32. Teo, J.; Lim, T.P.; Hsu, L.Y.; Tan, T.Y.; Sasikala, S.; Hon, P.Y.; Kwa, A.L.; Apisarnthanarak, A. Extensively drug-resistant *Acinetobacter baumannii* in a Thai hospital: A molecular epidemiologic analysis and identification of bactericidal Polymyxin B-based combinations. *Antimicrob. Resist. Infect. Control.* **2015**, *4*, 2. [CrossRef]
33. Adams Mark, D.; Nickel Gabrielle, C.; Bajaksouzian, S.; Lavender, H.; Murthy, A.R.; Jacobs Michael, R.; Bonomo Robert, A. Resistance to Colistin in *Acinetobacter baumannii* Associated with Mutations in the PmrAB Two-Component System. *Antimicrob. Agents Chemother.* **2009**, *53*, 3628–3634. [CrossRef]
34. Lee, C.R.; Lee, J.H.; Park, M.; Park, K.S.; Bae, I.K.; Kim, Y.B.; Cha, C.J.; Jeong, B.C.; Lee, S.H. Biology of *Acinetobacter baumannii*: Pathogenesis, Antibiotic Resistance Mechanisms, and Prospective Treatment Options. *Front. Cell Infect. Microbiol.* **2017**, *7*, 55. [CrossRef]
35. Ayoub Moubareck, C.; Hammoudi Halat, D. Insights into *Acinetobacter baumannii*: A Review of Microbiological, Virulence, and Resistance Traits in a Threatening Nosocomial Pathogen. *Antibiotics* **2020**, *9*, 119. [CrossRef]
36. McConnell, M.J.; Actis, L.; Pachon, J. *Acinetobacter baumannii*: Human infections, factors contributing to pathogenesis and animal models. *FEMS Microbiol. Rev.* **2013**, *37*, 130–155. [CrossRef] [PubMed]
37. Costa, A.R.; Monteiro, R.; Azeredo, J. Genomic analysis of *Acinetobacter baumannii* prophages reveals remarkable diversity and suggests profound impact on bacterial virulence and fitness. *Sci. Rep.* **2018**, *8*, 15346. [CrossRef] [PubMed]
38. Elhosseiny, N.M.; Attia, A.S. *Acinetobacter*: An emerging pathogen with a versatile secretome. *Emerg. Microbes Infect.* **2018**, *7*, 33. [CrossRef] [PubMed]

39. Liu, C.C.; Kuo, H.Y.; Tang, C.Y.; Chang, K.C.; Liou, M.L. Prevalence and mapping of a plasmid encoding a type IV secretion system in *Acinetobacter baumannii*. *Genomics* **2014**, *104*, 215–223. [CrossRef] [PubMed]

40. Morris, F.C.; Dexter, C.; Kostoulias, X.; Uddin, M.I.; Peleg, A.Y. The Mechanisms of Disease Caused by *Acinetobacter baumannii*. *Front. Microbiol.* **2019**, *10*, 1601. [CrossRef] [PubMed]

41. Palmieri, M.; D'Andrea, M.M.; Pelegrin, A.C.; Perrot, N.; Mirande, C.; Blanc, B.; Legakis, N.; Goossens, H.; Rossolini, G.M.; van Belkum, A. Abundance of Colistin-Resistant, OXA-23- and ArmA-Producing *Acinetobacter baumannii* Belonging to International Clone 2 in Greece. *Front. Microbiol.* **2020**, *11*, 668. [CrossRef]

42. Abdulzahra, A.T.; Khalil, M.A.F.; Elkhatib, W.F. First report of colistin resistance among carbapenem-resistant *Acinetobacter baumannii* isolates recovered from hospitalized patients in Egypt. *N. Microbes N. Infect.* **2018**, *26*, 53–58. [CrossRef] [PubMed]

43. Nies, D.H. Efflux-mediated heavy metal resistance in prokaryotes. *FEMS Microbiol. Rev.* **2003**, *27*, 313–339. [CrossRef]

44. Nies, D.H. The cobalt, zinc, and cadmium efflux system CzcABC from *Alcaligenes eutrophus* functions as a cation-proton antiporter in *Escherichia coli*. *J. Bacteriol.* **1995**, *177*, 2707–2712. [CrossRef]

45. Gheorghe, I.; Barbu, I.C.; Surleac, M.; Sarbu, I.; Popa, L.I.; Paraschiv, S.; Feng, Y.; Lazar, V.; Chifiriuc, M.C.; Otelea, D.; et al. Subtypes, resistance and virulence platforms in extended-drug resistant *Acinetobacter baumannii* Romanian isolates. *Sci. Rep.* **2021**, *11*, 13288. [CrossRef] [PubMed]

46. Smith, C.A.; Antunes, N.T.; Stewart, N.K.; Frase, H.; Toth, M.; Kantardjieff, K.A.; Vakulenko, S. Structural Basis for Enhancement of Carbapenemase Activity in the OXA-51 Family of Class D beta-Lactamases. *ACS Chem. Biol.* **2015**, *10*, 1791–1796. [CrossRef] [PubMed]

47. Gerson, S.; Betts Jonathan, W.; Lucaßen, K.; Nodari Carolina, S.; Wille, J.; Josten, M.; Göttig, S.; Nowak, J.; Stefanik, D.; Roca, I.; et al. Investigation of Novel *pmrB* and *eptA* Mutations in Isogenic *Acinetobacter baumannii* Isolates Associated with Colistin Resistance and Increased Virulence In Vivo. *Antimicrob. Agents Chemother.* **2019**, *63*, e01586-18. [CrossRef] [PubMed]

48. Sun, B.; Liu, H.; Jiang, Y.; Shao, L.; Yang, S.; Chen, D. New Mutations Involved in Colistin Resistance in *Acinetobacter baumannii*. *mSphere* **2020**, *5*, e00895-19. [CrossRef]

49. Balows, A. *Manual of Cinical Microbiology*, 8th ed.; Murray, P.R., Baron, E.J., Jorgenson, J.H., Pfaller, M.A., Yolken, R.H., Eds.; ASM Press: Washington, DC, USA, 2003; Volume 2, ISBN 1-555810255-4.

50. CLSI. *Performance Standards for Antimicrobial Susceptibility Testing*, 31st ed.; CLSI. Supplement M100-ED31; Clinical and Laboratory Standards Institute: Wayne, PA, USA, 2021; ISBN 978-1-68440-104-8.

51. Benov, L. Effect of growth media on the MTT colorimetric assay in bacteria. *PLoS ONE* **2019**, *14*, e0219713. [CrossRef]

52. Hundie, G.B.; Woldemeskel, D.; Gessesse, A. Evaluation of Direct Colorimetric MTT Assay for Rapid Detection of Rifampicin and Isoniazid Resistance in *Mycobacterium tuberculosis*. *PLoS ONE* **2016**, *11*, e0169188. [CrossRef]

53. Salvà Serra, F.; Svensson Stadler, L.; Busquets, A.; Jaén-Luchoro, D.; Gomila, M. A Protocol for Extraction and Purification of High-Quality and Quantity Bacterial DNA Applicable for Genome Sequencing: A Modified Version of the Marmur Procedure. *Protoc. Exch.* **2018**. [CrossRef]

54. Krueger, F. Trim Galore: A Wrapper Tool around Cutadapt and FastQC to Consistently Apply Quality and Adapter Trimming to FastQ Files. Available online: http://www.bioinformatics.babraham.ac.uk/projects/trim_galore/ (accessed on 27 June 2019).

55. Andrews, S. FastQC: A Quality Control Tool for High Throughput Sequence Data. Available online: https://www.bioinformatics.babraham.ac.uk/projects/fastqc/ (accessed on 4 October 2018).

56. Bankevich, A.; Nurk, S.; Antipov, D.; Gurevich, A.A.; Dvorkin, M.; Kulikov, A.S.; Lesin, V.M.; Nikolenko, S.I.; Pham, S.; Prjibelski, A.D.; et al. SPAdes: A new genome assembly algorithm and its applications to single-cell sequencing. *J. Comput. Biol.* **2012**, *19*, 455–477. [CrossRef] [PubMed]

57. Walker, B.J.; Abeel, T.; Shea, T.; Priest, M.; Abouelliel, A.; Sakthikumar, S.; Cuomo, C.A.; Zeng, Q.; Wortman, J.; Young, S.K.; et al. Pilon: An integrated tool for comprehensive microbial variant detection and genome assembly improvement. *PLoS ONE* **2014**, *9*, e112963. [CrossRef] [PubMed]

58. Li, H.; Handsaker, B.; Wysoker, A.; Fennell, T.; Ruan, J.; Homer, N.; Marth, G.; Abecasis, G.; Durbin, R.; Genome Project Data Processing, S. The Sequence Alignment/Map format and SAMtools. *Bioinformatics* **2009**, *25*, 2078–2079. [CrossRef] [PubMed]

59. Davis, J.J.; Wattam, A.R.; Aziz, R.K.; Brettin, T.; Butler, R.; Butler, R.M.; Chlenski, P.; Conrad, N.; Dickerman, A.; Dietrich, E.M.; et al. The PATRIC Bioinformatics Resource Center: Expanding data and analysis capabilities. *Nucleic Acids Res.* **2020**, *48*, D606–D612. [CrossRef]

60. Gurevich, A.; Saveliev, V.; Vyahhi, N.; Tesler, G. QUAST: Quality assessment tool for genome assemblies. *Bioinformatics* **2013**, *29*, 1072–1075. [CrossRef]

61. Wick, R.R.; Schultz, M.B.; Zobel, J.; Holt, K.E. Bandage: Interactive visualization of de novo genome assemblies. *Bioinformatics* **2015**, *31*, 3350–3352. [CrossRef]

62. Brettin, T.; Davis, J.J.; Disz, T.; Edwards, R.A.; Gerdes, S.; Olsen, G.J.; Olson, R.; Overbeek, R.; Parrello, B.; Pusch, G.D.; et al. RASTtk: A modular and extensible implementation of the RAST algorithm for building custom annotation pipelines and annotating batches of genomes. *Sci. Rep.* **2015**, *5*, 8365. [CrossRef]

63. Bortolaia, V.; Kaas, R.S.; Ruppe, E.; Roberts, M.C.; Schwarz, S.; Cattoir, V.; Philippon, A.; Allesoe, R.L.; Rebelo, A.R.; Florensa, A.F.; et al. ResFinder 4.0 for predictions of phenotypes from genotypes. *J. Antimicrob. Chemother.* **2020**, *75*, 3491–3500. [CrossRef]

64. Alcock, B.P.; Raphenya, A.R.; Lau, T.T.Y.; Tsang, K.K.; Bouchard, M.; Edalatmand, A.; Huynh, W.; Nguyen, A.V.; Cheng, A.A.; Liu, S.; et al. CARD 2020: Antibiotic resistome surveillance with the comprehensive antibiotic resistance database. *Nucleic Acids Res.* **2020**, *48*, D517–D525. [CrossRef]

65. Sayers, E.W.; Beck, J.; Brister, J.R.; Bolton, E.E.; Canese, K.; Comeau, D.C.; Funk, K.; Ketter, A.; Kim, S.; Kimchi, A.; et al. Database resources of the National Center for Biotechnology Information. *Nucleic Acids Res.* **2020**, *48*, D9–D16. [CrossRef] [PubMed]

66. Li, J.; Tai, C.; Deng, Z.; Zhong, W.; He, Y.; Ou, H.Y. VRprofile: Gene-cluster-detection-based profiling of virulence and antibiotic resistance traits encoded within genome sequences of pathogenic bacteria. *Brief Bioinform.* **2018**, *19*, 566–574. [CrossRef]

67. Arndt, D.; Grant, J.R.; Marcu, A.; Sajed, T.; Pon, A.; Liang, Y.; Wishart, D.S. PHASTER: A better, faster version of the PHAST phage search tool. *Nucleic Acids Res.* **2016**, *44*, W16–W21. [CrossRef] [PubMed]

68. Antipov, D.; Hartwick, N.; Shen, M.; Raiko, M.; Lapidus, A.; Pevzner, P.A. plasmidSPAdes: Assembling plasmids from whole genome sequencing data. *Bioinformatics* **2016**, *32*, 3380–3387. [CrossRef] [PubMed]

69. Jolley, K.A.; Bliss, C.M.; Bennett, J.S.; Bratcher, H.B.; Brehony, C.; Colles, F.M.; Wimalarathna, H.; Harrison, O.B.; Sheppard, S.K.; Cody, A.J.; et al. Ribosomal multilocus sequence typing: Universal characterization of bacteria from domain to strain. *Microbiology* **2012**, *158*, 1005–1015. [CrossRef]

70. Jolley, K.A.; Bray, J.E.; Maiden, M.C.J. Open-access bacterial population genomics: BIGSdb software, the PubMLST.org website and their applications. *Wellcome Open Res.* **2018**, *3*, 124. [CrossRef]

71. Stamatakis, A. RAxML version 8: A tool for phylogenetic analysis and post-analysis of large phylogenies. *Bioinformatics* **2014**, *30*, 1312–1313. [CrossRef]

72. Davis, J.J.; Gerdes, S.; Olsen, G.J.; Olson, R.; Pusch, G.D.; Shukla, M.; Vonstein, V.; Wattam, A.R.; Yoo, H. PATtyFams: Protein Families for the Microbial Genomes in the PATRIC Database. *Front. Microbiol.* **2016**, *7*, 118. [CrossRef]

73. Rambaut, A. FigTree, a Graphical Viewer of Phylogenetic Trees. Available online: http://tree.bio.ed.ac.uk/software/figtree/ (accessed on 1 November 2006).

74. Vallenet, D.; Nordmann, P.; Barbe, V.; Poirel, L.; Mangenot, S.; Bataille, E.; Dossat, C.; Gas, S.; Kreimeyer, A.; Lenoble, P.; et al. Comparative analysis of Acinetobacters: Three genomes for three lifestyles. *PLoS ONE* **2008**, *3*, e1805. [CrossRef]

75. Adams, M.D.; Goglin, K.; Molyneaux, N.; Hujer, K.M.; Lavender, H.; Jamison, J.J.; MacDonald, I.J.; Martin, K.M.; Russo, T.; Campagnari, A.A.; et al. Comparative genome sequence analysis of multidrug-resistant *Acinetobacter baumannii. J. Bacteriol.* **2008**, *190*, 8053–8064. [CrossRef]

76. Hamidian, M.; Wick, R.R.; Hartstein, R.M.; Judd, L.M.; Holt, K.E.; Hall, R.M. Insights from the revised complete genome sequences of *Acinetobacter baumannii* strains AB307-0294 and ACICU belonging to global clones 1 and 2. *Microb. Genom.* **2019**, *5*, e000298. [CrossRef]

77. Zhou, H.; Zhang, T.; Yu, D.; Pi, B.; Yang, Q.; Zhou, J.; Hu, S.; Yu, Y. Genomic analysis of the multidrug-resistant *Acinetobacter baumannii* strain MDR-ZJ06 widely spread in China. *Antimicrob. Agents Chemother.* **2011**, *55*, 4506–4512. [CrossRef] [PubMed]

78. Mustapha, M.M.; Li, B.; Pacey, M.P.; Mettus, R.T.; McElheny, C.L.; Marshall, C.W.; Ernst, R.K.; Cooper, V.S.; Doi, Y. Phylogenomics of colistin-susceptible and resistant XDR *Acinetobacter baumannii. J. Antimicrob. Chemother.* **2018**, *73*, 2952–2959. [CrossRef] [PubMed]

79. Smith, M.G.; Gianoulis, T.A.; Pukatzki, S.; Mekalanos, J.J.; Ornston, L.N.; Gerstein, M.; Snyder, M. New insights into *Acinetobacter baumannii* pathogenesis revealed by high-density pyrosequencing and transposon mutagenesis. *Genes Dev.* **2007**, *21*, 601–614. [CrossRef] [PubMed]

80. Barbe, V.; Vallenet, D.; Fonknechten, N.; Kreimeyer, A.; Oztas, S.; Labarre, L.; Cruveiller, S.; Robert, C.; Duprat, S.; Wincker, P.; et al. Unique features revealed by the genome sequence of *Acinetobacter* sp. ADP1, a versatile and naturally transformation competent bacterium. *Nucleic Acids Res.* **2004**, *32*, 5766–5779. [CrossRef] [PubMed]

81. Li, H.; Durbin, R. Fast and accurate short read alignment with Burrows-Wheeler transform. *Bioinformatics* **2009**, *25*, 1754–1760. [CrossRef] [PubMed]

82. Garrison, E.; Marth, G. Haplotype-based variant detection from short-read sequencing. *arXiv* **2012**, arXiv:1207.3907.

83. Madeira, F.; Park, Y.M.; Lee, J.; Buso, N.; Gur, T.; Madhusoodanan, N.; Basutkar, P.; Tivey, A.R.N.; Potter, S.C.; Finn, R.D.; et al. The EMBL-EBI search and sequence analysis tools APIs in 2019. *Nucleic Acids Res.* **2019**, *47*, W636–W641. [CrossRef]

 *antibiotics*

Article

Role of *dipA* and *pilD* in *Francisella tularensis* Susceptibility to Resazurin

Kendall Souder [1], Emma J. Beatty [1], Siena C. McGovern [1], Michael Whaby [1], Emily Young [1], Jacob Pancake [1], Daron Weekley [1], Justin Rice [1], Donald A. Primerano [2], James Denvir [2], Joseph Horzempa [1] and Deanna M. Schmitt [1,*]

[1] Department of Biomedical Sciences, West Liberty University, West Liberty, WV 26074, USA; kssouder@westliberty.edu (K.S.); ejbeatty@westliberty.edu (E.J.B.); scmcgovern@westliberty.edu (S.C.M.); whaby@musc.edu (M.W.); emily.young@westliberty.edu (E.Y.); jppancake2121@gmail.com (J.P.); dmweekley@westliberty.edu (D.W.); jcrice1@westliberty.edu (J.R.); joseph.horzempa@westliberty.edu (J.H.)

[2] Department of Biomedical Sciences, Joan C. Edwards School of Medicine, Marshall University, Huntington, WV 25755, USA; primeran@marshall.edu (D.A.P.); denvir@marshall.edu (J.D.)

* Correspondence: deanna.schmitt@westliberty.edu; Tel.: +1-304-336-8576

Abstract: The phenoxazine dye resazurin exhibits bactericidal activity against the Gram-negative pathogens *Francisella tularensis* and *Neisseria gonorrhoeae*. One resazurin derivative, resorufin pentyl ether, significantly reduces vaginal colonization by *Neisseria gonorrhoeae* in a mouse model of infection. The narrow spectrum of bacteria susceptible to resazurin and its derivatives suggests these compounds have a novel mode of action. To identify potential targets of resazurin and mechanisms of resistance, we isolated mutants of *F. tularensis* subsp. *holarctica* live vaccine strain (LVS) exhibiting reduced susceptibility to resazurin and performed whole genome sequencing. The genes *pilD* (FTL_0959) and *dipA* (FTL_1306) were mutated in half of the 46 resazurin-resistant (RZR) strains sequenced. Complementation of select RZR LVS isolates with wild-type *dipA* or *pilD* partially restored sensitivity to resazurin. To further characterize the role of *dipA* and *pilD* in resazurin susceptibility, a *dipA* deletion mutant, Δ*dipA*, and *pilD* disruption mutant, FTL_0959d, were generated. Both mutants were less sensitive to killing by resazurin compared to wild-type LVS with phenotypes similar to the spontaneous resazurin-resistant mutants. This study identified a novel role for two genes *dipA* and *pilD* in *F. tularensis* susceptibility to resazurin.

Keywords: *Francisella tularensis*; resazurin; DipA; PilD; tularemia; antimicrobial; antibiotic; resistance

Citation: Souder, K.; Beatty, E.J.; McGovern, S.C.; Whaby, M.; Young, E.; Pancake, J.; Weekley, D.; Rice, J.; Primerano, D.A.; Denvir, J.; et al. Role of *dipA* and *pilD* in *Francisella tularensis* Susceptibility to Resazurin. *Antibiotics* **2021**, *10*, 992. https://doi.org/10.3390/antibiotics10080992

Academic Editor: Teresa V. Nogueira

Received: 22 July 2021
Accepted: 14 August 2021
Published: 17 August 2021

Publisher's Note: MDPI stays neutral with regard to jurisdictional claims in published maps and institutional affiliations.

1. Introduction

The Centers for Disease Control and Prevention (CDC) estimates that there are nearly three million new cases of antibiotic resistant bacterial infections annually, resulting in 35,000 deaths and billions of dollars in health care costs in the United States [1]. The rise in antibiotic resistance has been attributed to over-prescription and improper use of commonplace antibiotics [2]. Despite the clinical threat posed by increasing antibiotic resistance, the development of new antibiotics has significantly slowed due to decreased profitability [3]. While there are over 40 antimicrobial agents in clinical trials, most belong to existing classes of antibiotics such as beta lactams and beta-lactamase inhibitors, tetracyclines, and aminoglycosides [4]. Each of these classes of antibiotics targets similar bacterial processes including cell wall synthesis and protein synthesis. The development of bacterial resistance against these new drugs in the future is likely since there is already selective pressure from current antibiotics. Therefore, new antibiotic targets should be explored to minimize the emergence of resistance and provide effective alternatives.

Phenoxazine-based compounds have been shown to have antimicrobial activity [5,6]. Actinomycin D, the most well-known phenoxazine-containing antibiotic, suppresses the growth of various Gram-positive and Gram-negative bacteria as well as *Mycobacterium tu-*

berculosis [7]. Other actinomycins, neo-actinomycins A and B, exhibit moderate antibacterial activity against methicillin-resistant *Staphylococcus aureus* (MRSA) and vancomycin-resistant *Enterococcus* (VRE) strains [8]. We previously showed the phenoxazine dye resazurin inhibits growth of the Gram-negative human pathogens *Neisseria gonorrhoeae* and *Francisella tularensis* [9,10]. Resorufin pentyl ether, an analog of resazurin, significantly reduces vaginal colonization by *N. gonorrhoeae* in a mouse model of infection [10]. However, the mechanism by which resazurin kills *F. tularensis* and *N. gonorrhoeae* has not been defined. Based on its structural similarity to other phenoxazine compounds, resazurin may function as a DNA intercalating agent [5]. Unlike other phenoxazine compounds, resazurin is non-toxic to eukaryotic cells, and its bactericidal activity is limited to only two Gram-negative bacterial species suggesting a novel mechanism of action [5,6,9].

In this study, we sought to identify *F. tularensis* genetic determinants involved in susceptibility to resazurin in hopes of elucidating the mode of action of this compound.

2. Results

2.1. Isolation of Resazurin-Resistant F. tularensis LVS Mutants

A common strategy bacteria employ to develop antimicrobial resistance is by mutating gene(s) associated with the mechanism of action of the antibiotic. Therefore, to identify potential targets of resazurin, we selected for mutants of *F. tularensis* LVS that were capable of growing in the presence of 22 µg/mL of resazurin. This concentration of resazurin is twenty-fold higher than the previously determined MIC of resazurin for *F. tularensis* LVS [9]. The genomes of forty-six spontaneous resazurin-resistant (RZR) mutants were sequenced and compared to wild-type LVS to identify genetic variants. Nonsynonymous mutations were identified in ten different protein-coding *F. tularensis* LVS genes with approximately 50% of the isolates possessing mutations in FTL_1306 (*dipA*) and FTL_0959 (*pilD*) (Table 1). The same *pilD* variant was observed in twenty-two of the RZR LVS isolates involving a single base pair substitution leading to a premature stop codon at position 187 of the protein sequence (Table 2). Fifteen different coding mutations were characterized in *dipA* that included single base pair substitutions, insertions, and deletions with many resulting in nonsense mutations (Table 2). Of note, one RZR LVS isolate, RZR46, contained a deletion of a single thymine residue at Position 2 in *dipA* resulting in loss of the start codon (Table 2). Western blot analysis confirmed this mutation abolished expression of DipA protein in RZR46 (data not shown). These data suggest the genes *dipA* and *pilD* play a role in *F. tularensis* LVS susceptibility to resazurin.

Table 1. Protein-coding genes containing nonsynonymous mutations in 46 resazurin-resistant *F. tularensis* isolates sequenced.

F. tularensis Gene Name	Number of Resazurin-Resistant Isolates Containing Gene Mutation
FTL_1306 (*dipA*)	24
FTL_0959 (*pilD*)	22
FTL_1666	5
FTL_0073	4
FTL_0358	2
FTL_0610	2
FTL_0132	1
FTL_0315	1
FTL_1634	1
FTL_1654	1

Table 2. Description of *dipA* and *pilD* mutations found in sequenced resazurin-resistant (Rzr) *F. tularensis* LVS isolates.

Mutation Type	Nucleotide Change	Protein Change
DipA		
Frameshift and start lost	1_19delATGAAAGCTAAATTTATAA	Start lost
Frameshift and start lost	2delT	Start lost
Stop gained	169G > T	Glu57 *
Stop gained	211G > T	Glu71 *
Stop gained	268G > T	Glu89 *
Stop gained	310G > T	Glu104 *
Frameshift and stop gained	342_346delGGTAG	Lys114Xfs6 *
Frameshift and stop gained	349_350delCA	Gln117Lysfs6 *
Frameshift and stop gained	350_353delAAAA	Lys119Valfs2 *
Frameshift and stop gained	387_396delTTTGGGCTAT	Asn129Lysfs8 *
Stop gained	578 C > A	Ser193 *
Stop gained	655C > T	Gln219 *
Stop gained	676G > T	Glu226 *
Stop gained	723T > A	Tyr240 *
Frameshift and stop gained	751_752insA	Tyr251 *
PilD		
Stop gained	532G > T	Gly178X

* indicates stop codon.

2.2. Characterizing the Role of pilD and dipA in F. tularensis Susceptibility to Resazurin

Almost all of the RZR LVS strains isolated possessed nonsynonymous mutations in multiple genes with some exceptions. The sole nonsynonymous mutation in RZR47 is in *pilD* while RZR46 contains a single nucleotide deletion resulting in loss in expression of the DipA protein (Table 2; data not shown). These findings suggest that loss of either *pilD* or *dipA* function may confer resazurin resistance with limited contributions/effects from other genes. To confirm the mutations in *dipA* and/or *pilD* were contributing to the reduced susceptibility of RZR strains to resazurin, we individually cloned wild-type copies of *dipA* and *pilD* into the *Francisella* vector pABST which contains the robust *groE* promotor of *F. tularensis* [11] and then introduced these new plasmids, pDipA and pPilD, into RZR46 and RZR47, respectively. Then, we tested the sensitivity of these strains to resazurin. We hypothesized that restoring expression of DipA and/or PilD in these RZR strains would alter their resistance phenotype. RZR46 and RZR47 containing pABST alone served as controls. The resazurin MICs for RZR46/pABST and RZR47/pABST were two-fold higher than the MIC for wild-type LVS (Table 3). Introduction of the pABST vector into the RZR strains did not alter their susceptibility to resazurin (data not shown). Complementation of RZR46 with *dipA* restored resazurin sensitivity back to wild-type with both RZR46/pDipA and LVS having MICs of 5.5 µg/mL (Table 3). In contrast, RZR47 containing the *pilD* expression construct has the same resazurin MIC as the vector control (Table 3). These data confirm a role for *dipA*, but not *pilD*, in *F. tularensis* LVS susceptibility to resazurin.

Table 3. Resazurin MIC for *dipA* and *pilD* complemented *F. tularensis* resazurin resistant strains.

Strain	MIC (µg/mL) [1]
LVS	5.5
Rzr 46/pABST	11
Rzr 47/pABST	11
Rzr 46/pDipA	5.5
Rzr 47/pPilD	11

[1] MICs are mode values from at least six independent determinations.

To individually investigate the role of *dipA* and *pilD* in resazurin susceptibility in a wild-type background, we generated a *pilD* disruption mutant (FTL_0959d) and a *dipA* deletion mutant (Δ*dipA*) in *F. tularensis* LVS. The MICs of both FTL_0959d and Δ*dipA* were two-fold higher than that of wild-type LVS and comparable to the MICs determined for the spontaneous resazurin-resistant mutants RZR46 and RZR47 (Tables 3 and 4). We next sought to evaluate the bactericidal activity of resazurin against FTL_0959d and Δ*dipA* over time. Treatment with resazurin resulted in a significant reduction in viable bacteria after 24 h compared to untreated controls for all strains tested (Figure 1). However, significantly more FTL_0959d and Δ*dipA* bacteria were recovered 24 h post resazurin treatment compared to wild-type LVS (Figure 1). These data suggest that neither DipA nor PilD are targets of resazurin, but might be involved in the uptake and metabolism of this compound by *F. tularensis* LVS.

Table 4. MIC of resazurin for *dipA* and *pilD F. tularensis* LVS mutants.

Strain	MIC (µg/mL) [1]
LVS	5.5
Δ*dipA*	11
FTL_0959d	11

[1] MICs are mode values from at least three independent determinations.

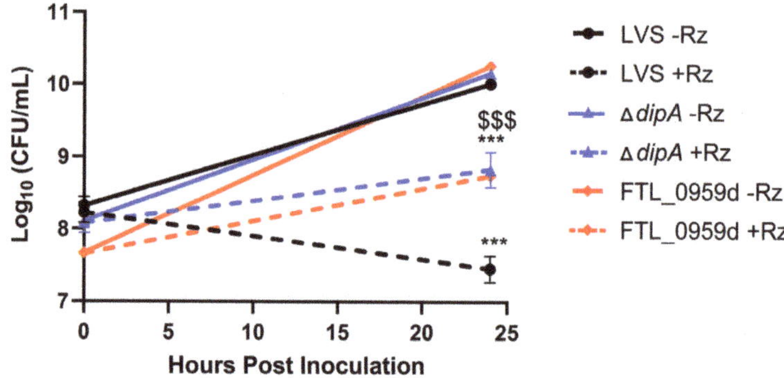

Figure 1. Reduced killing of FTL_0959d and Δ*dipA* by resazurin compared to wild-type LVS. Bacteria were cultivated in tryptic soy broth supplemented with 0.1% cysteine HCl (TSBc) in the presence or absence of resazurin (Rz, 1.375 µg/mL) for 24 h. Cultures were then diluted and plated to determine the number of viable *F. tularensis* LVS bacteria 24 h post inoculation. Data shown are mean ± SEM from three individual experiments. The limit of detection was 100 CFU per ml. Statistically significant differences in growth post-inoculation were determined by two-way ANOVA followed by Tukey's multiple comparisons test (***, $p < 0.001$ comparing +Rz to −Rz for each strain; $$$, $p < 0.001$ comparing FTL_0959d and Δ*dipA* to LVS 24 h. post Rz treatment).

3. Discussion

The clinical threat posed by antibiotic resistant bacterial infections highlights the need to identify and apply novel antimicrobial therapies [2]. Resazurin exhibits antibacterial activity against *F. tularensis* and *N. gonorrhoeae*, but its mode of action has not been defined [9,10]. Here, we screened for resazurin-resistant mutants of *F. tularensis* LVS to identify genetic determinants of resazurin susceptibility. Whole genome sequencing of RZR isolates identified nonsynonymous mutations in ten different protein-coding genes. None of the genes identified in this screen were previously described to play a role in *F. tularensis* resistance to other clinically relevant antibiotics such as aminoglycosides, tetracyclines and fluoroquinolones [12,13]. Therefore, these data suggest resazurin has a unique mechanism of action.

In particular, two genes, *dipA* and *pilD*, were mutated in the majority of the RZR LVS clones isolated and sequenced. Complementation of RZR46 with *dipA* restored sensitivity to resazurin comparable to wild-type LVS while expression of *pilD* in RZR47 did not.

However, both a laboratory generated *pilD* disruption mutant and *dipA* deletion mutant had higher MICs compared to wild-type LVS similar to the spontaneous RZR mutants. The failure to restore resazurin sensitivity in RZR47 upon complementation with wild-type *pilD* may be due to a dominant negative effect of the original *pilD* mutation or another intergenic/undetected variant in RZR47 may be contributing to the resazurin resistance. Regardless, less killing of FTL_0959d and Δ*dipA* was observed over 24 h following resazurin treatment compared to wild-type LVS confirming a role for both *dipA* and *pilD* in resazurin susceptibility.

DipA is a 353 amino acid protein predicted to form a surface complex with other outer membrane proteins to facilitate translocation of virulence factors that interact with host cells [14]. Most of the *dipA* mutations characterized in this study occurred within the Sel1-like repeat domains (amino acids 95–170 and 192–263) which are known to be required for the biological function of DipA [14]. These structural domains provide a scaffold for protein–protein interactions [15]. DipA has been shown to associate with FopA, a *F. tularensis* outer membrane protein with noted homology to the OmpA porin [14]. OmpA has been shown to play a role in antibiotic resistance in other Gram-negative pathogens including *Escherichia coli* and *Acinetobacter baumannii* [16,17]. Given the correlation between DipA mutations and resazurin resistance in this study, it is possible that DipA plays a role in the stability or function of FopA allowing for uptake of resazurin. Without functional DipA, less resazurin may be taken up by *F. tularensis* LVS resulting in increased resistance to resazurin. Interestingly, none of the resazurin resistant strains isolated and sequenced in this study contained mutations in the LVS gene encoding FopA. The role of DipA and FopA in uptake of resazurin by *F. tularensis* is currently being investigated.

The PilD protein was also shown to contribute to *F. tularensis* susceptibility to resazurin. PilD is an inner membrane peptidase responsible for processing major and minor (pseudo) pilins involved in the formation of Type IV pili and assembly of a Type II protein secretion system [18,19]. In *P. aeruginosa*, PilD has also been shown to play a role in the export of select enzymes including alkaline phosphatase, phospholipase C, elastase, and exotoxin A [20]. Although the exact function of PilD in *F. tularensis* has not been fully investigated and defined, it is possible PilD may play a role in the processing and/or export of a protein that is a target of resazurin. Mutation of *pilD* would thus affect processing or export of the antibiotic target rendering *F. tularensis* less susceptible to resazurin. The exact role PilD plays in *F. tularensis* susceptibility to resazurin is under further investigation.

As *F. tularensis* is not the only bacterium susceptible to resazurin, we wondered whether *pilD* (FTL_0959) and *dipA* (FTL_1306) homologs were present in *N. gonorrhoeae* and played similar roles as described here. A BLAST search [21] against the National Center for Biotechnology Information database of non-redundant protein sequences identified an A24 family peptidase (WP_003689814.1) and a SEL1-like repeat protein (MBS5742101.1) in *Neisseria* that were only 43% and 26% identical to *F. tularensis* PilD and DipA, respectively. The low homology between these proteins suggests that PilD and DipA have unique roles in *F. tularensis* susceptibility to resazurin. It also remains unknown whether DipA and PilD function in the same pathway to mediate *F. tularensis* killing by resazurin. The proposed function of DipA in transport and PilD in pilin processing suggests that they function separately from one another, but the relationship between DipA and PilD in mediating susceptibility to resazurin is currently being explored. Overall, mutations in *pilD* and *dipA* failed to confer complete resistance to resazurin. The growth of both FTL_0959d and Δ*dipA* was still inhibited by resazurin and the MICs for these mutants were only two-fold higher than wild-type LVS. An alternative method to identify potential antibiotic targets from the one used in this study is high-throughput transposon sequencing (Tn-Seq) [22–24]. In this approach, a library of LVS transposon mutants would be generated and screened for resazurin resistance [25]. Then, next-generation sequencing of the transposon insertion would be performed and the insertion sites would be mapped to the LVS genome. Further understanding of resazurin's mode of action and resistance mechanisms will help guide the development of resazurin derivatives with more potent activity as well as new therapeutic strategies to combat *Francisella* infections.

4. Materials and Methods

4.1. Bacterial Strains and Growth Conditions

Bacterial strains used in this study are listed in Table 5. For cultivation of *F. tularensis* LVS strains, frozen stock cultures were plated on chocolate II agar and incubated at 37 °C with 5% CO_2 for 2–4 days. These bacteria were then used to inoculate TSBc (trypticase soy broth (BD Biosciences) containing 0.1% L-cysteine hydrochloride monohydrate (Fisher)). *Escherichia coli* 5-α (New England Biolabs) bacteria cultivated on LB agar were used to inoculate LB broth. All broth cultures were incubated at 37 °C with agitation. When required, kanamycin (35 μg/mL for *E. coli*; 10 μg/mL for *F. tularensis*) and hygromycin (250 μg/mL) were supplemented into the media.

Table 5. Description of strains, plasmids, and primers used in this study.

Plasmid, Primer, or Strain	Description or Sequence	Source or Reference
F. tularensis strains		
LVS	*F. tularensis* subsp. holarctica live vaccine strain	Karen Elkins
RZR46	Spontaneous resazurin-resistant LVS mutant containing deletion of single thymine residue at position 2 in *dipA*	This study
RZR47	Spontaneous resazurin-resistant LVS mutant containing 532G > T substitution in *pilD*	This study
Δ*dipA*	LVS *dipA* deletion mutant	This study
FTL_0959d	LVS *pilD* disruption mutant	This study
E. coli strains		
DH5α	*fhuA2* Δ*(argF-lacZ)U169 phoA glnV44 Φ80* Δ*(lacZ)M15 gyrA96 recA1 relA1 endA1 thi-1 hsdR17*	NEB [1]
Plasmids		
pABST	pFNLTP8 with *F. tularensis* LVS groE promoter	[11]
pDipA	pABST containing *dipA* under control of *groE* promoter	This study
pPilD	pABST containing *dipA* under control of *groE* promoter	This study
pJH1	pMQ225 with the I-SceI restriction site	[26]
pGUTS	pGRP with I-SceI under the control of FGRp	[26]
pJH1ΔdipA	pJH1 with the left and right 500-bp flanking regions of *dipA* cloned into the PstI-SphI site of pJH1	This study
pJH1-0959d	pJH1 with the central 470 base pair region of FTL_0959	This study
Primers		
1F_dipA	5′-CATGCTGCAGGTAGTGTTTGGATCTTTTGTATTAGCAG -3′	IDT [2] This study
2R_dipA	5′-ATAATAAGTCGACGGTACCACCGGTAATTATAAATTTAGCTTTCATCTATTTCTCCTG-3′	IDT This study
3F_dipA	5′-ACCGGTGGTACCGTCGACTTATTATTACTACCAAAGAGCAGCAAAG-3′	IDT This study
4R_dipA	5′-CATGGCATGCCATCACCACATTTAGAACTATTGGC-3′	IDT This study
F_0959d	5′-CATGGGATCCGGGATTGATAAGCAACAATCTCCAC-3′	IDT This study
R_0959d	5′-CATTGCATGCCCAAAGCCTTCTTTACCTGTAAGG-3′	IDT This study
dipA_F	5′-CATGGAATTC GTGACGCTTGATGTTTTTGTATTGCAGG-3′	IDT This study
dipA_R	5′-CATGGCTAGCGCGCTATTTAGAAGTCACCGCATTTTG-3′	IDT This study
pilD_F	5′-ATCGGAATTCATGTTGTTATATATCGAGTGCCAAATAAAC-3′	IDT This study
pilD_R	5′-ATCACATATGTTATACATAGATATTATCTTTAGTCAGTAG ATAAAAAAATGTTG-3′	IDT This study
pJH1_conf	5′-CTGATTTAATCTGTATCAGGCTGAAAATC-3′	IDT This study

[1] NEB—New England Biotechnologies. [2] IDT—Integrated DNA Technologies.

4.2. Selection of Resazurin-Resistant (RZR) LVS Mutants and Whole Genome Sequencing

Mutants resistant to resazurin were obtained by plating suspensions containing approximately 5×10^8 CFU of *F. tularensis* LVS bacteria on chocolate II agar supplemented with 11 μg/mL resazurin sodium salt (Acros Organics, dissolved in water) [9]. This concentration of resazurin is ten-fold higher than the previously determined minimal inhibitory concentration (MIC) of resazurin [9]. Colonies that formed on plates containing 11 μg/mL resazurin were then replica-plated onto 22 μg/mL resazurin. Forty-six resazurin-resistant (RZR) LVS clones that grew in the presence of 22 μg/mL resazurin were selected for further analysis. Genomic DNA was isolated using the Bacterial Genomic DNA Isolation Kit (Norgen Biotek Corp., Thorold, ON, Canada) per the manufacturer's instructions. Sequencing libraries were prepared using Nextera XT DNA Library Preparation Kits (Illumina, San Diego, CA, USA) and individual libraries were indexed with barcodes using Nextera XT v2 Index Kits (Illumina, San Diego, CA, USA) in the Marshall University (MU) Genomics Core Facility. Library quality and size distribution was assessed on an Agilent Bioanalyzer equipped with High Sensitivity DNA Chips. Libraries were quantitated by Qubit fluorimetry. High throughput sequencing (2 × 100 bp paired-end) was performed on an Illumina HiSeq1500 in Rapid Run mode in the Genomics Core. This approach allowed for approximately 150× coverage of each individual bacterial genome.

Sequencing reads were trimmed to remove Illumina adapters and low quality base calls using Trimmomatic version 0.36 [27]. Trimmed reads were aligned to the *Francisella tularensis* reference genome (NCBI accession number AM23362.1) using BWA version 0.7.12 [28] and sorted and indexed using picard tools version 2.9.4. Variant calling was performed following the GATK version 3.8.0 pipeline [29]. Briefly, duplicates were marked with the MarkDuplicates tool and per-sample variants identified using HaplotypeCaller. Variants were then called across the entire sample set using the GenotypeGVCFs tool with the sample-ploidy option set to one. Variant calls with quality score below 20 were filtered out using the VariantFiltration tool. Variants were exported to tab-delimited format using VariantsToTable. After the conclusion of these GATK pipeline steps, variants were annotated to determine their effects on the protein sequence using snpEff version 4.3q [30].

4.3. Construction of dipA and pilD Complemented Strains and Mutants

Primers and plasmids used in this study are listed in Table 5. To complement RZR LVS mutants with either wild-type *dipA* or *pilD*, *F. tularensis* LVS chromosomal DNA extracted from stationary-phase broth cultures using the Bacterial Genomic DNA Isolation Kit (Norgen Biotek Corp.) served as a template for amplification of *dipA* (FTL_1306) and *pilD* (FTL_0959). Primer pairs dipA_F/dipA_R and pilD_F/pilD_R were used to PCR-amplify *dipA* and *pilD*, respectively. The dipA amplicon was cloned into the EcoRI-NheI site of pABST [24] to generate the plasmid, pDipA. The pilD amplicon was cloned into the EcoRI-NdeI site of pABST to create the plasmid, pPilD. Both pDipA and pPilD were introduced into select *F. tularensis* LVS resazurin-resistant mutant strains by electroporation and selection on chocolate II agar supplemented with 10 µg/mL kanamycin.

The *pilD* disruption mutant (FTL_0959d) was generated using an unstable, integrating suicide vector, pJH1, as previously reported [31,32]. An internal 470 base pair region of FTL_0959 (*pilD*) was amplified by PCR using the primers F_0959d and R_0959d. This amplicon was digested with BamHI and SphI and then cloned into pJH1 that had been digested with these same enzymes, which produced pJH1-0959d. This disruption construct was mobilized into *F. tularensis* LVS by triparental mating [31]. Hygromycin-resistant colonies were screened by PCR using the primers pJH1-conf and 1F_pilD which are specific for the vector-portion of pJH1-FTL_0959 and pilD respectively, to confirm disruption of the *pilD* gene.

A *F. tularensis* LVS *dipA* (FTL_1306) deletion mutant was generated using pJH1 and the I-SceI endonuclease as described previously [26]. To begin, the *dipA* deletion construct pJH1-ΔdipA was generated using splicing by overlap-extension PCR. Regions (~500 bp) upstream and downstream of the *dipA* sequence targeted for deletion were amplified by PCR using the primer pairs 1F_dipA with 2R_dipA and 3F_dipA with 4R_dipA, respectively. The resulting amplicons contained regions of overlap and served as template DNA for a second PCR reaction using primers 1F_dipA and 4R_dipA. The resulting 1176 bp fragment was digested with PstI and SphI and then ligated into pJH1, which had been digested with the same enzymes, to produce pJH1-ΔdipA. This vector was then transferred into *F. tularensis* LVS by tri-parental mating [31]. Merodiploid strains were recovered and transformed with pGUTS by electroporation to allow for expression of I-SceI which causes a double-stranded break forcing recombination and allelic replacement [26]. Colonies resistant to kanamycin were screened by PCR for deletion of *dipA* using primers 1F_dipA and 4R_dipA. To cure pGUTS, the *F. tularensis* Δ*dipA* strains were repeatedly sub-cultured in TSBc, diluted, and plated to a density of 100 to 300 colony forming units CFU per chocolate II agar plate. Plates were incubated for at least 3 days at 37 °C, 5% CO_2 and colonies that formed were replica plated onto chocolate II agar supplemented with and without 10 µg/mL kanamycin to select against colonies harboring pGUTS. Those colonies sensitive to kanamycin were isolated and again tested for sensitivity to this antibiotic. The resulting LVS strain was named Δ*dipA*. Western blot analysis using an anti-DipA antibody [14] confirmed loss in expression of the DipA protein in LVS Δ*dipA* (data not shown).

4.4. Agar Dilution Susceptibility Testing

Minimum inhibitory concentrations (MICs) were determined by the agar dilution method using chocolate II agar according to CLSI guidelines [33]. Briefly, *F. tularensis* LVS bacteria were cultivated overnight in TSBc, diluted to an OD600 of 0.3 (approximately 5×10^8 CFU/mL), and then diluted to 1×10^6 CFU/mL. Ten microliters of this suspension (1×10^4 CFU) were plated onto chocolate II agar containing a series of two-fold dilutions of resazurin. Following incubation (37 °C with 5% CO_2 for 48–72 h), the MIC reported for each *F. tularensis* strain was the lowest concentration of resazurin that completely inhibited the growth of bacteria on the agar plate. No strain differed by more than one dilution in 3–5 tests.

4.5. Time Kill Assays

F. tularensis LVS overnight broth cultures were used to inoculate TSBc containing 1.375 µg/mL resazurin. Immediately following inoculation and 24 h later, cultures were serially diluted and plated onto chocolate II agar. Plates were incubated at 37 °C, 5% CO_2 and individual colonies were enumerated. The limit of detection was 100 CFU/mL.

4.6. Statistical Analyses

Data were analyzed for significant differences using GraphPad Prism software (Graph-Pad Software Inc., La Jolla, CA, USA). The statistical tests are indicated in the figure legend.

Author Contributions: Conceptualization, D.M.S.; methodology, J.H., D.A.P., J.D. and D.M.S.; resources, K.S., E.J.B., S.C.M., M.W., E.Y., J.P., D.W., J.H., D.A.P., J.D. and D.M.S.; data curation, D.A.P., J.D. and D.M.S.; writing—original draft preparation, K.S. and D.M.S.; writing—review and editing, K.S., E.J.B., S.C.M., M.W., E.Y., J.P., D.W., J.R., J.H., D.A.P., J.D. and D.M.S.; visualization, K.S., E.J.B., S.C.M., M.W., J.P., D.W., J.R. and D.M.S.; supervision, D.M.S.; project administration, D.M.S.; funding acquisition, D.A.P., J.H. and D.M.S. All authors have read and agreed to the published version of the manuscript.

Funding: This work was funded by a pilot grant from the West Virginia IDeA Network of Biomedical Research Excellence (WV-INBRE) program which is supported by a grant from the National Institute of General Medical Sciences of the National Institutes of Health (P20GM103434). Whole genome sequencing was carried out by the Marshall University Genomics Core Facility, which is supported by the WV-INBRE grant (P20GM103434), the COBRE ACCORD grant (1P20GM121299) and the West Virginia Clinical and Translational Science Institute (WV-CTSI) grant (2U54GM104942). The Research Resource ID (RRID) citation for the MU Genomics Core is RRID:SCR_018885. In addition, a portion of this work was funded by the National Heart Lung and Blood Institute of the National Institutes of Health (1R15HL147135).

Data Availability Statement: All sequencing data have been submitted to the Sequence Read Archive at the National Center for Biotechnology Information (NCBI) and are accessible via project number PRJNA748943.

Acknowledgments: The authors thank Karen Elkins for providing LVS and Jean Celli for the rat anti-FTT0369c antibody to detect expression of DipA.

Conflicts of Interest: The authors acknowledge that Joseph Horzempa has been awarded a patent (US20150258103A1) on the antimicrobial activity of the compounds used in this manuscript. However, the authors declare that the research was conducted in the absence of any commercial or financial relationships that could be construed as a potential conflict of interest.

References

1. Centers for Disease Control and Prevention. *Antibiotic Resistance Threats in the United States, 2019*; Department of Health and Human Services, CDC: Atlanta, GA, USA, 2019.
2. Ventola, C.L. The antibiotic resistance crisis: Causes and threats. *Phys. Ther. J.* **2015**, *40*, 277–283.
3. Plackett, B. Why big pharma has abandoned antibiotics. *Nature* **2020**, *586*, S50–S52. [CrossRef]
4. Hutchings, M.; Truman, A.; Wilkinson, B. Antibiotics: Past, present and future. *Curr. Opin. Microbiol.* **2019**, *51*, 72–80. [CrossRef] [PubMed]
5. Sridhar, B.T.; Girish, K.; Channu, B.C.; Thimmaiah, K.N.; Kumara, M.N. Antibacterial activity of phenoxazine derivatives. *J. Chem. Pharm. Res.* **2015**, *7*, 1074–1079.
6. Onoabedje, E.A.; Ayogu, J.I.; Odoh, A.S. Recent Development in Applications of Synthetic Phenoxazines and Their Related Congeners: A Mini-Review. *ChemistrySelect* **2020**, *5*, 8540–8556. [CrossRef]
7. Praveen, V.; Tripathi, C.K.M. Studies on the production of actinomycin-D by Streptomyces griseoruber—A novel source. *Lett. Appl. Microbiol.* **2009**, *49*, 450–455. [CrossRef]
8. Wang, Q.; Zhang, Y.; Wang, M.; Tan, Y.; Hu, X.; He, H.; Xiao, C.; You, X.; Wang, Y.; Gan, M. Neo-actinomycins A and B, natural actinomycins bearing the 5H-oxazolo[4,5-b]phenoxazine chromophore, from the marine-derived *Streptomyces* sp. IMB094. *Sci. Rep.* **2017**, *7*, 3591. [CrossRef]
9. Schmitt, D.M.; O'Dee, D.M.; Cowan, B.N.; Birch, J.W.M.; Mazzella, L.K.; Nau, G.J.; Horzempa, J. The use of resazurin as a novel antimicrobial agent against Francisella tularensis. *Front. Cell. Infect. Microbiol.* **2013**, *3*, 93. [CrossRef]
10. Schmitt, D.M.; Connolly, K.L.; Jerse, A.E.; Detrick, M.S.; Horzempa, J. Antibacterial activity of resazurin-based compounds against Neisseria gonorrhoeae in vitro and in vivo. *Int. J. Antimicrob. Agents* **2016**, *48*, 367–372. [CrossRef]
11. Robinson, C.M.; Kobe, B.N.; Schmitt, D.M.; Phair, B.; Gilson, T.; Jung, J.-Y.; Roberts, L.; Liao, J.; Camerlengo, C.; Chang, B.; et al. Genetic engineering of Francisella tularensis LVS for use as a novel live vaccine platform against Pseudomonas aeruginosa infections. *Bioengineered* **2015**, *6*, 82–88. [CrossRef]
12. Biot, F.V.; Bachert, B.A.; Mlynek, K.D.; Toothman, R.G.; Koroleva, G.I.; Lovett, S.P.; Klimko, C.P.; Palacios, G.F.; Cote, C.K.; Ladner, J.T.; et al. Evolution of Antibiotic Resistance in Surrogates of Francisella tularensis (LVS and Francisella novicida): Effects on Biofilm Formation and Fitness. *Front. Microbiol.* **2020**, *11*, 593542. [CrossRef] [PubMed]
13. Kassinger, S.J.; Van Hoek, M.L. Genetic Determinants of Antibiotic Resistance in Francisella. *Front. Microbiol.* **2021**, *12*, 644855. [CrossRef]
14. Chong, A.; Child, R.; Wehrly, T.D.; Rockx-Brouwer, D.; Qin, A.; Mann, B.J.; Celli, J. Structure-Function Analysis of DipA, a Francisella tularensis Virulence Factor Required for Intracellular Replication. *PLoS ONE* **2013**, *8*, e67965. [CrossRef]
15. Mittl, P.R.E.; Schneider-Brachert, W. Sel1-like repeat proteins in signal transduction. *Cell. Signal.* **2007**, *19*, 20–31. [CrossRef]
16. Smani, Y.; Fàbrega, A.; Roca, I.; Sánchez-Encinales, V.; Vila, J.; Pachón, J. Role of OmpA in the Multidrug Resistance Phenotype of Acinetobacter baumannii. *Antimicrob. Agents Chemother.* **2014**, *58*, 1806–1808. [CrossRef] [PubMed]
17. Viveiros, M.; Dupont, M.; Rodrigues, L.; Couto, I.; Davin-Regli, A. Antibiotic Stress, Genetic Response and Altered Permeability of E. coli. *PLoS ONE* **2007**, *2*, e365. [CrossRef] [PubMed]
18. Forsberg, Å.; Guina, T. Type II Secretion and Type IV Pili of Francisella. *Ann. N. Y. Acad. Sci.* **2007**, *1105*, 187–201. [CrossRef]
19. Nguyen, Y.; Harvey, H.; Sugiman-Marangos, S.; Bell, S.D.; Buensuceso, R.N.C.; Junop, M.S.; Burrows, L.L. Structural and Functional Studies of the Pseudomonas aeruginosa Minor Pilin, PilE. *J. Biol. Chem.* **2015**, *290*, 26856. [CrossRef]
20. Strom, M.S.; Nunn, D.; Lory, S. Multiple roles of the pilus biogenesis protein PilD: Involvement of PilD in excretion of enzymes from Pseudomonas aeruginosa. *J. Bacteriol.* **1991**, *173*, 1175–1180. [CrossRef]
21. Altschul, S.F.; Madden, T.L.; Schäffer, A.A.; Zhang, J.; Zhang, Z.; Miller, W.; Lipman, D.J. Gapped BLAST and PSI-BLAST: A new generation of protein database search programs. *Nucleic Acids Res.* **1997**, *25*, 3389–3402. [CrossRef]
22. Willcocks, S.; Huse, K.K.; Stabler, R.; Oyston, P.C.; Scott, A.; Atkins, H.S.; Wren, B.W. Genome-wide assessment of antimicrobial tolerance in Yersinia pseudotuberculosis under ciprofloxacin stress. *Microb. Genom.* **2019**, *5*, e000304. [CrossRef]
23. Gallagher, L.A.; Shendure, J.; Manoil, C. Genome-scale identification of resistance functions in Pseudomonas aeruginosa using Tn-seq. *MBio* **2011**, *2*, e00315-10. [CrossRef]
24. Vitale, A.; Pessi, G.; Urfer, M.; Locher, H.H.; Zerbe, K.; Obrecht, D.; Robinson, J.A.; Eberl, L. Identification of Genes Required for Resistance to Peptidomimetic Antibiotics by Transposon Sequencing. *Front. Microbiol.* **2020**, *11*, 1681. [CrossRef]
25. Schmitt, D.M.; O'Dee, D.M.; Horzempa, J.; Carlson, P.E., Jr.; Russo, B.C.; Bales, J.M.; Brown, M.J.; Nau, G.J. A Francisella tularensis live vaccine strain that improves stimulation of antigen-presenting cells does not enhance vaccine efficacy. *PLoS ONE* **2012**, *7*, e31172. [CrossRef]
26. Horzempa, J.; Shanks, R.M.Q.; Brown, M.J.; Russo, B.C.; O'Dee, D.M.; Nau, G.J. Utilization of an unstable plasmid and the I-SceI endonuclease to generate routine markerless deletion mutants in Francisella tularensis. *J. Microbiol. Methods* **2010**, *80*, 106–108. [CrossRef] [PubMed]
27. Bolger, A.M.; Lohse, M.; Usadel, B. Genome analysis Trimmomatic: A flexible trimmer for Illumina sequence data. *Bioinformatics* **2014**, *30*, 2114–2120. [CrossRef] [PubMed]

28. Li, H.; Durbin, R. Fast and accurate short read alignment with Burrows-Wheeler transform. *Bioinformatics* **2009**, *25*, 1754–1760. [CrossRef]
29. Van der Auwera, G.A.; Carneiro, M.O.; Hartl, C.; Poplin, R.; Del Angel, G.; Levy-Moonshine, A.; Jordan, T.; Shakir, K.; Roazen, D.; Thibault, J.; et al. From FastQ data to high confidence variant calls: The GenomeAnalysis Toolkit best practices pipeline. *Curr. Protoc. Bioinform.* **2013**, *43*, 11.10.1–11.10.33. [CrossRef]
30. Cingolani, P.; Platts, A.; Wang, L.L.; Coon, M.; Nguyen, T.; Wang, L.; Land, S.J.; Lu, X.; Ruden, D.M. A program for annotating and predicting the effects of single nucleotide polymorphisms, SnpEff SNPs in the genome of Drosophila melanogaster strain w^{1118}; iso-2; iso-3. *Fly* **2012**, *6*, 80–92. [CrossRef] [PubMed]
31. Horzempa, J.; Carlson, P.E., Jr.; O'Dee, D.M.; Shanks, R.M.Q.; Nau, G.J. Global transcriptional response to mammalian temperature provides new insight into Francisella tularensis pathogenesis. *BMC Microbiol.* **2008**, *8*, 172. [CrossRef]
32. Shanks, R.M.Q.; Caiazza, N.C.; Hinsa, S.M.; Toutain, C.M.; O'Toole, G.A. Saccharomyces cerevisiae-based molecular tool kit for manipulation of genes from gram-negative bacteria. *Appl. Environ. Microbiol.* **2006**, *72*, 5027–5036. [CrossRef] [PubMed]
33. Clinical and Laboratory Standards Institute. *Methods for Dilution Antimicrobial Susceptibility Tests for Bacteria That Grow Aerobically*, 12th ed.; Clinical and Laboratory Standards Institute: Wayne, PA, USA, 2018.

Article

Co-Existence of Certain ESBLs, MBLs and Plasmid Mediated Quinolone Resistance Genes among MDR *E. coli* Isolated from Different Clinical Specimens in Egypt

Salwa Mahmoud Masoud [1], Rehab Mahmoud Abd El-Baky [1,2,*] , Sherine A. Aly [3] and Reham Ali Ibrahem [1]

[1] Department of Microbiology and Immunology, Faculty of Pharmacy, Minia University, Minia 61519, Egypt; salwa.mahmoud@mu.edu.eg (S.M.M.); reham.ali@mu.edu.eg (R.A.I.)
[2] Department of Microbiology and Immunology, Faculty of Pharmacy, Deraya University, Minia 11566, Egypt
[3] Department of Medical Microbiology and Immunology, Faculty of Medicine, Assiut University, Assiut 71516, Egypt; s-aly71@windowslive.com
* Correspondence: rehab.mahmoud@mu.edu.eg; Tel.: +20-010-9248-7412

Abstract: The emergence of multi-drug resistant (MDR) strains and even pan drug resistant (PDR) strains is alarming. In this study, we studied the resistance pattern of *E. coli* pathogens recovered from patients with different infections in different hospitals in Minia, Egypt and the co-existence of different resistance determinants. *E. coli* was the most prevalent among patients suffering from urinary tract infections (62%), while they were the least isolated from eye infections (10%). High prevalence of MDR isolates was found (73%) associated with high ESBLs and MBLs production (89.4% and 64.8%, respectively). bla_{TEM} (80%) and bla_{NDM} (43%) were the most frequent ESBL and MBL, respectively. None of the isolates harbored bla_{KPC} and bla_{OXA-48} carbapenemase like genes. Also, the fluoroquinolone modifying enzyme gene *aac-(6′)-Ib-cr* was detected in 25.2% of the isolates. More than one gene was found in 81% of the isolates. Azithromycin was one of the most effective antibiotics against MDR *E. coli* pathogens. The high MAR index of the isolates and the high prevalence of resistance genes, indicates an important public health concern and high-risk communities where antibiotics are abused.

Keywords: MDR *E. coli*; ESBLs; MBLs; MAR index

Citation: Masoud, S.M.; Abd El-Baky, R.M.; Aly, S.A.; Ibrahem, R.A. Co-Existence of Certain ESBLs, MBLs and Plasmid Mediated Quinolone Resistance Genes among MDR *E. coli* Isolated from Different Clinical Specimens in Egypt. *Antibiotics* **2021**, *10*, 835. https://doi.org/10.3390/antibiotics10070835

Academic Editor: Teresa V. Nogueira

Received: 3 June 2021
Accepted: 4 July 2021
Published: 9 July 2021

Publisher's Note: MDPI stays neutral with regard to jurisdictional claims in published maps and institutional affiliations.

1. Introduction

Escherichia coli, belongs to the family Enterobacteriaceae, is the most common human gastrointestinal commensal as well as important etiological agent of many hospital and community-acquired infections. Pathogenic strains are capable of causing a wide variety of diseases including diarrhea, dysentery, overwhelming sepsis, and the hemolytic-uremic syndrome and neonatal meningitis. *E. coli* can be sorted into intestinal or extraintestinal according to the site of infection [1].

Antibiotics have been the most successful form of chemotherapy developed in the 20th century, saving human lives every day [2]. The evolution of pathogens resistant to antibiotics limits their clinical use, making such infections difficult to control. The antimicrobial resistance (AMR) can be of chromosomal or mobile genetic elements origin [3]. The most common resistance mechanism is the production of the β-lactamase hydrolytic enzymes, which specifically have an inactivated β-lactam ring so that they cannot inhibit the bacterial transpeptidases [4].

β-lactamases are classified into four classes. Serine classes (A, C and D) have serine residue at the hydrolysis active sites. Metallo- β-lactamases (MBLs) (class B) in which the hydrolytic action is promoted by one or two zinc ions at the active site [5]. Class A enzymes include bla_{TEM} which is the first identified plasmid-encoded β-lactamase; bla_{SHV} which has similar activity to bla_{TEM}; bla_{CTX-M} (cefotaximase) and bla_{KPC} which confers carbapenem resistance [6]. Class A Extended-spectrum β-lactamase (ESBL) producing strains (bla_{TEM}, bla_{SHV} and bla_{CTX-M} types) are of the most clinically significant pathogens which can resist all β-lactam drugs including monobactams [5,6]. The most clinically significant class B enzymes are bla_{VIM}, bla_{IMP} and bla_{NDM}. MBLs are a group of carbapenemases that resist most β-lactam drugs except the monobactams. Monobactams (e.g., aztreonam) are intrinsically stable to MBLs, but their susceptibility to other serine β-lactamases which are often co-expressed with the MBL limit their usage against MBL expressing strains [7]. Another group that able to hydrolyze carbapenems in addition to other β-lactams are class D β-lactamases (e.g., bla_{OXA-48} like enzymes) [8].

Another example for enzymatic inactivation of antibiotics is the enzymatic modification at different -OH or $-NH_2$ groups of aminoglycosides. As a result of the induced steric and/or electrostatic interactions, the modified antibiotic is unable to bind to the target RNA. They can be nucleotidyltransferases (ANTs), phosphotransferases (APHs), or acetyltransferases (AACs) [4]. In addition, the enzyme variant *aac(6′)-Ib-cr* has two amino acid changes that allow the enzyme to inactivate quinolones as well [9,10].

Being plasmid-encoded, hydrolyzing enzymes are likely to be transmissible and widespread. As a single plasmid may encode more than one enzyme, a strain may express many different enzymes, as each one deactivates a different antibiotic [5]. As a result, MDR, or even PDR strains, arising and returning to the pre-antibiotic era has become a nightmare for medical professionals.

The present study aimed to report the resistance pattern of *E. coli* pathogens, detect the co-existence of different resistance determinants and their correlations to the resistance of *E. coli* pathogens of different infection origins, which would help in identifying local effective therapeutic options and infection control.

2. Results

2.1. Prevalence of E. coli Among Samples

In the present study, 200 (47%) *E. coli* pathogens were isolated from 425 patients suffering from different infections attending three hospitals in EL-Minia, Egypt. The highest prevalence was among urinary tract infections (62%) while it was lowest among eye infections (10%) (Table 1). Among the three hospitals, *E. coli* isolates were most prevalent in Minia University Hospital samples (51.37%), followed by Minia General Hospital (40.8%) (Table 2).

Table 1. Prevalence of *E. coli* among different clinical samples.

Infection	No. of Samples	No. of *E. coli* Isolates	*E. coli* (%) *
Wound infections (burns, diabetic foot, surgery wound, cuts)	150	66	44%
Urinary tract infection	100	62	62%
Gastro-enteritis	50	24	48%
Blood	75	39	52%
Chest infection	20	4	20%
Ear infection	20	4	20%
Eye	10	1	10%
Total	425	200	47%

* Percent of *E. coli* were correlated to the number of samples of each infection.

Table 2. Distribution of *E. coli* among samples collected from different hospitals.

Hospitals	No. of Samples	E. coli	
		No.	% *
Minia University Hospitals	290	149	51.37
Minia Chest Hospital	20	4	20
Minia General Hospital	115	47	40.8
Total	425	200	

* Percent of *E. coli* were correlated to the number of collected samples from each hospital.

2.2. Antibiotic Resistance of the E. coli Pathogens

The antibiotic susceptibility was tested in 27 antibiotics that cover most of the available antibiotics in the Egyptian market. Table S1 (Supplementary Data) indicates the different used antibiotics and their different targets. The test revealed that 73% of *E. coli* were MDR. The pathogens were approximately totally resistant to Amoxycillin/clavulanic (97.5%), cephalothin (97%) and cefadroxil (93%). Also, high resistance levels were observed for Ceftazidime (78%) and Aztreonam (65.5%). Imipenem was the most effective antibiotic (20%), followed by Azithromycin (26%) (Figure 1). Our supplementary spread sheet indicates the resistance patterns of the isolates. One hundred MDR isolates were subjected for further investigation.

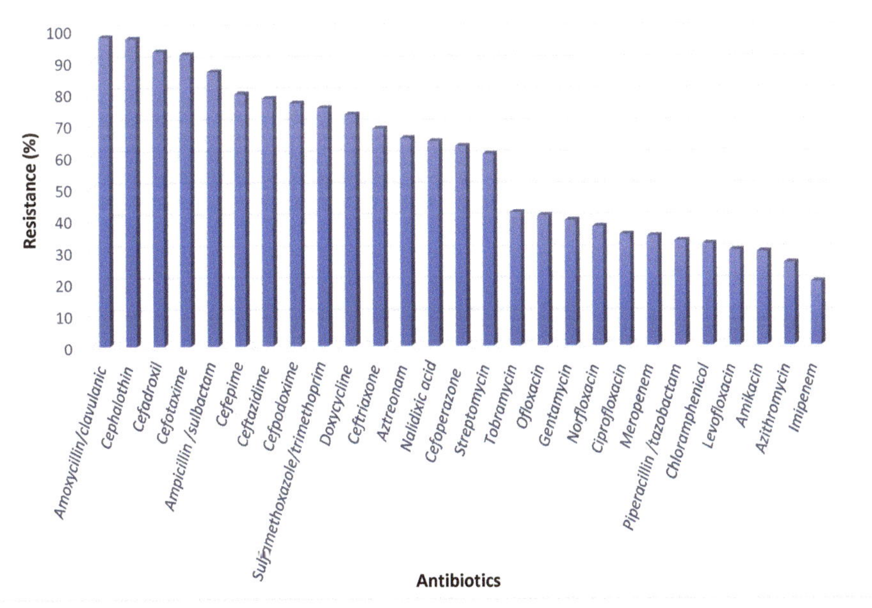

Figure 1. Antibiotic resistance of the total *E. coli* isolates.

2.3. Serotyping of the Intestinal E. coli

Out of the selected 100 isolates, 20 were isolated from stool. Since *E. coli* normally inhabit the intestine, stool isolates were serotyped to confirm its pathogenicity. Out of 20 intestinal *E. coli* isolates 15 (75%) isolates were diarrheagenic *E. coli* (DEC). Three isolates (20%) were identified as Enterohaemorrhagic *E. coli* (EHEC) O157:H7. Different O serotypes were observed as O115, O158, O55, O126, O125 and O86a. The identified pathotypes are listed in Table S3 (Supplementary Data). Untyped five isolates were excluded from the further testing so that 95 isolates were further tested phenotypically and genotypically.

2.4. Multiple Antibiotic Resistance Index MDR E. coli Pathogens

The multiple antibiotic resistance index (MARI) ratio between the number of antibiotics that an isolate is resistant to and the total number of antibiotics the organism is exposed to, have been calculated for 95 MDR *E. coli*. It was found that 98.9% of isolates have showed MAR index higher than 0.2, indicating high risk communities where antibiotics are abused. However, there was a statistically significant difference in MARI mean among *E. coli* isolates of different sources (*p* value < 0.05). Eye and blood isolates showed highest MARI mean of 0.82 and 0.74, respectively. On the other hand, stool samples had the lowest MARIs (Figure 2).

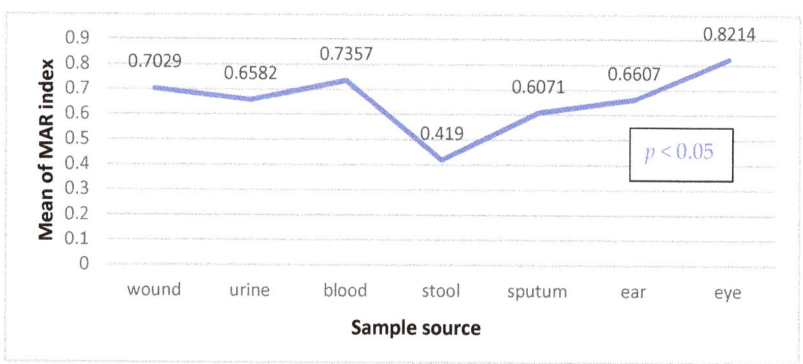

Figure 2. MAR index mean values ($\bar{x}$), *p* value calculated by One-way ANOVA test.

2.5. Phenotypic Characteristics

ESBL production phenotypically tested by combined disk test (CDT). It was found that 89.4% (85/95) of the tested strains were ESBL producers. Carbapenemase production was tested by Modified Hodge test (MHT), then carbapenemase producers were tested for MBL using combined-disk synergy test. MBL producers accounted for 50.50% of the isolates (64.8% of carbapenemase producers). Positive significant association was found between ESBL and MBL phenotypes (*p* = 0.001) as all MBL producers were ESBL producers. Statistically significant difference in distribution of ESBL and MBL producers between different infection groups was observed (*p* < 0.001). Regardless of the eye infection, MBL producers were mostly frequent in UTIs (71%) while no intestinal *E. coli* was reported as MBL producers (Table 3).

Table 3. Distribution of ESBLs and MBLs producing isolates among clinical specimens.

Type of Infection	β-Lactam Resistant Isolates	ESBLs		MBL	
		No.	% *	No.	% *
Wound infections	25	25	100%	12	48%
Urinary tract infection	28	28	100%	20	71.4%
Gastro-enteritis	15	5	33.3%	0	0%
Blood	20	20	100%	12	46.67%
Respiratory infection	4	4	100%	2	50%
Ear infection	2	2	100%	1	50%
Eye	1	1	100%	1	100%
Total	95	85	89.4% **	48	50.5% **

* percent was correlated to the total number of samples of each infection type; ** percent was correlated to total number of samples. Significant if $p \leq 0.05$.

2.6. Antimicrobial Resistance of ESBLs and MBLs Producers

Resistance patterns of ESBL-producers revealed that ESBL producers were highly resistant to β-lactam antibiotics such as amoxycillin/clavulanic (99%), cefadroxil (95%), ceftazidime (85%) and meropenem (69%). MBL producers showed higher resistance rates to same antibiotics (100%, 97.9%, 93.75% and 93.75%, respectively). Azithromycin and chloramphenicol were the most effective drugs against ESBLs and MBLs producers. Table S4 (Supplementary Data) indicates the antibiogram of ESBLs and MBLs producers.

2.7. Prevalence of Resistance Genotypes Among the Tested Isolates

There was a statistically significant difference in the distribution of the different genotypes between different infections. The most prevalent gene was the bla_{TEM} (80%) followed by bla_{SHV}, bla_{CTX-M} and bla_{NDM} (54.7%, 42% and 44.2%, respectively). All isolates were negative for bla_{KPC} or bla_{oxa-48} genes. Also, *the aac-(6′)-Ib-cr gene was observed in 26.3% of the tested pathogens* (Table 4).

Table 4. Distribution of detected MBLs and ESBLs genotypes.

Type of Infection	β-Lactam Resistant Isolates N	bla_{NDM} N (%) *	bla_{TEM} N (%) *	bla_{CTX-M} N (%) *	bla_{SHV} N (%) *	bla_{IMP} N (%) *	aac-(6′)-Ib-cr N (%) *
Wound infections	25	14 (56%)	23 (92%)	14 (56%)	15 (60%)	12 (48%)	9 (36%)
UTI	28	17 (60.7%)	27 (96.4%)	11 (39.2%)	12 (42.8%)	16 (57%)	6 (21%)
Gastro-enteritis	15	1 (7.6%)	7 (46.6%)	0 (0%)	0 (0%)	0 (0%)	0 (0%)
Blood	20	7 (35%)	20 (100%)	13 (65%)	20 (100%)	7 (35%)	10 (50%)
Chest infection	4	1 (25%)	3 (75%)	0 (0%)	4 (100%)	0 (0%)	0 (0%)
Ear infection	2	1 (50%)	2 (100%)	1 (50%)	0 (0%)	0 (0%)	0 (0%)
Eye	1	1 (100%)	1 (100%)	1 (100%)	1 (100%)	0 (0%)	0 (0%)
Total	95	42 (44.2%)	76 (80%)	40 (42%)	52 (54.7%)	35 (36.8%)	25 (26.3%)
p value **		0.11	<0.001	0.011	<0.001	0.011	0.042

* percent was correlated to the total number of β-lactam resistant isolates in each type of infection. ** Significant *p* value at $p \leq 0.05$.

2.8. Genotypic-Phenotypic Agreement of the Tested Genes

Out of 48 MBL phenotypic positive samples, 38 (79%) isolates were confirmed genotypically as MBL producers. It was found that 15.78% of isolates were phenotypically positive and harbored both bla_{NDM} and bla_{IMP} (Figure 3). Furthermore, there was a significant decrease in MAR index when isolates were both bla_{NDM} and bla_{IMP} negative (*p* value < 0.001). The bla_{NDM} producers had higher MAR index than those which were only bla_{IMP} producers. However, isolates harbored both bla_{NDM} and bla_{IMP} was observed resisting higher number of antibiotics (Figure 4). Moreover, positive correlations between MBLs phenotype, genotypes and carbapenem resistance were observed. Statistically significant correlations between detected MBL genotypes and meropenem resistance were observed. The MBLs phenotypes were more significantly associated with bla_{NDM} than bla_{IMP} (Table 5).

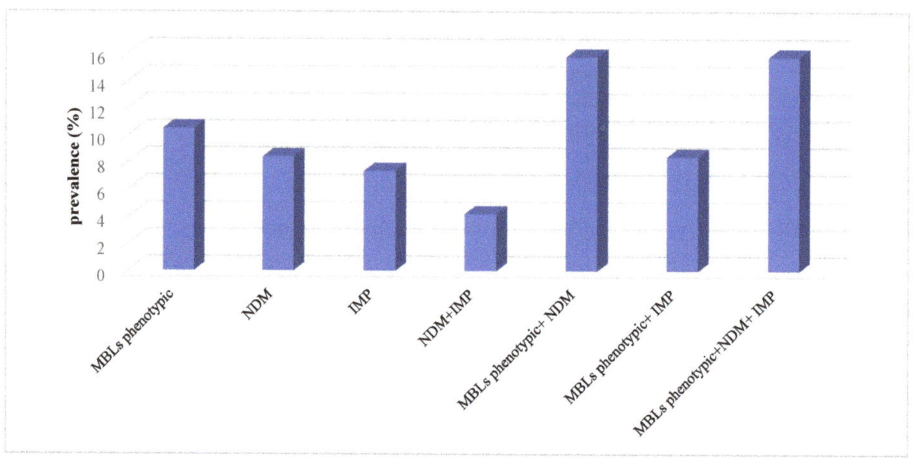

Figure 3. Phenotypic and genotypic agreement of MβL tested genes.

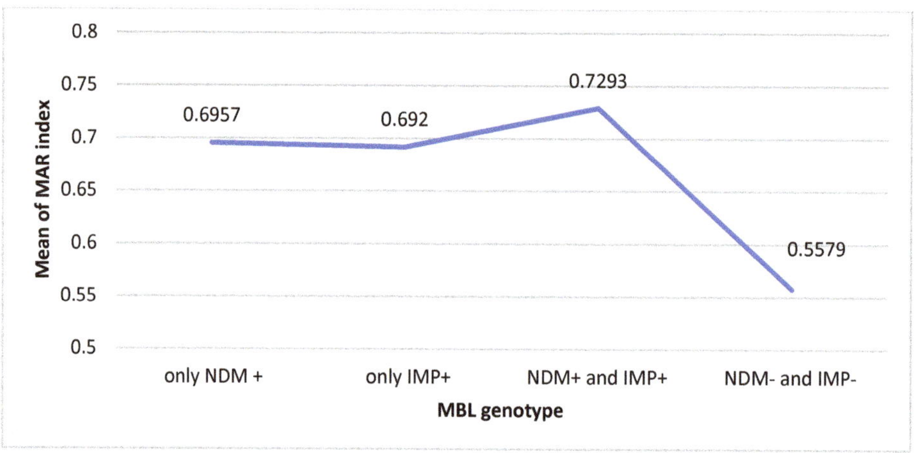

Figure 4. Distribution of MAR index mean values ($\bar{x}$) among the detected MBLs genotypes.

Table 5. Relation between MBLs phenotype, detected genotype and carbapenem resistance across the *E. coli* isolates.

	MBLs Phenotype	bla_{NDM}	bla_{IMP}	Imipenem Resistance	Meropenem Resistance
MBLs phenotype	1	0.372 **	0.232 *	0.237 *	0.465 **
bla_{NDM}		1	0.155	0.054	0.212 *
bla_{IMP}			1	0.122	0.275 **
Imipenem resistance				1	0.292 **
Meropenem resistance					1

* Correlation is significant at the 0.05 level (2-tailed). ** Correlation is significant at the 0.01 level (2-tailed). *p* values calculated by Fisher's exact test.

ESBLs phenotypes showed strong positive correlation with the presence of bla_{TEM}. However, all correlations were significant at the 0.01 level. Among the detected ESBL genes, presence of bla_{TEM} and bla_{CTX-M} types had the upper hand on the isolates' resistance followed by $aac(6')Ib-cr$ gene (Table 6).

Table 6. Correlation matrix (r^2) between phenotypes, genotypes and antibiotic resistance across the *E. coli* isolates.

	ESBLs Production	MBLs Production	bla_{NDM}	bla_{IMP}	bla_{TEM}	bla_{CTX-M}	bla_{SHV}	$aac(6')1b-cr$
MAR index	0.611 **	0.342 **	0.330 **	0.289 **	0.366 **	0.365 **	0.251 *	0.360 **
ESBLs phenotype	1	0.347 **	0.305 **	0.262 *	0.696 **	0.293 **	0.377 **	0.205 *

* Correlation is significant at the 0.05 level (2-tailed). ** Correlation is significant at the 0.01 level (2-tailed). *p* values calculated by Fisher's exact test.

The current study identified that $aac(6')Ib-cr$ gene was mainly related to aminoglycoside antibiotics than fluoroquinolones. It was found that 47.2% of amikacin resistant isolates and 32.4% of ciprofloxacin resistant isolates harbored $aac(6')-Ib-cr$ gene. The presence of $aac(6')-Ib-cr$ gene was least correlated to ofloxacin resistance. Significant moderate positive correlation was observed between $aac(6')-Ib-cr$ and the resistance to amikacin and tobramycin, *p* values < 0.001 and 0.003, respectively (Table 7).

Table 7. Correlation between $aac(6')Ib-cr$, aminoglycoside and fluoroquinolone resistance.

Antibiotics	Number of Resistant Isolates	Number of aac(6')Ib-cr Positive Isolates (%) *	Person Correlation (r^2)	*p* Value
Streptomycin	61	19 (31)	0.135	0.193
Tobramycin	43	17 (39.5)	0.301	0.003 **
Gentamycin	43	13 (30.2)	0.044	0.670
Amikacin	36	17 (47.2)	0.374	<0.01 ***
Ofloxacin	39	9 (23)	0.018	0.866
Norfloxacin	42	14 (33.3)	0.1450	0.162
Ciprofloxacin	37	12 (32.4)	0.152	0.142

* percent correlated to no. of resistant isolates of each antibiotic. *p* values were calculated by Fisher's exact test. ** *p* value is significant at 0.05 level (2-tailed), *** *p* value is significant at 0.01 level (2-tailed).

2.9. Association of Different Resistance Genotypes

Most isolates harbored more than one resistance gene (81%). The resistance frequency has significantly increased with the increased number of the co-existed genes ($p < 0.01$). The most frequent association was of the five genes bla_{NDM}, bla_{IMP}, bla_{TEM}, bla_{CTX-M} and bla_{SHV} (8.4%) (Table 8).

The correlation matrix of the detected genes indicated overall positive correlations. The strongest and most significant correlation was observed between bla_{CTX-M} and bla_{SHV} ($r = 0.519$). Moreover, $aac(6')Ib-cr$ gene was significantly associated with bla_{CTX-M} (Table 9).

Studying the association of bla_{NDM} gene with class A ESBL genes (bla_{TEM}, bla_{CTX-M} and bla_{SHV}) among the bla_{NDM} positive isolates indicated that the association of the bla_{NDM} with the three ESBLs (bla_{TEM} + bla_{SHV} + bla_{CTX-M}) genes was the highest, accounting for 40.47% of the isolates harboring bla_{NDM} (Figure 5).

Furthermore, the spectrum of antibiotics to which the isolates were resistance is significantly increased (*p* value < 0.01) with the number of positive genes. The isolate harbored five genes showed the highest mean of antibiotic resistance (21 antibiotics), as indicted in Figure 6. As indicated in Table S5 (Supplementary Data) the resistance of β-lactam drugs is significantly associated with presence of higher number of genes. Also, significant moderate correlation of number of positive genes with ciprofloxacin, norfloxacin and aminoglycoside were observed.

Table 8. Co-existence of different genotypes.

	Number of Isolates (%)
One gene	9 (9.5)
bla_{TEM}	8 (8.5)
bla_{SHV}	1 (1)
Two genes	20 (21.1)
blaTEM, aac(6′)Ib-cr	1 (1)
blaIMP, blaTEM	5 (5.2)
blaTEM, blaSHV	5 (5.2)
blaNDM, blaTEM	7 (7.3)
blaTEM, blaCTX-M	2 (2.1)
Three genes	21 (22.1)
blaIMP, blaTEM, blaSHV	4 (4.2)
blaIMP, blaTEM, blaCTX-M	1 (1)
blaNDM, blaTEM, aac(6′)Ib-cr	2 (2.1)
blaTEM, blaCTX-M, blaSHV	5 (5.2)
blaTEM, blaSHV, aac(6′)Ib-cr	1 (1)
blaNDM, blaCTX-M, blaSHV	1 (1)
blaNDM, blaIMP, blaTEM	2 (2.1)
blaNDM, blaTEM, blaSHV	5 (5.2)
Four genes	21(22.1)
blaTEM, blaCTX-M, blaSHV, aac(6′)Ib-cr	5 (5.2)
blaNDM, blaIMP, blaTEM, aac(6′)Ib-cr	3 (3.1)
blaNDM, blaTEM, blaCTX-M, blaSHV	3 (3.1)
blaIMP, blaTEM, blaCTX-M, blaSHV	3 (3.1)
blaIMP, blaTEM, blaSHV, aac(6′)Ib-cr	1 (1)
blaNDM, blaIMP, blaTEM, blaCTX-M	1 (1)
blaNDM, blaTEM, blaSHV, aac(6′)Ib-cr	1 (1)
blaNDM, blaCTX-M, blaSHV, aac(6′)Ib-cr	1 (1)
blaNDM, blaTEM, blaCTX-M, aac(6′)Ib-cr	3 (3.1)
Five genes	12 (12.6)
blaNDM, blaIMP,, blaTEM, blaCTX-M, blaSHV	8 (8.5)
blaNDM, blaIMP, blaTEM, blaCTX-M, aac(6′)Ib-cr	2 (2.1)
blaIMP, blaTEM, blaCTX-M, blaSHV, aac(6′)Ib-cr	2 (2.1)
Six genes	3 (3.2)
Total	95

Percentages were correlated to the total number of isolates.

Table 9. Correlation matrix (r^2) between the different genotypes.

	bla_{NDM}	bla_{IMP}	bla_{TEM}	bla_{CTX-M}	bla_{SHV}	aac(6′)1b-cr
bla_{NDM}	1	0.155	0.211 *	0.185	0.086	0.190
bla_{IMP}		1	0.290 **	0.233 *	0.081	0.089
bla_{TEM}			1	0.196	0.227 *	0.155
bla_{CTX-M}				1	0.519 **	0.265 **
bla_{SHV}					1	0.159
aac(6′)Ib-cr						1

* Correlation is significant at the 0.05 level (2-tailed). ** Correlation is significant at the 0.01 level (2-tailed). p values were calculated by Fisher's exact test.

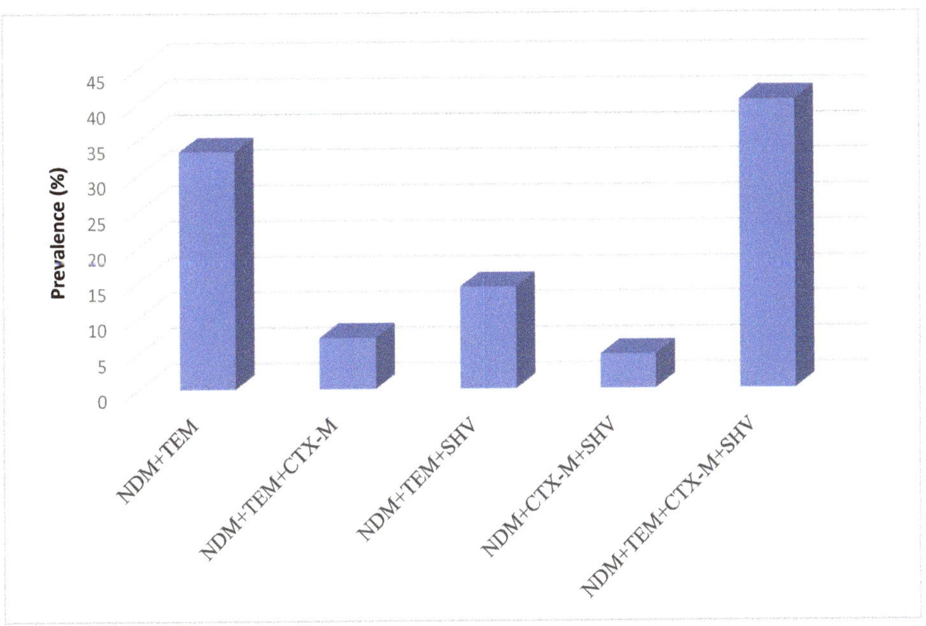

Figure 5. *bla*$_{NDM}$ association with class A ESBLs genes.

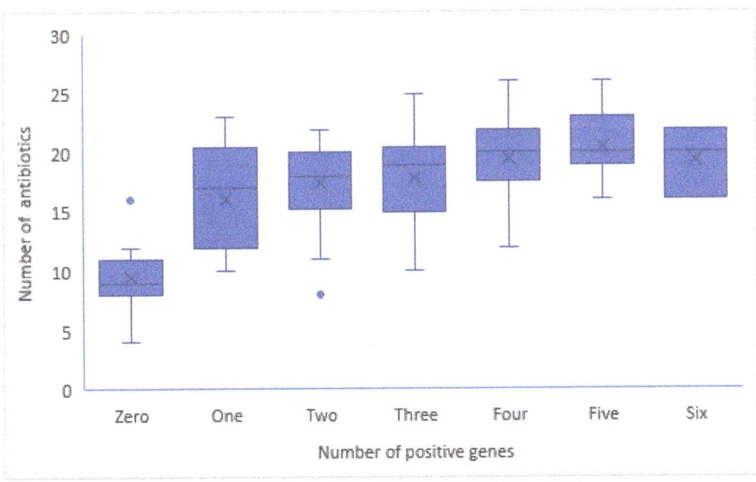

Figure 6. Compatibility of number of detected genes and number of antibiotics to which isolates were resistant among 95 isolates.

3. Discussion

The last two decades have witnessed a conspicuous increase in the number of infections caused by multi-drug resistant strains of *E. coli*, and this has impacted the outcomes of different infections [11].

The present study demonstrated that the prevalence of *E. coli* pathogens isolated from patients suffering from different infections in El-Minya hospitals accounted for 47%. This result was in accordance with results reported by Amer et al. (45%) [12] and Fam et al. (56%) [13].

Among extraintestinal infections, *E. coli* was the most common among urine isolates (62%) followed by blood infections (52%). Likewise, in Saudi Arabia Alanazi et al. [14] reported that *E. coli* was isolated from 60.24% of the urine samples and in Greece Koupetori et al. [15] reported high incidence of *E. coli* accounting for 48% of blood isolates. In contrary, many studies showed lower *E. coli* incidence [16,17]. On the other hand, *E. coli* was least isolated from eye infections (10%). This was higher than results reported in USA by Miller et al. (5.9%) [18].

Moreover, intestinal *E. coli* was serotyped to ensure its pathogenicity. Diarrheagenic *E. coli* (DEC) serotypes accounted for 54.1% which was considered very high in comparison to results obtained by Zhou et al. (7.9%) [19]. *E. coli* O157:H7 accounted for 12.5% of total DEC serotypes. Comparably, in the same region of study Abd El Gany et al. [20] reported incidence rate of 15.72%. The variations between the different studies may be ascribed to many socioeconomical, demographical and geographical factors. It was obviously noted that *E. coli* were relatively high in the present study compared to other studies indicates poor hygienic attitudes that correlated to the mentioned factors.

Antimicrobial resistance (AMR) emphasizes an overwhelming health and economic burden in both developed and developing countries. As Resistance narrows the therapeutic options leading to increased morbidity and mortality [21]. Our results showed high prevalence of MDR *E. coli* (73%) which were higher than results obtained by Siwakoti et al. [22] and Abdelaziz et al. [23] (28% and 60%, respectively).

Although carbapenem resistance is considered low, it is higher than previous studies done at the same government [9,20,24]. This may be attributed to the availability and the usage of the drugs when the studies were held. Concerning our results, the drugs were more available and highly used but in previous studies there was a shortage in many antibiotics inside the hospitals. Moreover, the resistance to meropenem was higher than imipenem which may be attributed to that meropenem is cheaper than imipenem, so it is commonly used while some other studies reported complete meropenem sensitivity [9,23].

The continuous spread of ESBLs and carbapenemase mediated resistance has dramatically increased in both hospital and community infections. It was found that 89.4% of the tested isolates were ESBL producers. Abd El-Baky et al. [24] in earlier study in our area reported that 46.8% of isolates were phenotypically ESBL producers. It seems very alerting as the prevalence is almost doubled in short period.

The incidence of the carbapenemase and MBLs producers accounted for 77.8% and 64.8%, respectively. These rates were very high when compared to results obtained by Ibrahim, et al. [25] who reported that carbapenemase and MBLs incidences were 37.6% and 46.3% respectively. A previous study in our area showed that 52.3% of *P. aeruginosa* were MBL producers [26]. The differences in prevalence may be due to strains and time variations but overall indicate high incidence of MBLs among bacteria in our area.

There was a significant difference in the distribution of detected genes among the different sample sources p values < 0.05 which was in agreement with many studies [27,28].

The most prevalent genotype was bla_{TEM} (80%) followed by bla_{SHV} (54.7%), bla_{NDM} (44.2%) and bla_{CTX-M} (42%). Similarly, bla_{TEM} was predominant in results reported by Mohamed et al. [28] and Maamoun et al. [29]. On the other hand, a study on *Escherichia coli* causing sepsis among Egyptian children reported that bla_{SHV} was the most common ESBL (61.22%), followed by bla_{TEM} (38.78%) and bla_{CTX-M} (20.41%) [30]. Furthermore, the higher incidence of bla_{TEM} gene reported by the current study or other studies in our region suggests that bla_{TEM} gene may be endemic. In contrast to our study, several studies in Asia reported that bla_{CTX-M} was the most frequent indicating that bla_{CTX-M} is a predominant genotype in Asia [31–33]. Also, reports from Qatar stated that bla_{CTX-M} type genes evolved through mutations in bla_{TEM} and bla_{SHV} genes and it is a recent endemic [34].

Similar to our study, studies in UK have reported 44% of isolates as bla_{NDM} producers, most of them were from urine samples [35]. The number of bla_{NDM} producers is increasing in Egypt which is reflected by many studies conducted in this area [30,36,37]. Lower bla_{IMP} incidence reported in previous studies in Egypt compared to this study (36.8%) suggesting an increasing rate of MBLs producers [36,37]. However, the current reported high prevalence of MBLs may be attributed to the ability of *E. coli* to acquire novel resistance genes through horizontal transfer or the increased use of carbapenems in the clinical treatment.

bla_{Oxa-48} like and bla_{KPC} were not detected in any isolate. Quite higher prevalence of bla_{Oxa-48} like and bla_{KPC} (38.46% and 23%, respectively) was reported in Bangladesh [38]. In accordance with the current study, bla_{KPC} wasn't detected in several previous studies in Egypt or detected in very low rate [37,38]. In addition, bla_{KPC} wasn't detected in countries such as Saudi Arabia [39] or those of the Arabian Peninsula [40]. These data confirmed that bla_{KPC} genes does not predominate in this geographical region, where it is frequently detected in the United States [41] and endemic in Israel [42]. The *aac(6′)-Ib-cr* gene prevailed in 26.3% of the isolates which were mostly isolated from wounds. This rate is lower than rates previously reported by Al-Agamy et al. [43] and Mohamed et al. [28]. The differences across studies may be attributed to differences in geographical locations, age groups, or clinical criteria.

Most of the isolates (81%) harbored more than one resistance determinant. Co-harboring of multiple ESBL genes was detected previously in Egypt [28,44] and some other countries; Burkina Faso [45], Qatar [33] and Iran [46].The co-existence of bla_{NDM}, bla_{IMP}, bla_{TEM}, bla_{CTX-M} and bla_{-SHV} was the most frequent, accounting for 8.5% of the isolates. There was a significant association between bla_{CTX-M} and bla_{SHV}, which agree with other studies [28,30,46]. There was a significant positive correlation between bla_{NDM} and bla_{IMP}. This was comparable to Zaki et al. [30] and Kamel et al. [47] where single *E. coli* isolate had more than one type of metallo β-lactamase. The association between *aac(6′)-Ib-cr* and bla_{CTX-M} genes was statistically significant, agreeing with previous studies [48–50]. None of the isolates harbored *aac(6′)Ib-cr* alone. Moreover, there was significant association of *aac(6′)-Ib-cr* gene with ESBL phenotypes. This may be due to the common presence of ESBL genes and PMQR genes on the same plasmid in *Enterobacteriaceae* [51]. Moreover, the *aac(6′)-Ib-cr* gene showed significant positive correlation with amikacin and tobramycin resistance in ESBL producers. Similarly, Mohamed et al. [28] reported significant association of ESBL genes with *aac(6′)-Ib-cr* gene that resulted in increased ciprofloxacin, gentamicin and amikacin resistance in ESBL producers.

The resistance rates were significantly increased in ESBL producers than non-producers that reported by several studies [28,52–54]. In agreement with our results, many studies reported higher resistance rate of MBL producers in comparison to MBLs non-producers. In accordance with the current work, previous studies have reported significant high resistance rates in MBL producers [55,56]. It was reported that bla_{TEM} and bla_{SHV} are important factors in increased resistance of ESBL *E. coli* producers to third-generation cephalosporin [57].

Finally, variations in rate and predominance of resistance genes between different countries and even among the same country institutions may be due to difference in locally prescribed antibiotics and if the infection control guidelines are followed or not in different health institutes. In Egypt, the high rate of ESBLs and MBLs is a reflection of the inappropriate use of antimicrobials due to the over counter availability of antibiotics without prescription and patients incompliance or the wide use of antibiotics in veterinary care and farms [58,59].

4. Material and Methods

4.1. Bacterial Isolates

Two hundred *E. coli* isolates were isolated from 425 patients attending different hospitals in El-Minia with different infections. All clinical samples were obtained as part of the routine hospital laboratory procedures. Samples were processed and cultured on trypticase soy agar (Lab M, Hewwood, UK) at 37 °C for overnight. *E. coli* colonies gave pink color on MacConkey agar and green metallic sheen on Eosin methylene blue (EMB) (lab M, Hewwood, UK). Colonies were further identified by regular microbiological biochemical tests [60].

4.2. Antimicrobial-Susceptibility Testing

The antimicrobial susceptibility of the isolates was tested by the Kirby-Bauer Disk Diffusion method [61]. The used antibiotics discs were ready cartilages purchased from Oxoid; Basingstoke, UK. The following antibiotic discs were used Cefpodoxime (10 μg), Streptomycin (10 μg), Aztreonam (30 μg), Ceftriaxone (30 μg), Gentamycin (10 μg), Amoxycillin/clavulanic (20/10 μg), Piperacillin/Tazobactam (100/10 μg), Ceftazidime (30 μg), Imipenem (10 μg), Meropenem (10 μg), Cefoperazone (75 μg), Doxycycline (30 μg), Ciprofloxacin (5 μg), Amikacin (30 μg), Nalidixic acid (30 μg), Cefotaxime (30 μg), Cefepime (30 μg), Ampicillin/sulbactam (10/10 μg), Norfloxacin (10 μg), Tobramycin (10 μg), Sulfamethoxazole/trimethoprim (23.75/1.25 μg), Chloramphenicol (30 μg). Isolates classified as sensitive, intermediate and resistant according to inhibition zones interpretation standards of Clinical Laboratory standards Institute (CLSI) 2018 (Table S2) [62].

4.3. Serotyping of Intestinal E. coli

Escherichia coli recovered from gastroenteritis infections were sent to the Animal Health Research Institute, Giza, Egypt to be serotyped. The isolates serotyped through detection of isolates agglutination with O and H antisera using the slide agglutination method according to the manufacturer instructions (Pro-Lab Diagnostics, Round Rock, TX, USA).

4.4. Phenotypic Detection of ESBLs and MBLs Production

Detection of ESBL in *E. coli* isolates was carried out by combined disc test (CDT). Isolates defined positive when the difference between the inhibition zones of cefotaxime and cefotaxime/clavulanic or ceftazidime and ceftazidime/clavulanic disks is ≥5 mm [62]. Carbapenemases were detected in carbapenem resistant isolates by Modified Hodge test (MHT). MHT positive isolates w ere further tested for MBL production using EDTA-combined disk synergy test. An increase in zone diameter of at least 7 mm around the imipenem–EDTA or meropenem–EDTA disks were recorded positive result [63].

4.5. Amplification of Resistance Genes

The DNA templet was extracted by available commercial kit QIAprep® Spin Miniprep Kit (QIAGEN, Germany) by following the manufacturer instruction. Resistance genes were detected using conventional PCR technique. Amplification was done using 25 μL PCR reaction mixture consisting of 12.5 μL master mix (BIOMATIk, Kitchener, Canada) 1 μL of each forward and reverse primers (BIOMATIk, Canada), 2 μL DNA template and 8.5 μL nuclease-free water. PCR cycling conditions are indicated in Table S5 (Supplementary Data).

4.6. Statistical Analysis

Data were analyzed using IBM SPSS version 20.0. First, normal distribution of data was tested by normality tests as Kolmogorov–Smirnov and Shapiro–Wilk *p*-values in addition to histograms. Descriptive analysis was done to analyze prevalence of *E. coli* isolates among different infections and hospitals, percentage of resistance and prevalence of ESBL and MBL producers and prevalence of the different genes. To compare differences in distribution between different groups chi-square (X^2) test was done but when more than 20% of cells were less than 5, Fisher's exact test was done to be more accurate. One-way ANOVA tests was done to compare mean values between different groups as MAR index mean values in different sample sources. Non parametric tests were used for non-parametric data as Kruskal–Wallis. To study associations between phenotypes, genotypes and resistance, correlations were established using Pearson's correlation coefficient (r^2) in bivariate. *p*-values are significant if they are ≤0.05.

5. Conclusions

High resistance reported in our study indicates poor awareness of the microbiological laboratory test importance, high empirical antimicrobial prescription and high patient incompliance. Moreover, the massive co-existence of the detected genes strongly supports

Antibiotics **2021**, *10*, 835

the presence of one or more circulating plasmids that harbors different resistance genes. Finally, the study highlighted the importance of continuous surveillance of the resistance trends and the direct need to strictly apply the infection control policies, implementing a national antimicrobial stewardship plan.

Supplementary Materials: The following are available online at https://www.mdpi.com/article/10.3390/antibiotics10070835/s1, Table S1: The different tested antibiotics and their different biological processes targets, Table S2: The different thresholds of inhibition zones according to CLSI (2018), Table S3: Different serotypes of the intestinal *E. coli*, Table S4: Resistance pattern of ESBLs and MBLs producers, Table S5: Correlation coefficient of the number of detected genes in isolates with the resistance of the antibiotics, Table S6: Primers used in the current study. References [63–70] are cited in Supplementary File.

Author Contributions: Formal analysis, S.M.M., S.A.A. and R.A.I.; Methodology, R.M.A.E.-B., S.M.M., S.A.A. and R.A.I.; Supervision, R.M.A.E.-B., S.A.A. and R.A.I.; Validation, S.A.A.; Visualization, R.A.I.; Writing—original draft, S.M.M.; Writing—review & editing, R.M.A.E.-B. and S.M.M. All authors have read and agreed to the published version of the manuscript.

Funding: This research did not receive any specific grant from funding agencies in the public, commercial, or not-for-profit sectors.

Institutional Review Board Statement: The study was conducted according to the guidelines of the Declaration of Helsinki and approved by the Commission on the Ethics of Scientific Research of Faculty of Pharmacy, Minia University (protocol code HV06/2020 and date of approval 17-6-2020).

Informed Consent Statement: Informed consent was obtained from all subjects involved in the study. Written informed consent has been obtained from the patient(s) to publish this paper.

Data Availability Statement: Not applicable.

Acknowledgments: The authors received no financial support for the research or publication of this article.

Conflicts of Interest: The authors declare no conflict of interest.

References

1. Abdelwahab, R.; Yasir, M.; Godfrey, R.E.; Christie, G.S.; Element, S.J.; Saville, F.; Hassan, E.A.; Ahmed, E.H.; Abu-Faddan, N.H.; Daef, E.A.J.V. Antimicrobial Resistance and Gene Regulation in Enteroaggregative Escherichia Coli from Egyptian Children with Diarrhoea: Similarities and Differences. *Virulence* **2021**, *12*, 57–74. [CrossRef] [PubMed]
2. Banin, E.; Hughes, D.; Kuipers, O.P. Bacterial Pathogens, Antibiotics and Antibiotic Resistance. *FEMS Microbiol. Rev.* **2017**, *41*, 450–452. [CrossRef]
3. Sultan, I.; Rahman, S.; Jan, A.T.; Siddiqui, M.T.; Mondal, A.H.; Haq, Q.M.R. Antibiotics, Resistome and Resistance Mechanisms: A Bacterial Perspective. *Front. Microbiol.* **2018**, *9*, 2066. [CrossRef]
4. Christaki, E.; Marcou, M.; Tofarides, A. Antimicrobial Resistance in Bacteria: Mechanisms, Evolution, and Persistence. *J. Mol. Evol.* **2020**, *88*, 26–40. [CrossRef] [PubMed]
5. Matlashewski, G.; Berghuis, A.; Sheppard, D.; Wainberg, M.A.; Gotte, M. *Handbook of Antimicrobial Resistance*; Springer: Berlin/Heidelberg, Germany, 2017.
6. Tooke, C.L.; Hinchliffe, P.; Bragginton, E.C.; Colenso, C.K.; Hirvonen, V.H.; Takebayashi, Y.; Spencer, J. β-Lactamases and β-Lactamase Inhibitors in the 21st Century. *J. Mol. Biol.* **2019**, *431*, 3472–3500. [CrossRef]
7. Reck, F.; Bermingham, A.; Blais, J.; Capka, V.; Cariaga, T.; Casarez, A.; Colvin, R.; Dean, C.R.; Fekete, A.; Gong, W.; et al. Optimization of Novel Monobactams with Activity against Carbapenem-Resistant *Enterobacteriaceae*—Identification of LYS228. *Bioorg. Med. Chem. Lett.* **2018**, *28*, 748–755. [CrossRef]
8. Bush, K.; Bradford, P.A. Interplay between β-Lactamases and New β-lactamase Inhibitors. *Nat. Rev. Genet.* **2019**, *17*, 295–306. [CrossRef] [PubMed]
9. Ramírez-Castillo, F.Y.; Moreno-Flores, A.C.; Avelar-González, F.J.; Márquez-Díaz, F.; Harel, J.; Guerrero-Barrera, A.L. An Evaluation of Multidrug-Resistant Escherichia Coli Isolates in Urinary Tract Infections from Aguascalientes, Mexico: Cross-Sectional Study. *Ann. Clin. Microbiol. Antimicrob.* **2018**, *17*, 34. [CrossRef]
10. Park, C.H.; Robicsek, A.; Jacoby, G.A.; Sahm, D.; Hooper, D.C. Prevalence in the United States of aac(6′)-Ib-cr Encoding a Ciprofloxacin-Modifying Enzyme. *Antimicrob. Agents Chemother.* **2006**, *50*, 3953–3955. [CrossRef]
11. Peralta, G.; Sanchez, M.B.; Garrido, J.C.; De Benito, I.; Cano, M.E.; Martínez-Martínez, L.; Roiz, M.P. Impact of Antibiotic Resistance and of Adequate Empirical Antibiotic Treatment in the Prognosis of Patients with *Escherichia coli* bacteraemia. *J. Antimicrob. Chemother.* **2007**, *60*, 855–863. [CrossRef]

12. Amer, R.; El-Baghdady, K.; Kamel, I.; El-Shishtawy, H. Prevalence of Extended Spectrum Beta- Lactamase Genes among Escherichia Coli and Klebsiella Pneumoniae Clinical Isolates. *Egypt. J. Microbiol.* **2019**. [CrossRef]
13. Fam, N.; Leflon-Guibout, V.; Fouad, S.; Aboul-Fadl, L.; Marcon, E.; Desouky, D.; El-Defrawy, I.; Abou-Aitta, A.; Klena, J.; Nicolas-Chanoine, M.-H. CTX-M-15-Producing Escherichia Coli Clinical Isolates in Cairo (Egypt), Including Isolates of Clonal Complex ST10 and Clones ST131, ST73, and ST405 in Both Community and Hospital Settings. *Microb. Drug Resist.* **2011**, *17*, 67–73. [CrossRef]
14. Alanazi, M.Q.; Alqahtani, F.Y.; Aleanizy, F.S. An Evaluation of E. Coli in Urinary Tract Infection in Emergency Department at KAMC in Riyadh, Saudi Arabia: Retrospective Study. *Ann. Clin. Microbiol. Antimicrob.* **2018**, *17*, 3. [CrossRef]
15. Koupetori, M.; Retsas, T.; Antonakos, N.; Vlachogiannis, G.; Perdios, I.; Nathanail, C.; Makaritsis, K.; Papadopoulos, A.; Sinapidis, D.; Giamarellos-Bourboulis, E.J.; et al. Bloodstream Infections and Sepsis in Greece: Over-Time Change of Epidemiology and Impact of de-escalation on final outcome. *BMC Infect. Dis.* **2014**, *14*, 272. [CrossRef]
16. Mohammed, M.A.; Alnour, T.M.; Shakurfo, O.M.; Aburass, M.M. Prevalence and antimicrobial resistance pat-tern of bacterial strains isolated from patients with urinary tract infection in Messalata Central Hospital, Libya. *Asian Pac. J. Trop. Med.* **2016**, *9*, 771–776. [CrossRef]
17. Musicha, P.; Cornick, J.E.; Bar-Zeev, N.; French, N.; Masesa, C.; Denis, B.; Kennedy, N.; Mallewa, J.; Gordon, M.A.; Msefula, C.L.; et al. Trends in Antimicrobial Resistance in Bloodstream Infection Isolates at a Large Urban Hospital in Malawi (1998–2016): A Surveillance Study. *Lancet Infect. Dis.* **2017**, *17*, 1042–1052. [CrossRef]
18. Miller, D. Update on the Epidemiology and Antibiotic Resistance of Ocular Infections. *Middle E. Afr. J. Ophthalmol.* **2017**, *24*, 30.
19. Zhou, Y.; Zhu, X.; Hou, H.; Lü, Y.; Yu, J.; Mao, L.; Mao, L.; Sun, Z. Characteristics of Diarrheagenic Escherichia Coli among Children under 5 Years of Age with Acute Diarrhea: A Hospital Based Study. *BMC Infect. Dis.* **2018**, *18*, 63. [CrossRef] [PubMed]
20. El Gany, S.A.; Gad, G.M.; Mousa, S.; Ibrahem, R. Characterization of Verotoxigenic E.Coli and Enteropathogenic E.Coli Isolated from Infants with Diarrhea in Combination with Antimicrobial Resistance Pattern in Minia, Egypt. *J. Adv. Biomed. Pharm. Sci.* **2020**, *3*, 101–109. [CrossRef]
21. Jit, M.; Ng, D.H.L.; Luangasanatip, N.; Sandmann, F.; Atkins, K.E.; Robotham, J.V.; Pouwels, K.B. Quantifying the Economic cost of Antibiotic Resistance and the Impact of Related Interventions: Rapid Methodological Review, Conceptual Framework and Recommendations for Future Studies. *BMC Med.* **2020**, *18*, 38. [CrossRef]
22. Siwakoti, S.; Subedi, A.; Sharma, A.; Baral, R.; Bhattarai, N.R.; Khanal, B. Incidence and Outcomes of Multidrug-Resistant Gram-negative Bacteria Infections in Intensive Care Unit from Nepal- a Prospective Cohort Study. *Antimicrob. Resist. Infect. Control.* **2018**, *7*, 114. [CrossRef] [PubMed]
23. Abdelaziz, S.M.; Aboshanab, K.M.; Yahia, I.S.; Yassien, M.A.; Hassouna, N.A.J.A. Correlation between the Antibiotic Resistance Genes and Susceptibility to Antibiotics among the Carbapenem-Resistant Gram-Negative Pathogens. *Antibiotics* **2021**, *10*, 255. [CrossRef]
24. Abd El-Baky, R.M.; Ibrahim, R.A.; Mohamed, D.S.; Ahmed, E.F.; Hashem, Z.S.J.I.; Resistance, D. Prevalence of Virulence Genes and Their Association with Antimicrobial Resistance Among Pathogenic E. Coli Isolated from Egyptian Patients with Different Clinical Infections. *Infect. Drug Resist.* **2020**, *13*, 1221. [CrossRef] [PubMed]
25. Ibrahim, Y.; Sani, Y.; Saleh, Q.; Saleh, A.; Hakeem, G. Phenotypic Detection of Extended Spectrum Beta Lactamase and Carbapenemase Co-Producing Clinical Isolates from Two Tertiary Hospitals in Kano, North West Nigeria. *Ethiop. J. Health Sci.* **2017**, *27*, 3–10. [CrossRef]
26. Farhan, S.M.; A Ibrahim, R.; Mahran, K.M.; Hetta, H.F.; El-Baky, R.M.A. Antimicrobial Resistance Pattern and Molecular Genetic Distribution of Metallo-β-Lactamases Producing *Pseudomonas aeruginosa* Isolated from Hospitals in Minia, Egypt. *Infect. Drug Resist.* **2019**, *12*, 2125–2133. [CrossRef]
27. Fazlul, M.; Deepthi, S.; Farzana, Y.; Najnin, A.; Rashid, M.; Munira, B.; Srikumar, S. Detection of Metal-lo-B-Lactamases-Encoding Genes among Clinical Isolates of Escherichia Coli in a Tertiary Care Hospital, Malaysia. *BMC Res. Notes* **2019**, *11*, 291–298.
28. Mohamed, E.S.; Khairy, R.M.M.; Abdelrahim, S.S. Prevalence and Molecular Characteristics of ESBL and AmpC β -Lactamase Producing *Enterobacteriaceae* Strains Isolated from UTIs in Egypt. *Antimicrob. Resist. Infect. Control.* **2020**, *9*, 198. [CrossRef]
29. El Sherif, R.H.; Maamoun, H.A.H. Molecular Characteristics of Extended-Spectrum Beta-Lactamases Among Gram-Negative Isolates Collected in Cairo University Hospital. *Comp. Haematol. Int.* **2012**, *22*, 733–739. [CrossRef]
30. Zaki, M.; El-Halaby, H.; Elmansoury, E.; Zeid, M.; Khaled, K.; Nomir, M. Genetic Study of Extended Spectrum Beta-Lactamase and Carbapenemase Producing Escherichia Coli Causing Sepsis among Egyptian Children. *Open Microbiol. J.* **2019**, *13*, 128–137. [CrossRef]
31. Myat, T.O.; Hannaway, R.F.; Zin, K.N.; Htike, W.W.; Win, K.K.; Crump, J.A.; Murdoch, D.R.; Ussher, J.E. ESBL-and Carbapenemase-Producing *Enterobacteriaceae* in Patients with Bacteremia, Yangon, Myanmar. *Emerg. Infect. Dis.* **2017**, *23*, 857. [CrossRef] [PubMed]
32. Abrar, S.; Ain, N.U.; Liaqat, H.; Hussain, S.; Rasheed, F.; Riaz, S. Distribution of blaCTX—M, blaTEM, blaSHV and blaOXA Genes in Extended-Spectrum-β-Lactamase-Producing Clinical Isolates: A Three-Year Multi-Center Study from Lahore, Pakistan. *Antimicrob. Resist. Infect. Control.* **2019**, *8*, 80. [CrossRef] [PubMed]
33. Eltai, N.O.; Al Thani, A.A.; Al-Ansari, K.; Deshmukh, A.S.; Wehedy, E.; Al-Hadidi, S.H.; Yassine, H.M. Molecular Characterization of Extended Spectrum β -Lactamases *Enterobacteriaceae* Causing Lower Urinary Tract Infection among Pediatric Population. *Antimicrob. Resist. Infect. Control.* **2018**, *7*, 90. [CrossRef] [PubMed]

34. Ahmed, M.A.S.; Bansal, D.; Acharya, A.; Elmi, A.A.; Hamid, J.M.; Ahmed, A.M.S.; Chandra, P.; Ibrahim, E.; Sultan, A.A.; Doiphode, S.; et al. Antimicrobial Susceptibility and Molecular Epidemiology of Extended-Spectrum Beta-Lactamase-Producing *Enterobacteriaceae* from Intensive Care Units at Hamad Medical Corporation, Qatar. *Clin. Microbiol. Rev.* **2016**, *5*. [CrossRef]
35. Kumarasamy, K.K.; Toleman, M.A.; Walsh, T.R.; Bagaria, J.; Butt, F.; Balakrishnan, R.; Chaudhary, U.; Doumith, M.; Giske, C.G.; Irfan, S.; et al. Emergence of A New Antibiotic Resistance Mechanism in India, Pakistan, and the UK: A Molecular, Biological, and Epidemiological Study. *Lancet Infect. Dis.* **2010**, *10*, 597–602. [CrossRef]
36. Amer, W.H.; Khalil, H.S.; Abd EL Wahab, M.A. Risk Factors, Phenotypic and Genotypic Characterization of Carbapenem Resistant *Enterobacteriaceae* in Tanta University Hospitals, Egypt. *Int. J. Infect. Control.* **2016**, *12*. [CrossRef]
37. Khalifa, H.; Soliman, A.M.; Ahmed, A.M.; Shimamoto, T.; Hara, T.; Ikeda, M.; Kuroo, Y.; Kayama, S.; Sugai, M.; Shimamoto, T. High Carbapenem Resistance in Clinical Gram-Negative Pathogens Isolated in Egypt. *Microb. Drug Resist.* **2017**, *23*, 838–844. [CrossRef]
38. Begum, N.; Shamsuzzaman, S. Emergence of Carbapenemase-Producing Urinary Isolates at A Tertiary Care Hospital in Dhaka, Bangladesh. *Tzu Chi Med. J.* **2016**, *28*, 94–98. [CrossRef]
39. Memish, Z.A.; Assiri, A.; Almasri, M.; Roshdy, H.; Hathout, H.; Kaase, M.; Gatermann, S.G.; Yezli, S. Molecular Characterization of Carbapenemase Production Among Gram-Negative Bacteria in Saudi Arabia. *Microb. Drug Resist.* **2015**, *21*, 307–314. [CrossRef]
40. Sonnevend, A.; Ghazawi, A.A.; Hashmey, R.; Jamal, W.; Rotimi, V.O.; Shibl, A.M.; Al-Jardani, A.; Al-Abri, S.S.; Tariq, W.U.Z.; Weber, S.; et al. Characterization of Carbapenem-Resistant *Enterobacteriaceae* with High Rate of Autochthonous Transmission in the Arabian Peninsula. *PLoS ONE* **2015**, *10*, e0131372. [CrossRef]
41. Sekyere, J.O.; Govinden, U.; Essack, S. The Molecular Epidemiology and Genetic Environment of Carbapenemases Detected in Africa. *Microb. Drug Resist.* **2016**, *22*, 59–68. [CrossRef] [PubMed]
42. Chen, L.; Anderson, D.; Paterson, D. Overview of the Epidemiology and the Threat of Klebsiella Pneumoniae Carbapenemases (KPC) Resistance. *Infect. Drug Resist.* **2012**, *5*, 133–141. [CrossRef]
43. Al-Agamy, M.H.; Aljallal, A.; Radwan, H.H.; Shibl, A.M. Characterization of Carbapenemases, ESBLs, and Plasmid-Mediated Quinolone Determinants in Carbapenem-Insensitive *Escherichia coli* and *Klebsiella pneumoniae* in Riyadh hospitals. *J. Infect. Public Health* **2018**, *11*, 64–68. [CrossRef]
44. Hassuna, N.A.; Khairalla, A.S.; Farahat, E.M.; Hammad, A.M.; Abdel-Fattah, M. Molecular Characterization of Extended-Spectrum β Lactamase- Producing E. coli Recovered from Community-Acquired Urinary Tract Infections in Upper Egypt. *Sci. Rep.* **2020**, *10*. [CrossRef]
45. Ouedraogo, A.-S.; Sanou, M.; Kissou, A.; Sanou, S.; Solaré, H.; Kaboré, N.F.; Poda, A.; Aberkane, S.; Bouzinbi, N.; Sano, I.; et al. High Prevalence of Extended-Spectrum ß-Lactamase producing *Enterobacteriaceae* among Clinical Isolates in Burkina Faso. *BMC Infect. Dis.* **2016**, *16*. [CrossRef] [PubMed]
46. Faghri, J.; Maleki, N.; Tahanasab, Z.; Mobasherizadeh, S.; Rezaei, A. Prevalence of CTX-M and TEM β-lactamases in *Klebsiella pneumoniae* Isolates from Patients with Urinary Tract Infection, Al-Zahra Hospital, Isfahan, Iran. *Adv. Biomed. Res.* **2018**, *7*, 10. [CrossRef]
47. Kamel, N.A.; El-Tayeb, W.N.; El-Ansary, M.R.; Mansour, M.T.; Aboshanab, K.M. Phenotypic Screening and Molecular Characterization of Carbapenemase-Producing Gram-Negative Bacilli Recovered from Febrile Neutropenic Pediatric Cancer Patients in Egypt. *PLoS ONE* **2018**, *13*, e0202119. [CrossRef]
48. Bado, I.; Gutiérrez, C.; Garcia-Fulgueiras, V.; Cordeiro, N.F.; Pirez, L.A.; Seija, V.; Bazet, C.; Rieppi, G.; Vignoli, R. CTX-M-15 in Combination with aac(6′)-Ib-cr is the Most Prevalent Mechanism of Resistance both in *Escherichia coli* and *Klebsiella pneumoniae*, including K. pneumoniae ST258, in an ICU in Uruguay. *J. Glob. Antimicrob. Resist.* **2016**, *6*, 5–9. [CrossRef]
49. Vignoli, R.; García-Fulgueiras, V.; Cordeiro, N.F.; Bado, I.; Seija, V.; Aguerrebere, P.; Laguna, G.; Araújo, L.; Bazet, C.; Gutkind, G.; et al. Extended-Spectrum β-lactamases, Transferable Quinolone Resistance, and Virulotyping in Extra-Intestinal E. coli in Uruguay. *J. Infect. Dev. Ctries.* **2016**, *10*, 43–52. [CrossRef] [PubMed]
50. Azargun, R.; Sadeghi, M.R.; Barhaghi, M.H.S.; Kafil, H.S.; Yeganeh, F.; Oskouee, M.A.; Ghotaslou, R. The Prevalence of Plasmid-mediated Quinolone Resistance and ESBL-Production in *Enterobacteriaceae* Isolated from Urinary tract Infections. *Infect. Drug Resist.* **2018**, *ume 11*, 1007–1014. [CrossRef]
51. Xue, G.; Li, J.; Feng, Y.; Xu, W.; Li, S.; Yan, C.; Zhao, H.; Sun, H. High Prevalence of Plasmid-Mediated Quinolone Resistance Determinants in *Escherichia coli* and *Klebsiella pneumoniae* Isolates from Pediatric Patients in China. *Microb. Drug Resist.* **2017**, *23*, 107–114. [CrossRef] [PubMed]
52. Mahato, S.; Mahato, A.; Yadav, J. Prevalence and identification of uropathogens in eastern Nepal and under-standing their antibiogram due to multidrug resistance and Esbl. *Asian Pac. J. Microbiol. Res.* **2018**, *2*, 09–17.
53. Alfola, M.M.H.R.; Kamel, Z.; Nada, M.G.E.D.; Rashed, L.A.; El-Awady, B.A. Phenotypic and Genotypic Characterization of ESBL-Producing *Escherichia coli* and *Klebsiella pneumonia* Isolates from Patient's Urine Specimens. *Int. Arab. J. Antimicrob. Agents* **2016**, *6*, 6. [CrossRef]
54. Abdel-Moaty, M.M.; Mohamed, W.S.; Abdel-All, S.M.; El-Hendawy, H.H. Prevalence and Molecular Epidemiology of Extended Spectrum β-Lactamase Producing *Escherichia coli* from Hospital and Community Settings in Egypt. *J. Appl. Pharm. Sci.* **2016**, *6*, 042–047. [CrossRef]

55. Nepal, K.; Pant, N.D.; Neupane, B.; Belbase, A.; Baidhya, R.; Shrestha, R.K.; Lekhak, B.; Bhatta, D.R.; Jha, B. Extended Spectrum Beta-Lactamase and Metallo Beta-Lactamase Production among *Escherichia coli* and *Klebsiella pneumoniae* Isolated from Different Clinical Samples in A Tertiary Care Hospital in Kathmandu, Nepal. *Ann. Clin. Microbiol. Antimicrob.* **2017**, *16*, 62. [CrossRef]
56. Javed, H.; Ejaz, H.; Zafar, A.; Rathore, A.W.; Haq, I.U. Metallo-Beta-Lactamase Producing *Escherichia coli* and *Klebsiella pneumoniae*: A Rising Threat for Hospitalized Children. *J. Pak. Med. Assoc.* **2016**, *66*, 1068–1072. [PubMed]
57. Wu, T.-L.; Siu, L.; Su, L.-H.; Lauderdale, T.; Lin, F.; Leu, H.-S.; Lin, T.-Y.; Ho, M. Outer membrane protein change combined with co-existing TEM-1 and SHV-1 β-lactamases lead to false identification of ESBL-producing Klebsiella pneu-moniae. *J. Antimicrob. Chemother.* **2001**, *47*, 755–761. [CrossRef] [PubMed]
58. WHO. *Egypt: Pharmaceutical Country Profile*; WHO: Geneva, Switzerland, 2011.
59. El-Shazly, D.A.; Nasef, S.; Mahmoud, F.F.; Jonas, D. Expanded Spectrum β–Lactamase Producing *Escherichia coli* Isolated from Chickens with Colibacillosis in Egypt. *Poult. Sci.* **2017**, *96*, 2375–2384. [CrossRef] [PubMed]
60. Tille, P. *Bailey & Scott's Diagnostic Microbiology-E-Book*; Elsevier Health Sciences: Amsterdam, The Netherlands, 2015.
61. Hudzicki, J. Kirby-Bauer Disk Diffusion Susceptibility Test Protocol. *Am. Soc. Microbiol.* **2009**, *3189*, 1–23.
62. CLSI. *Performance Standards for Antimicrobial Susceptibility Testing*, 28th ed.; CLSI supplement M100; Clinical and Laboratory Standards Institute: Wayne, PA, USA, 2018.
63. Lee, K.; Chong, Y.; Shin, H.; Kim, Y.; Yong, D.; Yum, J. Modified Hodge and EDTA-Disk Synergy Tests to Screen Metallo-β-Lactamase-Producing Strains of Pseudomonas and Acinetobactet Species. *Clin. Microbiol. Infect.* **2001**, *7*, 88–91. [CrossRef]
64. Nordmann, P.; Poirel, L.; Carrër, A.; Toleman, M.A.; Walsh, T.R. How to detect NDM-1 producers. *J. Clin. Microbiol.* **2011**, *49*, 718–721. [CrossRef] [PubMed]
65. Stürenburg, E.; Kühn, A.; Mack, D.; Laufs, R. A novel extended-spectrum β-lactamase CTX-M-23 with a P167T substitution in the active-site omega loop associated with ceftazidime resistance. *J. Antimicrob. Chemother.* **2004**, *54*, 406–409. [CrossRef]
66. Ghorbani-Dalini, S.; Kargar, M.; Doosti, A.; Abbasi, P.; Sarshar, M. Molecular epidemiology of ESBL genes and multi-drug resistance in diarrheagenic Escherichia coli strains isolated from adults in Iran. *Iran. J. Pharm. Res.* **2015**, *14*, 1257. [PubMed]
67. Ellington, M.J.; Kistler, J.; Livermore, D.M.; Woodford, N. Multiplex PCR for rapid detection of genes encoding acquired metallo-β-lactamases. *J. Antimicrob. Chemother.* **2007**, *59*, 321–322. [CrossRef]
68. Gröbner, S.; Linke, D.; Schütz, W.; Fladerer, C.; Madlung, J.; Autenrieth, I.B.; Witte, W.; Pfeifer, Y.J. Emergence of carbapenem-non-susceptible extended-spectrum β-lactamase-producing Klebsiella pneumoniae isolates at the university hospital of Tübingen, Germany. *J. Med. Microbiol.* **2009**, *58*, 912–922. [CrossRef] [PubMed]
69. Poirel, L.; Héritier, C.; Tolün, V.; Nordmann, P. Emergence of oxacillinase-mediated resistance to imipenem in Klebsiella pneumoniae. *J. Antimicrob. Agents. Chemother.* **2004**, *48*, 15–22. [CrossRef]
70. Jiang, Y.; Zhou, Z.; Qian, Y.; Wei, Z.; Yu, Y.; Hu, S.; Li, L. Plasmid-mediated quinolone resistance determinants qnr and aac (6′)-Ib-cr in extended-spectrum β-lactamase-producing Escherichia coli and Klebsiella pneumoniae in China. *J. Antimicrob. Chemother.* **2008**, *61*, 1003–1006. [CrossRef] [PubMed]

Article

Effect of Titanium Dioxide Nanoparticles on the Expression of Efflux Pump and Quorum-Sensing Genes in MDR *Pseudomonas aeruginosa* Isolates

Fatma Y. Ahmed [1], Usama Farghaly Aly [2], Rehab Mahmoud Abd El-Baky [1,3,*] and Nancy G. F. M. Waly [1]

[1] Department of Microbiology and Immunology, Faculty of Pharmacy, Minia University, Minia 61519, Egypt; fatma_yousef@s-mu.edu.eg (F.Y.A.); nancy.gamil1@mu.edu.eg (N.G.F.M.W.)

[2] Department of Pharmaceutics, Faculty of Pharmacy, Minia University, Minia 61519, Egypt; us_farghaly@mu.edu.eg

[3] Department of Microbiology and Immunology, Faculty of Pharmacy, Deraya University, Minia 11566, Egypt

* Correspondence: rehab.mahmoud@mu.edu.eg; Tel.: +20-10-9248-7412

Abstract: Most of the infections caused by multi-drug resistant (MDR) *P. aeruginosa* strains are extremely difficult to be treated with conventional antibiotics. Biofilm formation and efflux pumps are recognized as the major antibiotic resistance mechanisms in MDR *P. aeruginosa*. Biofilm formation by *P. aeruginosa* depends mainly on the cell-to-cell communication quorum-sensing (QS) systems. Titanium dioxide nanoparticles (TDN) have been used as antimicrobial agents against several microorganisms but have not been reported as an anti-QS agent. This study aims to evaluate the impact of titanium dioxide nanoparticles (TDN) on QS and efflux pump genes expression in MDR *P. aeruginosa* isolates. The antimicrobial susceptibility of 25 *P. aeruginosa* isolates were performed by Kirby–Bauer disc diffusion. Titanium dioxide nanoparticles (TDN) were prepared by the sol gel method and characterized by different techniques (DLS, HR-TEM, XRD, and FTIR). The expression of efflux pumps in the MDR isolates was detected by the determination of MICs of different antibiotics in the presence and absence of carbonyl cyanide m-chlorophenylhydrazone (CCCP). Biofilm formation and the antibiofilm activity of TDN were determined using the tissue culture plate method. The effects of TDN on the expression of QS genes and efflux pump genes were tested using real-time polymerase chain reaction (RT-PCR). The average size of the TDNs was 64.77 nm. It was found that TDN showed a significant reduction in biofilm formation (96%) and represented superior antibacterial activity against *P. aeruginosa* strains in comparison to titanium dioxide powder. In addition, the use of TDN alone or in combination with antibiotics resulted in significant downregulation of the efflux pump genes (*MexY, MexB, MexA*) and QS-regulated genes (*lasR, lasI, rhll, rhlR, pqsA, pqsR*) in comparison to the untreated isolate. TDN can increase the therapeutic efficacy of traditional antibiotics by affecting efflux pump expression and quorum-sensing genes controlling biofilm production.

Keywords: MDR *P. aeruginosa*; titanium dioxide nanoparticles; biofilm; efflux pumps; quorum-sensing system; real-time polymerase chain reaction

Citation: Ahmed, F.Y.; Aly, U.F.; Abd El-Baky, R.M.; Waly, N.G.F.M. Effect of Titanium Dioxide Nanoparticles on the Expression of Efflux Pump and Quorum-Sensing Genes in MDR *Pseudomonas aeruginosa* Isolates. *Antibiotics* **2021**, *10*, 625. https://doi.org/10.3390/antibiotics10060625

Academic Editor: Teresa V. Nogueira

Received: 23 April 2021
Accepted: 19 May 2021
Published: 24 May 2021

Publisher's Note: MDPI stays neutral with regard to jurisdictional claims in published maps and institutional affiliations.

1. Introduction

According to the World Health Organization (WHO), *P. aeruginosa* is one of the most virulent and resistant bacteria for which new antimicrobials are urgently needed [1]. *P. aeruginosa* is characterized by bacterial resistance syndrome, as nearly all recognized antimicrobial resistance pathways can be found in this strain [2].

One of the important mechanisms of antibiotic resistance in *P. aeruginosa* is the exclusion of antibiotics through multidrug resistance (MDR) efflux systems, especially those that belong to the resistance-nodulation-division (RND) family. MexAB-OprM, MexXY and MexCD-OprJ are considered the main cause of intrinsic and acquired multidrug resistance [3].

Biofilm formation by *P. aeruginosa* is another mechanism of antibiotic resistance, since biofilm cells are much more resistant to antibiotics than planktonic cells [4]. The quorum-sensing system (QS) regulates the formation of biofilms and the expression of many extracellular virulence factors in many bacterial pathogens such as *P. aeruginosa* [5]. *P. aeruginosa* has four types of QS schemes, including *rhl*, *las*, *PQS*, and integrated QS (IQS). The *rhl*, *las* and *PQS* systems have been widely tested, while the IQS system was recently detected in *P. aeruginosa* [6].

Although numerous antimicrobial drugs are available commercially, they often lack efficacy against multidrug resistant (MDR) microorganisms, which is a great challenge to the health care teams [7]. Nowadays, the substitution of conventional antimicrobials with modern technology to overcome antimicrobial resistance is under trials. Nanotechnology showed promising results in many studies dealing with antibiotic resistance. Many nanostructures involving metallic particles have been created to counteract microbial pathogens [8]. Among them, titanium dioxide nanoparticles have attracted the attention of researchers due to their oxidative and hydrolytic properties [9].

Nanoparticles can exert their antibacterial activity via different mechanisms such as the following: interactions with DNA and/or the bacterial cell wall; inhibition of biofilm development; or the formation of reactive oxygen species (ROS). Nanoparticles, according to previous studies, have different modes of action than antibiotics, which may enhance their effect on MDR bacteria. [10]. This study aims to evaluate the impact of titanium dioxide nanoparticles (TDN) on the activity of the tested antimicrobials, QS and efflux pump genes expression in MDR *P. aeruginosa* isolates.

2. Results

2.1. Antimicrobial Resistance Pattern of P. aeruginosa Isolates

Twenty-five *P. aeruginosa* isolates were recovered from different clinical specimens, including three ear discharge specimens, thirteen urine, and nine wound exudate samples. Their resistance patterns were detected by the disc diffusion method against six different antimicrobial agents that are commonly used to treat *P. aeruginosa* infections.

The antimicrobial susceptibility of *P. aeruginosa* strains was detected by using the Kirby–Bauer disk diffusion method according to criteria provided by the Clinical & Laboratory Standards Institute (CLSI 2018). The *P. aeruginosa* strains were multidrug resistant (MDR) as they were resistant to three or more of the tested antibiotics (Table 1).

2.2. Synthesis of Titanium Dioxide Nanoparticles and Characterization

Titanium dioxide nanoparticles were prepared by the sol gel method. Dynamic light scattering was used to detect the size distribution of the formed nanoparticles on a scale that ranges from submicron to one nanometer. The average size of the formed TDN was 64.77 ± 0.14 nm (Figure 1A). The HR-TEM of TDN showed that most of the prepared TDN was spheres with asymmetrical edges and slight aggregations (Figure 1B).

X-ray diffraction (XRD) spectrum was performed to detect the phase and crystallinity of the formed TDN (Figure 1C). The XRD patterns showed that TDN was amorphous and had brookite polymorphs that were previously stated by the International Centre for Diffraction Data (ICDD card No. 015-0875).

The chemical composition of the synthesized TDN was analyzed by Fourier transform infrared spectrometry (FTIR). The FTIR spectra of TDN showed a small broad band at 1624 cm^{-1}, which is characteristic to the O–Ti–O bond, a broad band at 3385 cm^{-1}, which is related to O–H stretching, and two small bands observed below 1000 cm^{-1} representing Ti–O–Ti vibrations (Figure 1D).

Table 1. Antibiotic resistance patterns of the tested *P. aeruginosa* strains.

No.	Type *	CF	IM	AK	CP	LV	CT
1	W	R	S	R	R	R	R
2	W	R	R	R	R	R	R
3	W	R	R	R	R	R	R
4	U	R	R	R	R	R	R
5	ER	R	R	R	R	R	R
6	U	R	R	R	R	I	R
7	U	R	R	R	R	R	R
8	U	R	R	R	R	I	R
9	U	R	R	R	R	I	R
10	U	R	S	R	S	I	R
11	U	R	R	R	R	I	R
12	W	R	I	R	R	R	R
13	W	R	R	R	R	R	R
14	W	R	R	R	R	R	R
15	U	R	S	R	S	I	R
16	U	R	I	R	I	I	R
17	U	R	R	R	R	R	R
18	W	R	S	R	S	R	R
19	W	R	R	I	R	S	R
20	U	R	S	I	R	S	R
21	ER	R	R	R	R	R	R
22	W	R	R	R	R	R	R
23	U	R	R	R	S	S	R
24	U	R	R	R	S	S	R
25	ER	R	R	I	S	S	R

* Type of clinical specimen; **U**: urine specimen; **W**: wound exudate; **ER**: ear discharge. **CP**: ciprofloxacin; **CF**: cefepime; **CT**: ceftriaxone; **AK**: amikacin; **IM**: imipenem; **LV**: levofloxacin. **R**: resistance; **I**: intermediate; **S**: sensitive.

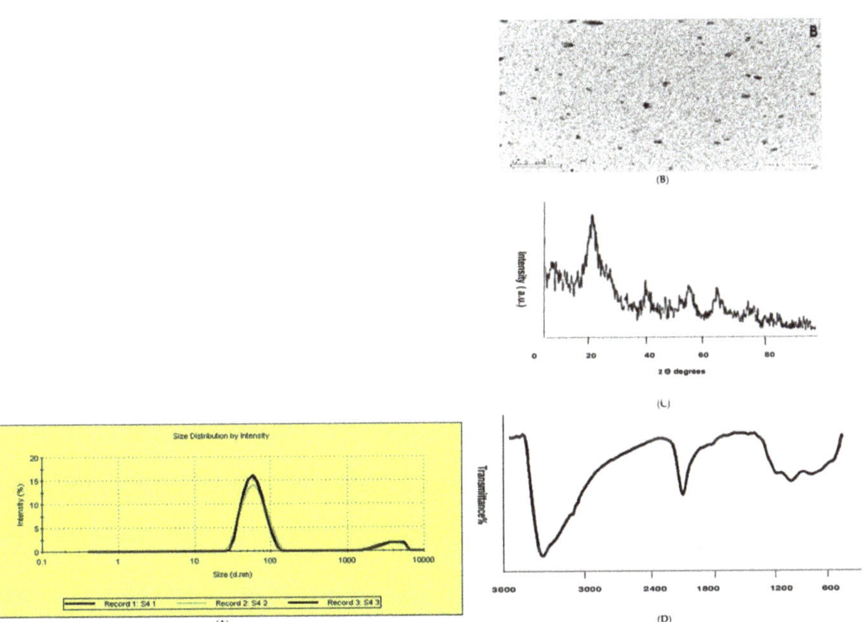

Figure 1. Characterization of titanium dioxide nanoparticles. (**A**) Dynamic light scattering (DLS). (**B**) High-resolution transmission electron microscopy. (**C**) X-ray differentiation. (**D**) Fourier transform infrared spectrometry.

2.3. Susceptibility of P. aeruginosa Isolates to TDN, TDP

The susceptibility of 25 MDR *P. aeruginosa* isolates to titanium dioxide powder (TDP) in bulk form and in nanoform was determined at different concentrations (1 μg/mL–1024 μg/mL) using the agar dilution method.

The *P. aeruginosa* strains showed high resistance to the titanium dioxide powder (TDP), as 11 isolates (44%) had MICs $\geq$ 1024 μg/mL and only 2 strains showed MICs $\leq$ 1 μg/mL. On the other hand, the titanium dioxide nanoparticles (TDN) showed excellent antimicrobial activity in comparison to TDP. As the MICs of TDN were ranged from 8 to 64 μg/mL, and three strains (12%) were sensitive to it (MIC less than 1 μg/mL) (Table 2).

Table 2. Distribution of minimum inhibitory concentrations of TDN, TDP against MDR *P. aeruginosa* isolates (25 isolates).

	<1	**1**	**2**	**4**	**8**	**16**	**32**	**64**	**128**	**256**	**512**	**$\geq$1024**
	Number of Isolates with MICs (μg/mL)											
TDN	3	0	0	0	1	6	1	14	0	0	0	0
TDP	2	0	0	0	0	0	2	1	6	2	1	11

TDN: titanium dioxide nanoparticles; **TDP**: titanium dioxide powder (bulk form).

2.4. Study of Efflux Pump System

To detect the expression of the efflux pump in MDR *P. aeruginosa* isolates, the effect of CCCP on the MICs of the tested antibiotics were determined. Table 3 showed that CCCP caused 4-fold decreases in the MICs of antibiotics in some isolates, which indicates the expression of efflux pumps in the highlighted isolates.

Table 3. Effect of addition of CCCP on antibiotic resistance pattern against 25 MDR *Pseudomonas aeruginosa* isolates.

No.	Et Br		CP		M		CT		AK	
	MIC1	MIC2	MIC1	MIC2	MIC1	MIC2	MIC1	MIC2	MIC1	MIC2
1	>1024	256	8	8	S	ND *	32	32	512	16
2	>1024	2	32	8	64	<1	1024	16	512	16
3	512	16	32	<1	512	<1	512	<1	128	<1
4	>1024	512	32	32	512	<1	>1024	16	>1024	16
5	>1024	512	32	2	512	<1	>1024	16	128	1
6	>1024	512	8	8	8	4	64	64	512	8
7	>1024	256	4	<1	16	8	128	128	512	16
8	1024	256	4	<1	16	<1	256	<1	64	<1
9	>1024	512	32	<1	512	<1	256	2	64	<1
10	512	128	S	ND *	S	ND *	256	<1	16	<1
11	>1024	512	8	8	16	<1	128	16	512	16
12	1024	512	16	<1	16	8	64	64	128	16
13	>1024	256	32	<1	16	8	>1024	8	1024	32
14	>1024	256	4	<1	16	<1	256	<1	128	<1
15	1024	256	S	ND *	S	ND *	256	128	16	<1
16	1024	512	8	<1	8	<1	32	<1	512	<1
17	>1024	64	8	<1	8	<1	256	4	512	<1
18	512	256	S	ND *	S	ND *	16	<1	512	<1
19	>1024	256	4	<1	64	<1	32	2	32	8
20	>1024	512	4	<1	S	ND *	32	<1	16	<1
21	1024	512	4	<1	32	2	32	8	64	<1

Table 3. *Cont.*

No.	Et Br		CP		M		CT		AK	
	MIC1	MIC2	MIC1	MIC2	MIC1	MIC2	MIC1	MIC2	MIC1	MIC2
22	>1024	512	4	<1	8	<1	32	<1	512	<1
23	>1024	512	S	ND *	16	<1	64	16	64	<1
24	>1024	256	S	ND *	32	<1	64	16	64	8
25	256	8	S	ND *	32	<1	256	8	16	<1

Et Br: ethidium bromide, **CP**: ciprofloxacin; **CT**: ceftriaxone; **AK**: amikacin; **M**: meropenem, **MIC1**: antibiotic alone, **MIC2**: antibiotic + CCCP, **ND** *: not determined, **S**: sensitive.

2.5. Effect of TDN on the Antimicrobial Susceptibility of the Tested P. aeruginosa

The addition of TDN resulted a in 2-fold MIC decrease or more in most of the tested isolates. As some of the antibiotics that showed no change in their MICs in the presence of CCCP showed more than a 2-fold decrease in their MICs in the presence of TDN, which suggested that TDN may have anti-efflux activity in addition to its other antimicrobial mechanisms (Table 4).

Table 4. Effect of the addition of TDN on MICs of the tested antibiotics.

No.	CP		M		CT		AK	
	MIC1	MIC2	MIC1	MIC2	MIC1	MIC2	MIC1	MIC2
1	8	0.5	S	S	32	1	512	1
2	32	0.5	64	16	1024	512	512	64
3	32	1	512	256	512	1	128	1
4	32	8	512	256	>1024	512	>1024	1024
5	32	16	512	128	>1024	256	128	128
6	8	0.5	8	4	64	1	512	1
7	4	4	16	8	128	16	512	4
8	4	0.5	16	16	256	16	64	4
9	32	2	512	256	256	32	64	4
10	S (<1)	S (<1)	S	S	256	32	16	1
11	8	0.5	16	16	128	16	512	1
12	16	0.5	16	16	64	1	128	1
13	32	0.5	16	16	>1024	256	1024	4
14	4	0.5	16	16	256	16	128	4
15	S	S	s	s	256	8	16	1
16	8	0.5	8	8	32	1	512	1
17	8	0.5	8	8	256	8	512	1
18	S (<1)	S (<1)	S (<1)	S (<1)	16	1	512	4
19	4	S	64	16	32	1	32	4
20	4	S	S	S	32	1	16	1
21	4	0.5	32	16	32	16	64	1
22	4	0.5	8	4	32	16	512	1
23	S (<1)	S (<1)	16	16	64	16	64	1
24	S (<1)	S (<1)	32	8	64	16	64	4
25	S (<1)	S (<1)	32	16	256	16	16	4

CP: ciprofloxacin; **CT**: ceftriaxone; **AK**: amikacin; **M**: meropenem, **MIC1**: antibiotic alone **MIC2**: antibiotic + TDN, **S**: sensitive.

2.6. Characterization of Biofilm Formation Using Tissue Culture Plate Method (TCP) or Microtitre Plate Test

Out of 25 isolates, 11 (44%) isolates were high biofilm producers, 7 (28%) isolates showed moderate biofilm formation, and 7 (28%) were non or weak biofilm producers (Table 5) according to the decrease in the optical density measured in the presence of TDN in comparison to O.D in the absence of TDN.

Table 5. Degree of biofilm formation in absence and presence of TDN among 25 MDR *P. aeruginosa* isolates.

Target	Degree of Biofilm Formation					
	High		Moderate		Non/Weak Biofilm	
	No.	%	No.	%	No.	% *
In absence of TDN	11	44	7	28	7	28
In presence of TDN	1	4	-	-	24	96

* Percentages were correlated to MDR *P. aeruginosa* isolates (25).

We tested the effect of the formed TDN at a sub-inhibitory concentration (4 μg/mL) on the biofilm formation of 25 MDR *P. aeruginosa* isolates. It was found that TDN showed a high significant inhibitory effect (96%) on biofilm formation.

2.7. Real Time PCR

The isolates (no.2) that showed complete resistance to all tested antibiotics, presented an active efflux pump and strong biofilm producer were selected for testing the effect of TDN on the relative expression of QS genes (*lasI, lasR, rhlI, rhlR, pqsA* and *pqsR*) and efflux pump genes (*MexY, MexB, MexA* and *oprM*) by using real-time polymerase chain reaction (RT-PCR).

A comparison of the expression ratios of the treated and the untreated isolates was performed for each gene, as shown in Table 6. The results showed overexpression of the efflux pump genes (*MexY, MexB, MexA*) in the untreated *P. aeruginosa* isolate, which agreed with the phenotypic results, confirming the importance of an efflux pump as one of the main resistance mechanisms in MDR *P. aeruginosa*. Furthermore, the expression ratios of the QS-regulated genes (*lasI, lasR, rhlI, rhlR, pqsA* and *pqsR*) were also high in the untreated sample.

Table 6. Relative quantitation of gene expression of treated and control *P. aeruginosa* isolate.

Sample	Ct	ΔCt	ΔΔCt	Fold Difference of Gene Expression
Target rpoD (Housekeeping Gene)		-	-	-
Ps *	37.203			
Ps1 **	34.0287			
Target lasI			−4.1318	17.607
Ps *	27.9	−9.253		
Ps1 **	20.637	-13.3911		
Target lasR			−3.5783	11.9447
Ps *	27.721	−9.4821		
Ps1 **	20.9683	−13.0604		
Target MexA			−5.599	48.47
Ps *	30.7584	−6.4446		
Ps1 **	21.984	−12.0438		
Target MexY			−7.458	175.86
Ps *	34.5908	−2.6122		
Ps1 **	25.9582	−10.0705		
Target Mex B			−6.0032	64.1428
Ps *	29.905	−7.2525		
Ps1 **	20.773	−13.2557		
Target OprM			3.937	0.0653
Ps *	34.9282	−2.2749		
Ps1 **	35.6917	1.663		
Target pqsA			−1.499	2.8279
Ps *	26.669	−10.5335		
Ps1 **	21.99	−12.033		

Table 6. *Cont.*

Sample	Ct	ΔCt	ΔΔCt	Fold Difference of Gene Expression
Target pqsR				
Ps *	27.9577	−9.2453	−3.9985	15.98
Ps$_1$ **	20.7849	−13.243		
Target rhlR				
Ps *	27.3754	−9.8277	−3.2054	9.2242
Ps1 **	20.9956	−13.0331		
Target rhlI				
Ps *	27.1394	−10.0637	−2.0059	4.0165
Ps1 **	21.959	−12.0696		

PS *: Treated isolate with TDN, **PS1 ****: control isolate, **Ct**: cycle threshold, **ΔCt**: target gene Ct of sample—housekeeping gene C_t of the same sample. **ΔΔCt**: ΔCt of treated strain−ΔCt of control.

The addition of TDN resulted in significant downregulation of the efflux pump genes (*MexY, MexB, MexA*) and the QS-regulated genes (*lasR, lasI, rhlI, rhlR, pqsA, pqsR*) in comparison to the untreated isolate. On other hand, a slight increase in the expression of *oprM* gene was shown.

3. Discussion

With the emergence of antibiotic-resistant bacteria like *P. aeruginosa*, the use of conventional antibiotics has contributed to the failure of treatments. Therefore, there is an urgent need for the introduction of new antimicrobial agents or the use of non-antimicrobial agents to increase the therapeutic activity of the current antibiotics [11].

The present study evaluated the antibacterial activity of TDN against MDR *P. aeruginosa* isolates. Multidrug resistant isolates represent one of the most important challenges in Egypt due to many factors including the misuse of antibiotics and biocides [12], which leads to the accumulation of antibiotic resistance and cross-resistance among antibiotics and the appearance of multidrug resistant (MDR), XRD, PDR *P. aeruginosa* [13].

The wide spread of MDR strains in Egypt in comparison to other countries warns us that strict antibiotic prescribing policies need to be implemented [14].

In the present study, twenty-five *P. aeruginosa* isolates recovered from different clinical specimens were considered as multidrug resistant isolates, as these strains were non-susceptible (resistant or intermediate) to one antimicrobial agent in three or more different antimicrobial classes [15].

The activity of TiO_2 nanoparticles is of interest to researchers due to the unique characterization of TiO_2 nanomaterial involving crystal size, shape, surface stability and structure [16]. Titanium dioxide nanoparticles (TDN) were prepared and characterized by DLS, HR-TEM, XRD, and FTIR. The average size of the titanium dioxide nanoparticles was 64.77 nm. The antimicrobial activities were due to the large surface area and their ability to penetrate the cell wall [17]. The prepared TDN was spheres with asymmetrical edges and slight aggregations. Nanoparticles have a strong tendency to agglomerate due to their large surface area. Typically, the agglomeration occurs due to the Van der Waals attraction forces among the nanoparticles [18]. The size and size distribution have an effect on the nanoparticle's properties and the possible applications. Similarly, the crystal structure of the NPs and the chemical composition of nanoparticles are extensively studied [19].

X-ray diffraction peaks indicate the small size and the amorphous structure (less crystalline) of the formed TDN. As the size of nanoparticles are inversely proportional to the peak width, increases in the peak width (broad peak) indicate a small size of the formed nanoparticles and the presence of material in nonorange [20], and that supported the data of DLS. FTIR spectroscopy is used to identify the functional groups that exist on nanoparticles. The spectrum represents a fingerprint of the nanoparticles [21].

FTIR confirms the formation of TDN. The characteristic peaks for nano TiO_2 were observed around 1624 cm^{-1}, which is characteristic of the O–Ti–O bond, a broad band at

3385 cm^{-1}, which is related to O–H stretching revealing the presence of water [22]; the two small bands for TiO$_2$ nanoparticles observed below 1000 cm^{-1} were due to Ti–O–Ti vibrations [17]. The absence of any band at 2900 cm^{-1} indicates complete removing of absolute ethanol from the samples [23].

The MICs of titanium dioxide in bulk and nanoform were detected against *P. aeruginosa* strains. It was clear that the titanium dioxide nanoparticles had superior antibacterial activity than the bulk form, which may be due to the small size of the nanoparticles and their large surface area [24]. In addition, the antibacterial activity of TiO$_2$ may be enhanced by exposure to UV light or reaction with water, leading to the generation of radical oxygen species (ROS), which have a potent oxidizing power that attacks microbial cells through various processes, such as lipid peroxidation of the cell membrane, inhibition of enzymes, damage of DNA, alteration of proteins, inhibition of enzymes, finally leading to cell death [25]. Many studies reported the antibacterial and antifungal activities of TDN [26]. However, Abdel-Fatah et al. [27] found that TiO$_2$ nanoparticles had no bactericidal activity against both Gram-negative and Gram-positive bacteria isolated in a study performed in Egypt. Also, [28] they noted that titanium dioxide NPs showed very low antibacterial activity against different bacterial species, including *P. aeruginosa*.

These variations among different studies may depend on the nanoparticles' size, concentration, particle shape, zeta potential and the tested pathogen [29]. Moreover, many environmental factors, including aeration, pH, and temperature, play a role and influence the bactericidal activity [30].

In this study, the MDR *P. aeruginosa* isolates showed a 4-fold reduction in the MICs of the tested antibiotics (ceftriaxone, ciprofloxacin, amikacin and meropenem) in the presence of CCCP. These results suggested that efflux pumps are the main mechanism of resistance in the tested isolates. Many studies from Egypt reported that the efflux pump was the main cause of the decrease in ciprofloxacin and meropenem resistance, as the MICs of ciprofloxacin and meropenem significantly reduced after the addition of CCCP [15,31].

Also, Abbas et al. [12] found that all the MDR isolates showed efflux pump activity as reported by using the ethidium bromide cartwheel method and confirmed with the presence of Mex AB-R genes by PCR.

Another study from Iran reported similar results, in which 65%, 71.5%, 60.5% and 66% of *P. aeruginosa* strains showed a significant decrease in the MICs of imipenem, cefepime, ciprofloxacin, and gentamicin, respectively, in the presence of CCCP. The inhibition of efflux pumps by CCCP in some isolates increased their sensitivity to different antibiotic classes and increased the accumulation of antibiotics that agree with our results [2]. Adabi et al. [32] noted that ciprofloxacin-resistant strains are mediated mainly by an efflux pump.

Titanium dioxide powder combined with antibiotics showed a significant decrease in the MICs of the tested strains by 2-fold or greater, which suggested its activity as an efflux inhibitor according to the method reported by Lamers et al. [33]. In addition, many studies have shown that combining TDN and antibiotics potentiates the antimicrobial action of different classes of antibiotics [34,35]. The combination of TDN with antibiotics increases the concentration of antibiotics at the bacterium–antibiotic interaction site, and the binding of bacteria to antibiotics [36]. Many studies reported that metal nanoparticles can affect proton motif force (PMF), which is essential for efflux pumps [37,38].

Chatterjee et al. [38] reported that copper nanoparticles increase the antimicrobial activity of the tested antibiotics due to the inhibition of the efflux pump of resistant *S. aureus* and *P. aeruginosa* because of copper nanoparticles on PMF.

Another significant factor leading to *P. aeruginosa* pathogenesis in clinical settings is biofilm formation [39]. Regarding biofilm results in the present study, 44%, 28%, and 28% of *P. aeruginosa* isolates were strong, moderate and weak biofilm producers, respectively.

Numerous previous studies reported various rates of biofilm formation by *P. aeruginosa* isolates. Elmaraghy et al. [40] from Egypt and Kamali et al. [41] from Iran, reported lower results than ours. While Abbas et al. [42] from Egypt reported higher results than ours.

Numerous experiments have shown the remarkable ability of metal nanoparticles to decompose thick biofilm barriers by different modes of action. The penetrating ability of metallic nanoparticles is often a valuable function for the prevention of biofilm infections [43]. In the current study, the antibiofilm activity of TDN was detected against 25 MDR *P. aeruginosa* strains. TDN showed a significant reduction in biofilm formation (96%), as TiO_2NPs target sulfhydryl (–SH) groups in the cell membrane, resulting in the creation of the S–TiO_2 bond, and this reaction suppresses the electron transport chain and enzymes essential for biofilm formation [24].

A recent study using TiO_2NPs with polyvinylpyrrolidone polymer (PVP) against pathogenic bacteria on catheters showed that PVP with titanium dioxide nanoparticle films had the capability to prevent biofilm formation by *S. aureus* (83.97%) and *E. coli* (65.3%) [44]. However, Polo et al. [45] reported that neither the photocatalytic treatment with TiO_2 film nor that with TiO_2 nanopowder had any effect on *P. aeruginosa* biofilms at all the interfaces investigated.

Quorum sensing regulates the expression of different virulence factors and the overall process of biofilm production by pathogenic bacteria, which attracted our interest to detect the anti-QS activity of TDN at sub-inhibitory concentrations (4 µg/mL). Our results showed that TDN not only reduced the biofilm formation of *P. aeruginosa* strains, but also reduced the expressions of QS-regulated genes (*lasR, lasI, rhlI, rhlR, pqsA, pqsR*).

Many researchers have recently began using nanotechnology for the development of nano-antimicrobials of the next generation, involving QS nano-inhibitors [46]. Silver nanoparticles were evaluated for the inhibition of QS-regulated virulence and biofilm formation in *P. aeruginosa*. AgNPs were able to decrease *LasIR-RhlIR* levels, inhibit biofilm formation, and significantly downregulate the expression of QS-regulated genes (*lasI, lasR, rhlI, rhlR*) [47]. Also, [48] found that glutathione-stabilized silver nanoparticles (GSH-Ag NPs) have antibiofilm activity in *P. aeruginosa* by reducing the expression the of *lasR, lasI* genes.

García-Lara et al. [49] found that ZnO nanoparticles decrease pyocyanin, elastase, and biofilm formation in *P. aeruginosa* strains; this indicates that ZnO nanoparticles may have QS inhibitor activity. It can be considered as an option for treatment of *P. aeruginosa* infections.

Our results also revealed that efflux pump genes (*MexY, MexB, MexA*) in an untreated *P. aeruginosa* sample were significantly overexpressed, especially the *MexY* gene. Nikaido and Pagès [50] reported that there is a noticeable relationship between MexXY-OprM overproduction in *P. aeruginosa* strains and the usage of different classes of antibiotics in treatment. It seems that MexXY-OprM has a significant benefit during antibiotic pressure in the medical setting and may play a vital role in the efflux of numerous antibiotics. Also, a direct association between the expression of the *MexA* and *MexB* genes and antibiotic resistance was reported by Dashtizadeh et al. [51], which was compatible with our findings.

In the presence of TDN, the efflux pump genes (*MexY, MexB,* and *MexA*) were found to be significantly downregulated. TDN may have efflux pump inhibitor activity by suppressing efflux pump genes. Abdolhosseini et al. [52] from Iran found that the efflux pump genes (*mexA* and *mexB*) in multidrug resistant *P. aeruginosa* strains were decreased by 6- and 2.75-fold, respectively, when exposed to Ag–TSC nanocomposite and ciprofloxacin. Also, Dolatabadi et al. [53] noted that both biosynthesized and commercial AgNPs downregulated the expression of the efflux pump gene *OxqAB* in resistant *Klebsiella pneumoniae* strains.

The inhibition of MDR efflux pumps by nanoparticles would be helpful in enhancing the therapeutic efficacy of traditional antibiotics by affecting quorum-sensing genes controlling biofilm production and other virulence factors [43].

4. Materials and Methods

4.1. Reagents

Titanium dioxide powder (TiO_2, 98%) was obtained from Loba Chemie, Mumbai, India. Meropenem, ceftriaxone, cefepime powder were obtained from Pharco B international, Egypt, ciprofloxacin from Amirya, Egypt and Amikacin from Amount, Egypt.

4.2. Bacterial Strains

In the current study, 25 *P. aeruginosa* strains were collected from different clinical specimens (urine, ear discharge, wound exudate) of patients attending Minia university hospital. *P. aeruginosa* strains were confirmed by using the traditional microbiological method and biochemical tests [54].

4.3. Antimicrobial Susceptibility Testing

Antimicrobial susceptibility of *P. aeruginosa* strains was performed by Kirby–Bauer disc diffusion method according to Clinical & Laboratory Standards Institute guidelines [55]. The used antibiotics were amikacin (AK, 30 μg), cefepime (FEP, 30 μg), levofloxacin (LEV, 10 μg), ceftriaxone (CRO, 30 μg), imipenem (IPM, 10 μg) and ciprofloxacin (CIP, 5 μg).

4.4. Synthesis of Titanium Dioxide Nanoparticles, Characterization

Titanium dioxide nanoparticles (TDN) were synthesized as described before by Ahmed et al. [56]. Dynamic light scattering (DLS) method, high-resolution transmission electron microscopy (HR-TEM), Fourier transform infrared spectroscopy and X-ray diffraction (XRD) were used for the characterization of titanium dioxide nanoparticles (TDN).

4.5. Preparation of TDN Suspension

Titanium dioxide nanoparticle suspensions were prepared by adding 50 mg of TDN to 5 mL of sterile MQ water followed by shaking using ultrasound for 30 min and autoclaving at 121 °C for 20 min [39].

4.6. Determination of Antibacterial Activity of Titanium Dioxide Powder (TDP) and Titanium Dioxide Nanoparticles (TDN)

The antibacterial activity of titanium dioxide powder (TDP) and titanium dioxide nanoparticles (TND) were detected by agar dilution method against 25 *P. aeruginosa* strains as previously described by [31].

Briefly, overnight cultures of 25 *P. aeruginosa* strains in a Mueller Hinton Broth (MHB) were adjusted to a cell density of 10^7 CFU/mL. Then, the bacterial culture spots were applied to the surface of Mueller Hinton Agar containing TDP and TDN with different concentrations (1 to 1024 μg/mL) using a multi-inoculator. The plates were incubated at 37 °C for 18 h and examined for growth. Spots showing no growth were defined as MIC.

4.7. Determination of Efflux Pumps Expression in Resistant Isolates

The MICs of ciprofloxacin, meropenem, amikacin, ceftriaxone were detected for 25 MDR *P. aeruginosa* isolates by agar dilution method in the presence and absence of efflux pump inhibitor carbonyl cyanide m-chlorophenylhydrazone (CCCP) (Sigma, San Jose, CA, USA) at a final concentration of 10 μM. A four-fold reduction in MIC or more of the tested antibiotics after adding CCCP is an indication for the presence of efflux pumps [31].

4.8. Effect of TDN on the Antimicrobial Susceptibility of the Tested P. aeruginosa

The MICs of ciprofloxacin, meropenem, amikacin, ceftriaxone either alone or in combination with TDN were detected for 25 MDR *P. aeruginosa* isolates by agar dilution method in the presence of titanium dioxide nanoparticles at sub-inhibitory concentrations. A two-fold or greater change in the MICs of the tested antibiotics in presence of TDN compared to MICs of antibiotics alone was reported as indicating a significant efflux pump inhibitor [33].

4.9. Biofilm Formation Assays

Biofilm formation of 25 *P. aeruginosa* strains was evaluated by the tissue culture plate assay method (TCP) as previously described by Christensen et al. [57]. About 1×10^7 CFU ml of the tested isolates were incubated in TCP for 24 h at 37 °C. Then, bacterial cultures were removed gently and washed with phosphate buffered saline. Crystal violet (0.1%)

was used to stain the adherent cells, followed by washing with PBS. Then, ethyl alcohol (95%) was applied followed by measuring absorbance at 630 nm and the results were interpreted according to Table 7. The assay was performed in triplicates for each isolate.

Table 7. Classification of bacterial adherence and biofilm formation by TCP method.

Biofilm Formation	Adherence	Mean OD Values
Non/Weak	Non/Weak	<0.120
Moderate	Moderate	0.120–0.240
High	Strong	>0.240

TDN at sub-MIC concentration was added to the tested isolates suspension in TCP and incubated for 24 h at 37 °C. Bacterial cultures were discarded and tissue culture plate wells were washed gently. The adhered cells were stained by crystal violet (0.1%) for 20 s, followed by washing using PBS. Finally, ethyl alcohol was added, and absorbance was measured as described before.

4.10. Gene Expression Using Real-Time PCR

The impact of TDN (4 μg/mL) on the relative expression of efflux pump genes (*MexY*, *MexB*, *MexA* and *OprM*) and quorum-sensing genes (*rhlR*, *lasI*, *lasR*, *rhlI*, *pqsA* and *pqsR*) in the chosen isolate (no.2) was evaluated using real-time polymerase chain reaction (RT-PCR).

The tested isolate was grown until the middle of the exponential phase in presence and absence of TDN. Cultures were pelleted, and mRNA was extracted using RNeasy Mini Kit (Qiagen, Hilden, Germany) according to the manufacturer's instructions. Then, cDNA was performed using High-Capacity cDNA Reverse Transcription Kit (Thermo Fisher, New York, NY, USA).

Real-time polymerase chain reaction (RT-PCR) was performed to determine the expression level of efflux and quorum-sensing genes using Quanti Tect SYBR Green RT-PCR Kit (Qiagen, Germany) according to the manufacturer's protocol. Synthesized cDNA, primers (Table 8) and master mix were mixed and transferred to Applied Biosystems Step One™ instrument.

Table 8. The sequence of the primers used in this study.

Gene	Primer Direction	Primer Sequence	Size of Amplified Product (bp)	Reference
rpoD	F R	5-GCGAGAGCCTCAAGGATAC-3 5-CGAACTGCTTGCCGACTT-3	131	(El-Shaer et al., 2016)
MexY	F R	5-CCGCTACAACGGCTATCCCT-3 5-AGCGGGATCGACCAGCTTTC-3	246	(Yoneda et al., 2005)
MexA	F R	5′ACCTACGAGCCGACTACCAGA-3′ 5′GTTGGTCACCAGGGCGCCTTC-3′	179	(Pourakbari et al., 2016)
MexB	F R	5-GTGTTCGGCTCGCAGTACTC-3 5-AACCGTCGGGATTGACCTTG-3	244	(Pourakbari et al., 2016)
OprM	F R	5-CCATGAGCCGCCAACTGTC-3 5-CCTGGAACGCCGTCTGGAT-3	205	(Pourakbari et al., 2016)
Las I	F R	5-CGCACATCTGGGAACTCA-3 5-CGGCACGACGATCATCATCT-3	176	(El-Shaer et al., 2016)
Las R	F R	5-CTGTGGATGCTCAAGGACTAC-3 5-AACTGGTCTTGCCGATGG-3	133	(El-Shaer et al., 2016)

Table 8. *Cont.*

Gene	Primer Direction	Primer Sequence	Size of Amplified Product (bp)	Reference
rhlI	F R	5-GTAGCGGGTTTGCGGATG-3 5-CGGCATCAGGTCTTCATCG-3	101	(El-Shaer et al., 2016)
rhlR	F R	5-GCCAGCGTCTTGTTCGG-3 5-CGGTCTGCCTGAGCCATC-3	160	(El-Shaer et al., 2016)
Pqs A	F R	5-GACCGGCTGTATTCGATTC-3 5-GCTGAACCAGGGAAAGAAC-3	74	(El-Shaer et al., 2016)
pqsR	F R	5-CTGATCTGCCGGTAATTGG-3 5-ATCGACGAGGAACTGAAGA-3	142	(El-Shaer et al., 2016)

The level of gene expression was relatively normalized to the expression of the housekeeping gene *rpoD*. The expression of genes in *P. aeruginosa* isolate cultivated with TDN was compared to their expression in the control cultures without TDN. Relative quantities of gene expression were calculated using the $2^{-\Delta\Delta Ct}$ method (Pfaffl, 2001).

5. Conclusions

Titanium dioxide nanoparticles can increase the therapeutic efficacy of traditional antibiotics by affecting efflux pump expression and quorum-sensing genes controlling biofilm production.

Author Contributions: Methodology, R.M.A.E.-B., U.F.A., N.G.F.M.W. and F.Y.A.; formal analysis, R.M.A.E.-B. and F.Y.A.; data curation, N.G.F.M.W.; writing—original draft preparation, R.M.A.E.-B., U.F.A., N.G.F.M.W. and F.Y.A.; writing—review and editing, R.M.A.E.-B. and U.F.A.; visualization, F.Y.A.; supervision, R.M.A.E.-B., U.F.A. and N.G.F.M.W. All authors have read and agreed to the published version of the manuscript.

Funding: This research did not receive any specific grant from funding agencies in the public, commercial, or not-for-profit sectors.

Institutional Review Board Statement: The study was conducted according to the guidelines of the Declaration of Helsinki, and approved by the Ethics committee of faculty of Medicine, Minia university.

Informed Consent Statement: Written informed consent has been obtained from the patient(s) to publish this paper.

Data Availability Statement: Data are applicable from authors.

Acknowledgments: The authors received no financial support for the research or publication of this article.

Conflicts of Interest: The authors declare no conflict of interest.

References

1. Awan, A.B.; Schiebel, J.; Böhm, A.; Nitschke, J.; Sarwar, Y.; Schierack, P.; Ali, A. Association of Biofilm Formation and Cytotoxic Potential with Multidrug Resistance in Clinical Isolates of Pseudomonas Aeruginosa. *EXCLI J.* **2019**, *18*, 79–90.
2. Talebi-Taher, M.; Majidpour, A.; Gholami, A.; Rasouli-Kouhi, S.; Adabi, M. Role of Efflux Pump Inhibitor in Decreasing Antibiotic Cross-Resistance of Pseudomonas Aeruginosa in a Burn Hospital in Iran. *J. Infect. Dev. Ctries.* **2016**, *10*, 600–604. [CrossRef]
3. Pourakbari, B.; Yaslianifard, S.; Yaslianifard, S.; Mahmoudi, S.; Keshavarz-Valian, S.; Mamishi, S. Evaluation of Efflux Pumps Gene Expression in Resistant Pseudomonas Aeruginosa Isolates in an Iranian Referral Hospital. *Iran. J. Microbiol.* **2016**, *8*, 249–256. [PubMed]
4. Lee, J.H.; Kim, Y.G.; Cho, M.H.; Lee, J. Zno Nanoparticles Inhibit Pseudomonas Aeruginosa Biofilm Formation and Virulence Factor Production. *Microbiol. Res.* **2014**, *169*, 888–896. [CrossRef] [PubMed]
5. Ugurlu, A.; Yagci, A.K.; Ulusoy, S.; Aksu, B.; Bosgelmez-Tinaz, G. Phenolic Compounds Affect Production of Pyocyanin, Swarming Motility and Biofilm Formation of Pseudomonas Aeruginosa. *Asian Pac. J. Trop. Biomed.* **2016**, *6*, 698–701. [CrossRef]

6.	Blach, S.; Zeuzem, S.; Manns, M.; Altraif, I.; Duberg, A.S.; Muljono, D.H.; Waked, I.; Alavian, S.M.; Lee, M.-H.; Negro, F. Global Prevalence and Genotype Distribution of Hepatitis C Virus Infection in 2015: A Modelling Study. *Lancet Gastroenterol. Hepatol.* **2017**, *2*, 161–176. [CrossRef]

7.	Singh, P.; Garg, A.; Pandit, S.; Mokkapati, V.R.S.S.; Mijakovic, I. Antimicrobial Effects of Biogenic Nanoparticles. *Nanomaterials* **2018**, *8*, 1009. [CrossRef] [PubMed]

8.	Rudramurthy, G.R.; Swamy, M.K.; Sinniah, U.R.; Ghasemzadeh, A. Nanoparticles: Alternatives against Drug-Resistant Pathogenic Microbes. *Molecules* **2016**, *21*, 836. [CrossRef] [PubMed]

9.	Arora, B.; Murar, M.; Dhumale, V. Antimicrobial Potential of TiO_2 Nanoparticles against MDR *Pseudomonas aeruginosa. J. Exp. Nanosci.* **2015**, *10*, 819–827. [CrossRef]

10.	Baptista, P.V.; McCusker, M.P.; Carvalho, A.; Ferreira, D.A.; Mohan, N.M.; Martins, M.; Fernandes, A.R. Nano-Strategies to Fight Multidrug Resistant Bacteria—"A Battle of the Titans". *Front. Microbiol.* **2018**, *9*, 1441. [CrossRef]

11.	Yang, Y.X.; Xu, Z.H.; Zhang, Y.Q.; Tian, J.; Weng, L.X.; Wang, L.H. A New Quorum-Sensing Inhibitor Attenuates Virulence and Decreases Antibiotic Resistance in Pseudomonas Aeruginosa. *J. Microbiol.* **2012**, *50*, 987–993. [CrossRef] [PubMed]

12.	Abbas, H.A.; El-Ganiny, A.M.; Kamel, H. Phenotypic and Genotypic Detection of Antibiotic Resistance of Pseudomonas Aeruginosa Isolated from Urinary Tract Infections. *Afr. Health Sci.* **2018**, *18*, 11–21. [CrossRef]

13.	Yayan, J.; Ghebremedhin, B.; Rasche, K. Antibiotic Resistance of Pseudomonas Aeruginosa in Pneumonia at a Single University Hospital Center in Germany over a 10-Year Period. *PLoS ONE* **2015**, *10*, e0139836. [CrossRef]

14.	Mohamed, A.; Abdelhamid, F. Antibiotic Susceptibility of Pseudomonas Aeruginosa Isolated from Different Clinical Sources. *Zagazig J. Pharm. Sci.* **2020**, *28*, 10–17.

15.	Kishk, R.M.; Abdalla, M.O.; Hashish, A.A.; Nemr, N.A.; El Nahhas, N.; Alkahtani, S.; Abdel-Daim, M.M.; Kishk, S.M. Efflux Mexab-Mediated Resistance in P. Aeruginosa Isolated from Patients with Healthcare Associated Infections. *Pathogens* **2020**, *9*, 471. [CrossRef] [PubMed]

16.	Mahdy, S.A.; Mohammed, W.H.; Emad, H.; Kareem, H.A.; Shamel, R.; Mahdi, S. The Antibacterial Activity of TiO_2 Nanoparticles. *J. Univ. Babylon* **2017**, *25*, 955–961.

17.	Saranya, K.S.; Padil, V.V.T.; Senan, C.; Pilankatta, R.; Saranya, K.; George, B.; Wacławek, S.; Černík, M. Green Synthesis of High Temperature Stable Anatase Titanium Dioxide Nanoparticles Using Gum Kondagogu: Characterization and Solar Driven Photocatalytic Degradation of Organic Dye. *Nanomaterials* **2018**, *8*, 1002. [CrossRef]

18.	Othman, S.H.; Rashid, S.A.; Ghazi, T.I.M.; Abdullah, N. Dispersion and Stabilization of Photocatalytic TiO_2 Nanoparticles in Aqueous Suspension for Coatings Applications. *J. Nanomater.* **2012**, *2012*. [CrossRef]

19.	Mourdikoudis, S.; Pallares, R.M.; Thanh, N.T. Characterization Techniques for Nanoparticles: Comparison and Complementarity Upon Studying Nanoparticle Properties. *Nanoscale* **2018**, *10*, 12871–12934. [CrossRef] [PubMed]

20.	Kantheti, P.; Alapati, P. Green Synthesis of TiO_2 Nanoparticles Using OcimumBasilicum Extract and Its Characterization. *Int. J. Chem. Stud.* **2018**, *6*, 670–674.

21.	Patra, J.K.; Baek, K.-H. Green Nanobiotechnology: Factors Affecting Synthesis and Characterization Techniques. *J. Nanomater.* **2014**, *2014*, 1–12. [CrossRef]

22.	Chatterjee, M.; Anju, C.; Biswas, L.; Kumar, V.A.; Mohan, C.G.; Biswas, R. Antibiotic Resistance in Pseudomonas Aeruginosa and Alternative Therapeutic Options. *Int. J. Med. Microbiol.* **2016**, *306*, 48–58. [CrossRef]

23.	Devi, R.; Venckatesh, R.; Sivaraj, R. Synthesis of Titanium Dioxide Nanoparticles by Sol-Gel Technique. *Int. J. Innov. Res. Sci. Eng. Technol.* **2014**, *3*, 15206–15211. [CrossRef]

24.	Abdulazeem, L.; AL-Amiedi, B.H.; Alrubaei, H.A.; AL-Mawlah, Y.H. Titanium Dioxide Nanoparticles as Antibacterial Agents against Some Pathogenic Bacteria. *Drug Invent. Today* **2019**, *12*.

25.	de Dicastillo, C.L.; Patiño, C.; Galotto, M.J.; Vásquez-Martínez, Y.; Torrent, C.; Alburquenque, D.; Pereira, A.; Escrig, J. Novel Hollow Titanium Dioxide Nanospheres with Antimicrobial Activity against Resistant Bacteria. *Beilstein. J. Nanotechnol.* **2019**, *10*, 1716–1725. [CrossRef]

26.	Manjunath, K.; Yadav, L.S.R.; Jayalakshmi, T.; Reddy, V.; Rajanaika, H.; Nagaraju, G. Ionic Liquid Assisted Hydrothermal Synthesis of TiO_2 Nanoparticles: Photocatalytic and Antibacterial Activity. *J. Mater. Res. Technol.* **2018**, *7*, 7–13. [CrossRef]

27.	Abdel-Fatah, W.I.; Gobara, M.M.; Mustafa, S.F.; Ali, G.W.; Guirguis, O.W. Role of Silver Nanoparticles in Imparting Antimicrobial Activity of Titanium Dioxide. *Mater. Lett.* **2016**, *179*, 190–193. [CrossRef]

28.	Kareem, P.A.; Alsammak, E.G.; Abdullah, Y.J.; Bdaiwi, Q.M. Estimation of Antibacterial Activity of Zinc Oxide, Titanium Dioxide, and Silver Nanoparticles against Multidrug-Resistant Bacteria Isolated from Clinical Cases in Amara City, Iraq. *Drug Invent. Today* **2019**, *11*, 2887–2890.

29.	Skandalis, N.; Dimopoulou, A.; Georgopoulou, A.; Gallios, N.; Papadopoulos, D.; Tsipas, D.; Theologidis, I.; Michailidis, N.; Chatzinikolaidou, M. The Effect of Silver Nanoparticles Size, Produced Using Plant Extract from Arbutus Unedo, on Their Antibacterial Efficacy. *Nanomaterials* **2017**, *7*, 178. [CrossRef]

30.	Beyth, N.; Houri-Haddad, Y.; Domb, A.; Khan, W.; Hazan, R. Alternative Antimicrobial Approach: Nano-Antimicrobial Materials. *Evid. Based Complementary Altern. Med.* **2015**, *2015*. [CrossRef]

31.	Gad, G.F.; El-Domany, R.A.; Zaki, S.; Ashour, H.M. Characterization of Pseudomonas Aeruginosa Isolated from Clinical and Environmental Samples in Minia, Egypt: Prevalence, Antibiogram and Resistance Mechanisms. *J. Antimicrob. Chemother.* **2007**, *60*, 1010–1017. [CrossRef] [PubMed]

32. Adabi, M.; Talebi-Taher, M.; Arbabi, L.; Afshar, M.; Fathizadeh, S.; Minaeian, S.; Moghadam-Maragheh, N.; Majidpour, A. Spread of Efflux Pump Overexpressing-Mediated Fluoroquinolone Resistance and Multidrug Resistance in Pseudomonas Aeruginosa by Using an Efflux Pump Inhibitor. *Infect. Chemother.* **2015**, *47*, 98. [CrossRef] [PubMed]

33. Lamers, R.P.; Cavallari, J.F.; Burrows, L.L. The Efflux Inhibitor Phenylalanine-Arginine Beta-Naphthylamide (PaβN) Permeabilizes the Outer Membrane of Gram-Negative Bacteria. *PLoS ONE* **2013**, *8*, e60666. [CrossRef]

34. Abdulrahman, N.B.; Nssaif, Z.M. Antimicrobial Activity of Zinc Oxide, Titanium Dioxide and Silver Nanoparticles against Mithicillin-Resistant Staphylococcus Aureus Isolates. *Tikrit J. Pure Sci.* **2018**, *21*, 49–53.

35. Kareem, P.A.; Alsammak, E.G. The Effect of Silver and Titanium Dioxide Nanoparticles on Klebsiella Pneumoniae Isolates Multi Resistant to Antibiotics and Observed by Scanning Electron Microscopy. *Cihan. Univ. Sci. J.* **2017**, 284–297. [CrossRef]

36. Allahverdiyev, A.M.; Kon, K.V.; Abamor, E.S.; Bagirova, M.; Rafailovich, M. Coping with Antibiotic Resistance: Combining Nanoparticles with Antibiotics and Other Antimicrobial Agents. *Expert Rev. Anti-Infect. Ther.* **2011**, *9*, 1035–1052. [CrossRef]

37. Dibrov, P.; Dzioba, J.; Gosink, K.K.; Häse, C.C. Chemiosmotic Mechanism of Antimicrobial Activity of Ag$^+$ in Vibrio Cholerae. *Antimicrob. Agents Chemother.* **2002**, *46*, 2668–2670. [CrossRef]

38. Chatterjee, A.K.; Chakraborty, R.; Basu, T. Mechanism of Antibacterial Activity of Copper Nanoparticles. *Nanotechnology* **2014**, *25*, 135101. [CrossRef]

39. Vincent, M.G.; John, N.P.; Narayanan, P.M.; Vani, C.; Murugan, S. In Vitro Study on the Efficacy of Zinc Oxide and Titanium Dioxide Nanoparticles against Metallo Beta-Lactamase and Biofilm Producing Pseudomonas Aeruginosa. *J. Appl. Pharm. Sci.* **2014**, *4*, 41.

40. Elmaraghy, N.; Abbadi, S.; Elhadidi, G.; Hashem, A.; Yousef, A. Virulence Genes in Pseudomonas Aeruginosa Strains Isolated at Suez Canal University Hospitals with Respect to the Site of Infection and Antimicrobial Resistance. *Int. J. Clin. Microbiol. Biochem. Technol.* **2019**, *2*, 8–19.

41. Kamali, E.; Jamali, A.; Ardebili, A.; Ezadi, F.; Mohebbi, A. Evaluation of Antimicrobial Resistance, Biofilm Forming Potential, and the Presence of Biofilm-Related Genes among Clinical Isolates of Pseudomonas Aeruginosa. *BMC Res. Notes* **2020**, *13*, 1–6. [CrossRef] [PubMed]

42. Abbas, H.A.; Serry, F.M.; EL-Masry, E.M. Combating Pseudomonas Aeruginosa Biofilms by Potential Biofilm Inhibitors. *Asian J. Res. Pharm. Sci.* **2012**, *2*, 2012.

43. Gupta, D.; Singh, A.; Khan, A.U. Nanoparticles as Efflux Pump and Biofilm Inhibitor to Rejuvenate Bactericidal Effect of Conventional Antibiotics. *Nanoscale Res. Lett.* **2017**, *12*, 1–6. [CrossRef]

44. Salman, J.A.S.; Al-Kadhemy, M.F.H.; Madhloom, S.A. Preparation of Titanium Dioxide Nanoparticles and Polyvinyl Pyrrolidone Polymer Films as Antibacterial, Antibiofilm against Pathogenic Bacteria on Different Surfaces. *MJS* **2017**, *36*, 132–144. [CrossRef]

45. Polo, A.; Diamanti, M.V.; Bjarnsholt, T.; Høiby, N.; Villa, F.; Pedeferri, M.P.; Cappitelli, F. Effects of Photoactivated Titanium Dioxide Nanopowders and Coating on Planktonic and Biofilm Growth of Pseudomonas Aeruginosa. *Photochem. Photobiol.* **2011**, *87*, 1387–1394. [CrossRef] [PubMed]

46. Singh, B.N.; Prateeksha, P.; Upreti, D.K.; Singh, B.R.; Defoirdt, T.; Gupta, V.K.; De Souza, A.O.; Singh, H.B.; Barreira, J.C.M.; Ferreira, I.C.F.R.; et al. Bactericidal, Quorum Quenching and Anti-Biofilm Nanofactories: A New Niche for Nanotechnologists. *Crit. Rev. Biotechnol.* **2017**, *37*, 525–540. [CrossRef] [PubMed]

47. Singh, B.R.; Singh, B.N.; Singh, A.; Khan, W.; Naqvi, A.H.; Singh, H.B. MycofabricatedBiosilver Nanoparticles Interrupt Pseudomonas Aeruginosa Quorum Sensing Systems. *Sci. Rep.* **2015**, *5*, 1–14. [CrossRef]

48. Mahnaie, M.P.; Mahmoudi, H. Effect of Glutathione-Stabilized Silver Nanoparticles on Expression of Las I and Las R of the Genes in Pseudomonas Aeruginosa Strains. *Eur. J. Med. Res.* **2020**, *25*, 1–11.

49. García-Lara, B.; Saucedo-Mora, M.; Roldán-Sánchez, J.A.; Pérez-Eretza, B.; Ramasamy, M.; Lee, J.; Coria-Jimenez, R.; Tapia, M.; Varela-Guerrero, V.; García-Contreras, R. Inhibition of Quorum-Sensing-Dependent Virulence Factors and Biofilm Formation of Clinical and Environmental P Seudomonas Aeruginosa Strains by Zno Nanoparticles. *Lett. Appl. Microbiol.* **2015**, *61*, 299–305. [CrossRef] [PubMed]

50. Nikaido, H.; Pagès, J.-M. Broad-Specificity Efflux Pumps and Their Role in Multidrug Resistance of Gram-Negative Bacteria. *FEMS Microbiol. Rev.* **2012**, *36*, 340–363. [CrossRef]

51. Dashtizadeh, Y.; Moattari, A.; Gorzin, A.A. Phenotypic and Genetically Evaluation of the Prevalence of Efflux Pumps and Antibiotic Resistance in Clinical Isolates of Pseudomonas Aeruginosa among Burned Patients Admitted to Ghotbodin Shirazi Hospital. *J. Microb. World* **2014**, *19*, 118–127.

52. Abdolhosseini, M.; Zamani, H.; Salehzadeh, A. Synergistic Antimicrobial Potential of Ciprofloxacin with Silver Nanoparticles Conjugated to Thiosemicarbazide against Ciprofloxacin Resistant Pseudomonas Aeruginosa by Attenuation of Mexa-B Efflux Pump Genes. *Biologia* **2019**, *74*, 1191–1196. [CrossRef]

53. Dolatabadi, A.; Noorbazargan, H.; Khayam, N.; Moulavi, P.; Zamani, N.; Asghari Lalami, Z.; Ashrafi, F. Ecofriendly Biomolecule-Capped Bifidobacterium Bifidum-Manufactured Silver Nanoparticles and Efflux Pump Genes Expression Alteration in Klebsiella Pneumoniae. *Microb. Drug Resist.* **2021**, *27*, 247–257. [CrossRef] [PubMed]

54. Abdelraheem, W.M.; Abdelkader, A.E.; Mohamed, E.S.; Mohammed, M.S. Detection of Biofilm Formation and Assessment of Biofilm Genes Expression in Different Pseudomonas Aeruginosa Clinical Isolates. *Meta Gene* **2020**, *23*, 100646. [CrossRef]

55. Wayne, P.A. Clinical and Laboratory Standards Institute: Performance Standards for Antimicrobial Susceptibility Testing: 20th Informational Supplement. *CLSI Document M100-S20*. Inform Suppl 2010. pp. 100–121. Available online: http://file.qums.ac.ir/repository/mmrc/CLSI2015.pdf (accessed on 20 April 2021).
56. Ahmed, F.Y.; Aly, U.F.; El-Baky, R.M.A.; Waly, N.G.F.M.; Youssef, F. Comparative Study of Antibacterial Effects of Titanium Dioxide Nanoparticles Alone and in Combination with Antibiotics on Mdr Pseudomonas Aeruginosa Strains. *Int. J. Nanomed.* **2020**, *15*, 3393.
57. Christensen, G.D.; A Simpson, W.; Younger, J.J.; Baddour, L.M.; Barrett, F.F.; Melton, D.M.; Beachey, E.H. Adherence of Coagulase-Negative Staphylococci to Plastic Tissue Culture Plates: A Quantitative Model for the Adherence of Staphylococci to Medical Devices. *J. Clin. Microbiol.* **1985**, *22*, 996–1006. [CrossRef] [PubMed]

*antibiotics* 

Article

Pan-Resistome Insights into the Multidrug Resistance of *Acinetobacter baumannii*

Diego Lucas Neres Rodrigues [1,†], Francielly Morais-Rodrigues [1,†], Raquel Hurtado [1], Roselane Gonçalves dos Santos [1], Daniela Camargos Costa [2], Debmalya Barh [1,3], Preetam Ghosh [4], Khalid J. Alzahrani [5], Siomar Castro Soares [6], Rommel Ramos [7], Aristóteles Góes-Neto [1], Vasco Azevedo [1,*,†] and Flávia Figueira Aburjaile [1,†]

1   Laboratory of Cellular and Molecular Genetics, Universidade Federal de Minas GeraisBelo Horizonte, Belo Horizonte 31270-901, MG, Brazil; dlnrodrigues@ufmg.br (D.L.N.R.); franrodriguesdacosta@ufmg.br (F.M.-R.); raquelgen1@gmail.com (R.H.); roselanegr@gmail.com (R.G.d.S.); dr.barh@gmail.com (D.B.); arigoesneto@icb.ufmg.br (A.G.-N.); faburjaile@gmail.com (F.F.A.)
2   FAMINAS-BH, Belo Horizonte 31744-007, MG, Brazil; daniela.costa@faminasbh.edu.br
3   Institute of Integrative Omics and Applied Biotechnology, Nonakuri West Bengal 721172, India
4   Department of Computer Science, Virginia Commonwealth University, Richmond, VA 23284, USA; preetam.ghosh@gmail.com
5   Department of Clinical Laboratories Sciences, College of Applied Medical Sciences, Taif University, P.O. Box 11099, Taif 21944, Saudi Arabia; ak.jamaan@tu.edu.sa
6   Department of Research Development and Technological Innovation, Universidade Federal do Triângulo Mineiro, Uberaba 38025-180, MG, Brazil; siomars@gmail.com
7   Faculty of Biotechnology, Universidade Federal de Pará, Belém 66075-110, PA, Brazil; rommelramos@ufpa.br
*   Correspondence: vasco@icb.ufmg.br
†   These authors contributed equally to this work.

Citation: Rodrigues, D.L.N.; Morais-Rodrigues, F.; Hurtado, R.; dos Santos, R.G.; Costa, D.C.; Barh, D.; Ghosh, P.; Alzahrani, K.J.; Soares, S.C.; Ramos, R.; et al. Pan-Resistome Insights into the Multidrug Resistance of *Acinetobacter baumannii*. *Antibiotics* **2021**, *10*, 596. https://doi.org/10.3390/antibiotics10050596

Academic Editor: Teresa V. Nogueira

Received: 29 March 2021
Accepted: 22 April 2021
Published: 18 May 2021

Publisher's Note: MDPI stays neutral with regard to jurisdictional claims in published maps and institutional affiliations.

Abstract: *Acinetobacter baumannii* is an important Gram-negative opportunistic pathogen that is responsible for many nosocomial infections. This etiologic agent has acquired, over the years, multiple mechanisms of resistance to a wide range of antimicrobials and the ability to survive in different environments. In this context, our study aims to elucidate the resistome from the *A. baumannii* strains based on phylogenetic, phylogenomic, and comparative genomics analyses. In silico analysis of the complete genomes of *A. baumannii* strains was carried out to identify genes involved in the resistance mechanisms and the phylogenetic relationships and grouping of the strains based on the sequence type. The presence of genomic islands containing most of the resistance gene repertoire indicated high genomic plasticity, which probably enabled the acquisition of resistance genes and the formation of a robust resistome. *A. baumannii* displayed an open pan-genome and revealed a still constant genetic permutation among their strains. Furthermore, the resistance genes suggest a specific profile within the species throughout its evolutionary history. Moreover, the current study performed screening and characterization of the main genes present in the resistome, which can be used in applied research to develop new therapeutic methods to control this important bacterial pathogen.

Keywords: antimicrobial; drug resistance; pan-genome; multilocus sequence typing; nosocomial infections

1. Introduction

Acinetobacter baumannii is a Gram-negative bacterium, aerobic, non-fermenting, catalase-positive coccobacillus with cosmopolitan distribution [1,2]. Most of the clinical cases involving this bacterial species are related to one or more of the following pathological conditions: severe pneumonia, meningitis, bacteremia, and erysipelas [1,3–5]. Although members of the genus *Acinetobacter* are ubiquitous, they are rarely isolated in the environment outside of hospitals, even during outbreaks [6].

This species has several intrinsic resistance mechanisms, such as (I) the presence of β-lactamases, which is responsible for the degradation of β-lactam drugs; (II) the presence of multiple drug efflux pumps that prevent the increase in the concentration of antimicrobials in the cytoplasm; (III) changes in the molecular pattern of proteins associated with plasma membrane; (IV) ribosomal methylation, which hinders the action of antimicrobials related to the regulation of protein translation processes, such as tigecyclines and quinolones; and (V) the presence of enzymes capable of degrading multiple antimicrobials [7–9].

The recommended treatment usually prescribed for infections with *A. baumannii* is based on β-lactam antibiotics, such as cephalosporins and carbapenems [10]. This class of antibiotics interferes with peptidoglycan biosynthesis and avoids forming the cell wall [11,12]. Nonetheless, over time, because of its high adaptation skills, strains capable of resisting the high concentrations of these antimicrobials have been detected [2,8]. Such cases of resistance can be classified into three categories: (i) Extensively Drug Resistant (XDR) refers to when it is resistant to more than three classes of antimicrobials; (ii) Multidrug Resistant (MDR), when it is resistant to almost all the antimicrobials except for two; or (iii) Pandrug Resistant (PDR), when it is resistant to all known antimicrobials.

Because of all these factors associated with the ability of *A. baumannii* to survive to adverse conditions (grow under a wide thermal range and in an environment with low concentrations of nutrients) and the resistance exhibited by *A. baumannii* generates numerous obstacles for the hospital treatment team, making it difficult to treat patients [1,4,6,9,13]. Some resistance mechanisms of protein origin have been previously evidenced, such as changes in the DNA-gyrase complex and an increase in the expression of the *ampC* that confers carbapenem resistance to *A. baumannii* [8].

In order to better clarify some of the resistance mechanisms present in the species genome, this study explores the genes occurring on the resistome of 206 complete genomes of strains of *A. baumannii* that are related to the resistance of this species by using multi-omic methodologies for comparative genomics, phylogenomics, and the pan-resistome of this species.

2. Results

2.1. Genomic Analysis and Geographic Distribution

Acinetobacter baumannii is a genetically diverse bacterial species and there is a variety of typing methods to identify genetic differences among the strains that could be associated with pathogenicity, epidemiological origin, dissemination, and evolutionary patterns [14]. Sequence type and phylogenetic analysis allow for the identification of genotype groups with a phylogenetic relationship and explore the diversity among the strains [15]. Similarity nucleotides and MLST analysis with geographical data can reveal a better knowledge of the epidemiological context and population structure among the strains around the world [16,17]. With the analysis of genomic similarity based on sequence alignment and geographic distribution, it is possible to infer bacterial clonality, considering that strains of bacterial species isolated from the same region tend to have the same genic repertoire. Even though events of gene drift and vertical gene transfer cannot be ruled out, genetic characteristics are generally conserved when dealing with isolated bacteria in the same site or nearby sites.

Numerous epidemiological studies of *A. baumannii* associate it with the presence of ST by local origin, as seen in the occurrence of ST 848 (CC 208) (Oxford scheme) carrying resistance gene to carbapenems in India [18], and likewise the frequent presence of ST15, ST25, ST79, and ST1 in South America [19,20]. A recent phylogeographical analysis of the Italian isolates belongs to the only clonal group ST78 (Pasteur scheme) [14].

The 206 *A. baumannii* complete NCBI genomes sequences were analyzed (see Supplementary Table S1). The genomes have sizes varying from 3.48 Mb to 4.43 Mb, with a genomic GC content of 39.05%. Considering that nearby isolated bacterial genomes tend to maintain the same genetic characteristics, the study of the geographical distribution of *A. baumannii* is an essential method for evaluating the conservation of the species in the global context.

It is important to note that all strains added to the study showed similarities greater than 95% based on the ANI results (see Supplementary Figure S1). This result corroborates the statement that all strains belong to the same species [21,22]. A total of five relevant clusters with high similarity ($\geq$98.5%) belonging mainly to specific STs (1, 2, 10, 79, and 437) were retrieved. This finding corroborates the conservation of genomes belonging to the same ST. Consequently, strains related to the same ST were expected to be isolated at locations to justify the high genomic similarity. Nevertheless, the geographic distribution of the strains according to the ST proved to be misplaced. Considering that different STs were isolated on distinct continents, possible factors that could justify this misplacing are microbial ubiquity and globalization (Figure 1). A higher number of deposited genomes belong to ST 2 (50% of the used dataset) and a more significant number of strains were isolated from the Asian continent (51.2% of the used dataset). These data do not corroborate the epidemiological information on the distribution of outbreaks caused by the bacterium *A. baumannii* [9,15,18]. Thus, this concludes that there is a more significant number of sequencing performed on the Asian and North American continents since epidemiological outbreaks have been reported in several developing countries over time (Argentina, Brazil, and South Africa). Furthermore, this pathogen has also reported outbreaks of infections on the European continent; however, the number of isolates from that continent is still much lower.

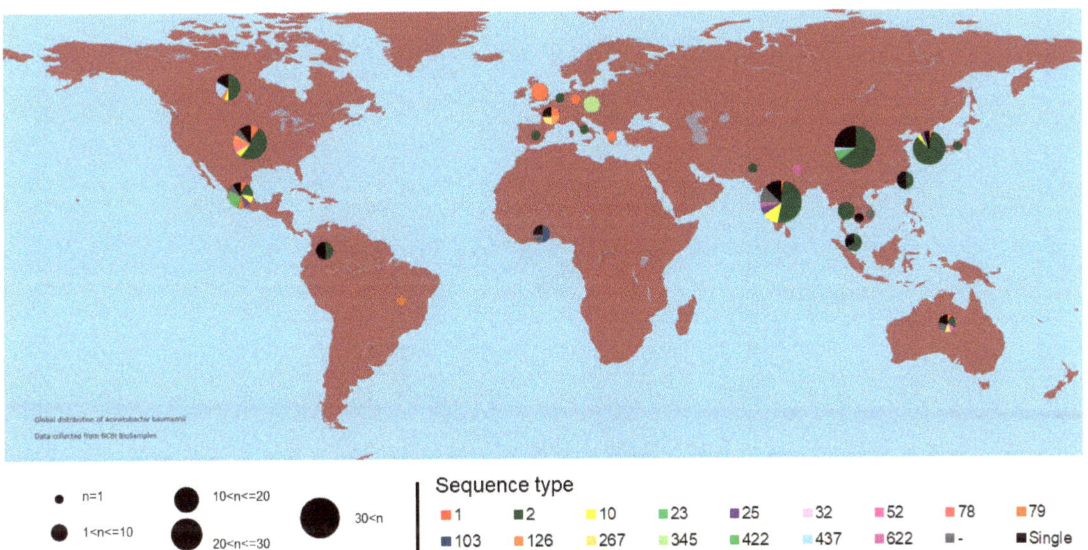

Figure 1. Graphical representation of the global distribution of isolation sites of different strains of *Acinetobacter baumannii* in a grouped way. The colors represent the sequence types of the strains in this study. The size of the circle indicates the number of isolated strains.

2.2. Phylogeny and Phylogenomics

Phylogenetically, all the *A. baumannii* strains were grouped in the same clade within the *Acinetobacter* genus, confirming the monophyly of this species (see Supplementary Figure S2). This result also points out that the *A. baumannii* strains are highly conserved within the species. It is also observed in different microbial species and is consistent with reports from the literature on phylogenetic analysis, indicating that the use of housekeeping genes to infer evolutionary history is a good qualifier of phylogenetic distance and epidemiology [23].

Three strains (FDAARGOS_494, FDAARGOS_493, and FDAARGOS_560), previously identified as *Acinetobacter* sp., were grouped together and inside the *A. baumannii* clade, strongly suggesting that they are, in fact, of this same species. This taxonomic re-classification has already occurred in other cases of bacterial species [24–26]. More phylogenomic studies, including tetranucleotide analyses, Average Nucleotide Identity (ANI), and the presence and absence of species-specific genes evaluation, are needed to confirm this hypothesis and assure taxonomic reclassification based on genomic data and theoretical background [24,27]. These three strains were not added to the subsequent analyses. The genomic similarity analysis integrated with a previous phylogenetic analysis was ideal for determining the exclusive addition of *A. baumannii* strains to the following in silico analysis, ensuring that the pan-genomic analyses were not skewed.

The *A. baumannii* strains were grouped according to their respective STs in the phylogenomic tree, using the core genome sequence (Figure 2). Nonetheless, in the phylogenomic analyses, the ST 2 strains (represented in green) formed paraphyletic clades, and, thus, these strains cannot be considered to be in the same group. The strains represented in gray do not have a defined ST, but they all grouped in the same clade, indicating the high similarity among them (see Supplementary Figure S1).

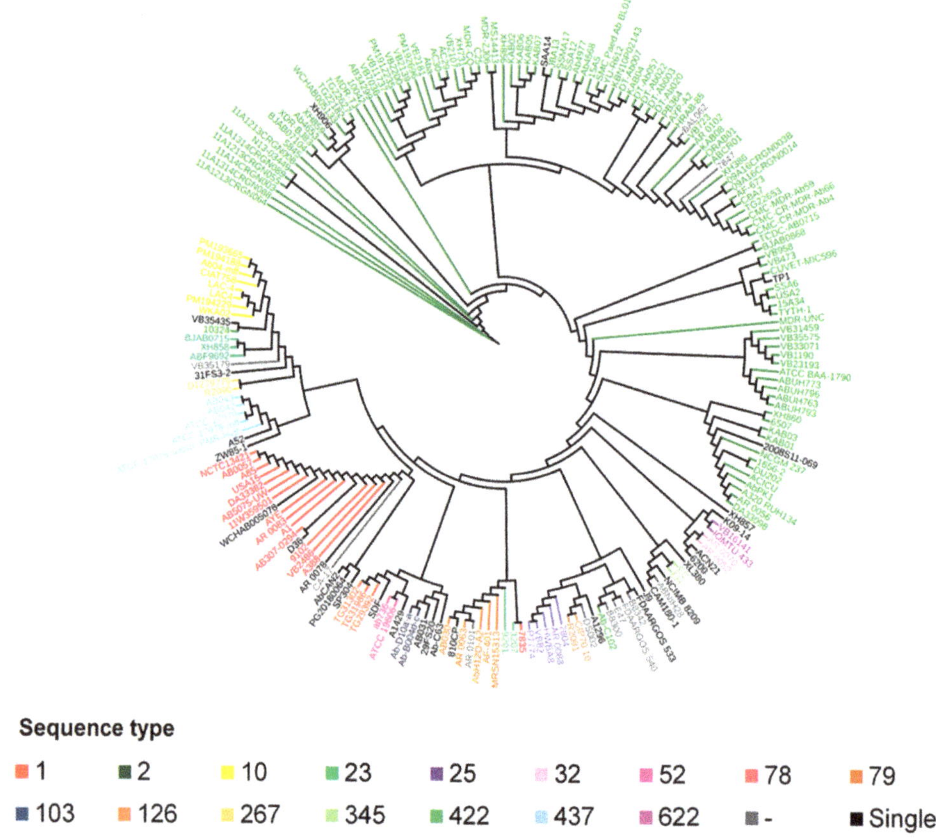

Figure 2. Phylogenomic tree based on the core genome of 206 *Acinetobacter baumannii* strains. The colors represent the grouping by sequence type. The method used was maximum likelihood with statistical support of 1000 bootstraps with 1999 genes present in the core genome.

2.3. Genomic Plasticity

During the analysis of genomic plasticity, a significant gap in the *A. baumannii* strains could be observed when visually compared. Even strains belonging to the same ST were not identical, although they were genomic, phylogenomically closer, and shared the same clade. This result suggests that the strains of this species are not very clonal and tend to have a high rate of gene permutation since there are many gaps between the genomes (Figure 3).

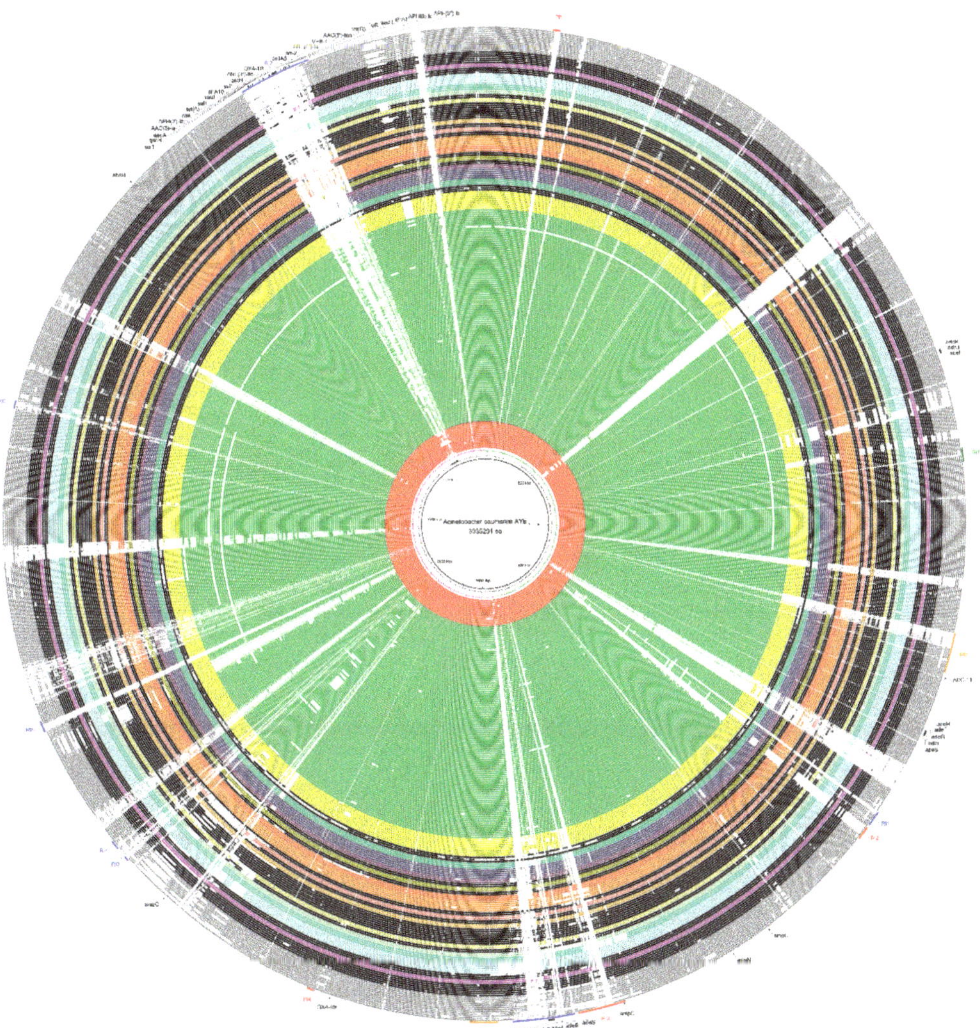

Figure 3. Representation of the circular genome of the *A. baumannii* AYE strain as a central genome. The compared strains were grouped and colored.

Comparative genomic analyses of the 206 *A. baumannii* genomes, using the AYE strain as a reference, showed the presence of 14 genomic islands (Figure 3). Among these 14 genomic islands, 4 were Pathogenicity islands, 2 were Metabolic islands, 1 was a Symbiotic island, and 7 were Resistance islands. Furthermore, 1 full-sized Resistance island (RI7 or

AbaR1) was identified within the AYE strain. This genomic region has a length of 96,878 nucleotides and contains the highest amount of resistance genes found in this species. There are 25 resistance genes within this island divided into efflux pumps and proteins with enzymatic activity.

The islands RI2 (80,220 bp) and RI7 (96,878 bp) were conserved within the species, which were more present within strains belonging to ST 1. Outside of this cluster, however, both islands were not entirely found. A similar result was observed in smaller islands, such as RI1 (20,317 bp), RI3 (6077 bp), RI4 (12,534 bp), RI5 (14,763 bp), and RI6 (10,374 bp), indicating that they are unstable regions within the genome.

There is a great number of genomic islands for the *A. baumannii* species, which reveals its high genomic plasticity. Although we identified a reduced number of type sequences and phylogenetically close strains, analyzing the complete genomes showed how all the strains are different in their gene content. This could be due to the horizontal acquisition of mobile genetic elements or gene duplication events.

2.4. Analysis of the Pan-Genome for Understanding This Species

There is an intensive effort to know the total repertoire of the *A. baumannii* species. Classically, the pan-genome assesses the total gene repertoire of a sample, population, or species. To this end, it considers subpartitions of the complete set, which are (I) a core genome consisting of genes shared by all the strains analyzed; (II) an accessory genome consisting of genes shared by two or more strains analyzed, but not all strains; (III) singletons (or exclusive genes), characterized to present exclusively in a single strain [28].

As a result, according to the Heaps' law, the pan-genome of *A. baumannii* remains open (α = 0.71), which by each newly added genome, the number of new genes will increase the genetic repertoire of the species. This result was obtained using the formula $n = a \times x^{1-\alpha}$, where n is the estimated size of the pan-genome for a given number of genomes, x is the number of genomes used, and α is a fitting parameter [28]. As a rule, when $0 < \alpha < 1$, the pan-genome is considered open. This fact also corroborates the high genomic plasticity already reported for this species, especially considering that this bacterium has an exceptional ability to obtain new gene content through transposable elements [14,29].

The pan-genome analysis revealed a total of 12,336 genes, of which 1999 genes are shared for all strains (complete genome sequences of *A. baumannii*), and 3920 were strain-specific genes. The accessory genome, except for single genes, is made up of 6417 genes. Figure 4 represents the development of the *A. baumannii* pan-genome. It is possible to observe that even using 206 genomes, the curve did not reach a point of stability or a plateau. This fact corroborates the alpha value found, as it also indicates an open pan-genome.

The different patterns of the presence of genes of the SDF strain can be observed in a detailed analysis. This strain is already known to be susceptible to antimicrobials and is the only representative of sequence type 17. Its accessory gene pattern differs from all the others and has about 362 unique genes, which contrasts with the pattern of the super-resistant AYE strain, which contains about 11 unique genes. This fact, combined with the distant phylogenomic position of the strain, shows how different the susceptible strain is from the others.

A more accurate analysis of the total pan-genome indicates the number of genes related to specific bacterial metabolic pathways. Such analysis is based on the KEGG database. It demonstrates a high number of core genes related to metabolic pathways intrinsic to microbial existence, such as energy metabolism (8.00%) and molecular translation (5.16%) (Figure 5). The accessory genes are related to amino acid metabolism (17.64%), carbohydrate metabolism (13.42%), and xenobiotics biodegradation and metabolism (7.51%). Most of the genes related to drug resistance are part of the accessory genome (2.24%), compared to their percentage represented in the core genome (1.89%). Similarly, genes related to infectious diseases are represented in the core genome (0.94%), accessory genome (2.24%), and strain-specific genes (2.82%).

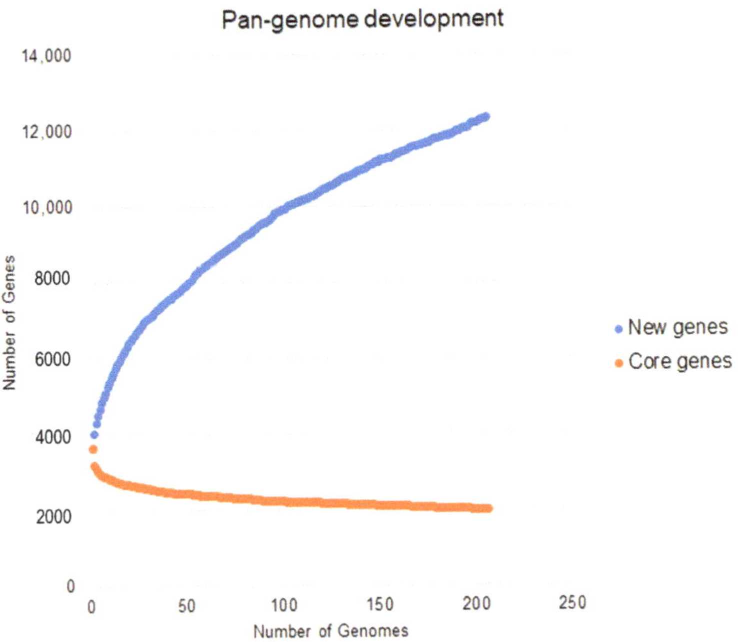

Figure 4. Development curve of the pan-genome of *Acinetobacter baumannii*.

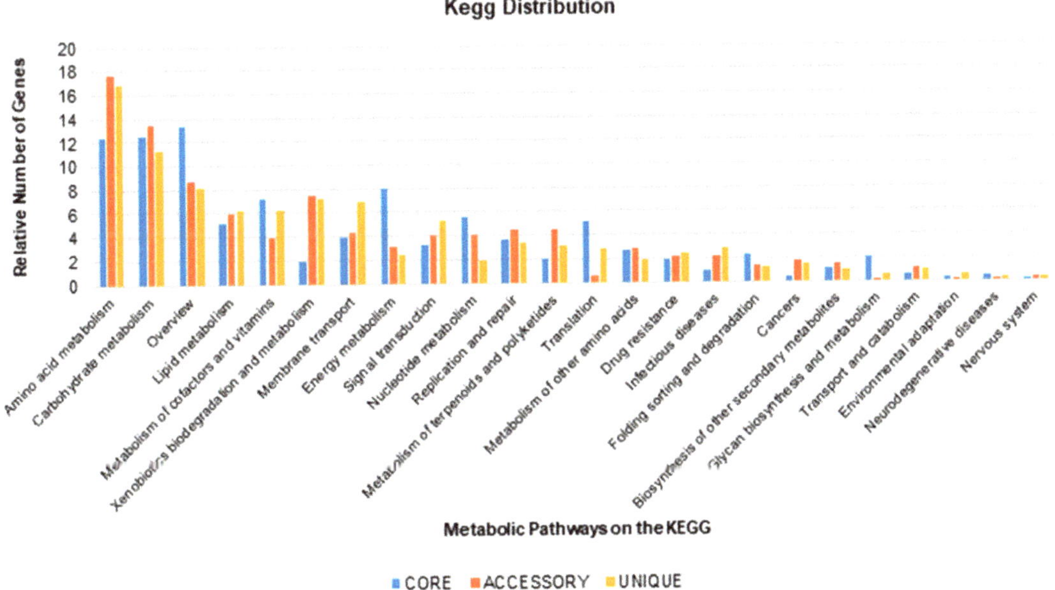

Figure 5. Graphical representation of the gene distribution by metabolic pathway within each subpartition of the total pan-genome. Only pathways with at least 0.1% of the genes represented in each subpartition of the pan-genome were considered.

As for genes related to adaptation to the environment, there is a very low gene repertoire associated with this process in the general pan-genome, with less than 1.0% of the total repertoire linked to such a pathway in any subdivision of the pan-genome.

2.5. Pan-Resistome Characterization of Acinetobacter Baumannii

A pan-resistome analysis contains analogous divisions applied to a pan-genomic analysis, but focused on microbial resistance factors [30]. Considering a similarity criterion greater than 70% and an E-value $< 5 \times 10^{-6}$, all the studied strains present a pan-resistome of 171 genes, and within that, a core resistome constituted of 9 genes is shown in Table 1 [11].

Table 1. Description of the genes present in the core resistome of the studied strains containing the mechanisms of action and antibiotics associated with these mechanisms [11].

Gene	Definition	Mechanism	Antibiotic
adeK	The outer membrane factor protein in the adeIJK multidrug efflux complex	Antibiotic efflux	Phenicol, rifamycin, penem, diaminopyrimidine, tetracycline, carbapenem, macrolide, lincosamide, floroquinolone, cephalosporin
adeJ	An RND efflux protein that acts as the inner membrane transporter of the AdeIJK efflux complex	Antibiotic efflux	Diaminopyrimidine, phenicol, tetracycline, rifamycin, carbapenem, penem, fluoroquinolone, macrolide, cephalosporin, lincosamide
adeI	The membrane fusion protein of the AdeIJK multidrug efflux complex	Antibiotic efflux	Phenicol, rifamycin, penem, diaminopyrimidine, tetracycline, carbapenem, macrolide, lincosamide, floroquinolone, cephalosporin
adeF	The membrane fusion protein of the multidrug efflux complex AdeFGH	Antibiotic efflux	Tetracycline, fluoroquinolone
adeG	The inner membrane transporter of the AdeFGH multidrug efflux complex.	Antibiotic efflux	Tetracycline, fluoroquinolone
adeL	A regulator of AdeFGH in *Acinetobacter baumannii*. AdeL mutations are associated with AdeFGH overexpression and multidrug resistance.	Antibiotic efflux	Tetracycline, fluoroquinolone
ampC	AmpC type beta-lactamases are commonly isolated from extended-spectrum cephalosporin-resistant Gram-negative bacteria.	Antibiotic inactivation	Cephalosporins
adeN	AdeN is a repressor of AdeIJK, an RND-type efflux pump in *Acinetobacter baumannii*. Its inactivation increases the expression of AdeJ.	Antibiotic efflux	Carbapenem, diaminopyrimidine, rifamycin, penem, tetracycline antibiotic, phenicol, lincosamide, fluoroquinolone, cephalosporin, macrolide
abeM	AbeM is a multidrug efflux pump found in *Acinetobacter baumannii*.	Antibiotic efflux	Acridine dye, fluoroquinolone antibiotic, triclosan

In these analyses, the strains that presented *ade*-type bombs were expected to have the complete gene repertoire to be functional. Nevertheless, this pattern was observed exclusively for the *ade*IJK efflux pump, as all the genomes presented the genes *adeI*, *adeJ*, and *adeK*. However, the same pattern was not observed for the other genes of the same family (see Supplementary Figure S3 and Supplementary Table S2). Similarly, to the genes capable of constituting the *ade*FGH pump, the presence only of the *adeF* and *adeG* genes was detected in all the strains. The gene *adeH* (the outer membrane factor protein in the *ade*FGH multidrug efflux complex) was not found in three strains (XDR-BJ83, ORAB01, and DS002), and, in theory, makes the activity of the pump unfeasible. Our study also identified an interesting protein present in all strains: the *ampC* enzyme. This is responsible for generating resistance to beta-lactams, specifically cephalosporin, and is thought to cause hydrolysis of the drug [31,32].

Analyzing the accessory portion of the resistome, an interesting distribution profile of specific genes was retrieved. The OXA-66 gene, responsible for coding a variant of beta-lactamase with action against penam, carbapenem, and cephalosporin, for example, was present in 99 strains, which is equivalent to approximately 48% of the dataset. Among these,

93 belonged to the ST 2. This fact makes this gene almost exclusive to strains belonging to ST 2. Regarding the other ST, only six strains had the OXA-66 gene, and they do not belong to ST 2, which are BAL062—ST unknown; SAA14—ST 187; XH857—ST 215; XH906—ST 922; 7847—ST unknown; TP1—ST 570.

A similar pattern was observed with the ADC-76 gene, responsible for encoding a beta-lactamase that caused cephalosporin inactivation and that was present in strains belonging exclusively to STs 23, 10, 85, 464, 575, and 639. The same was true for the OXA-68 gene, identified only in strains belonging to STs 23 and 10 but not present in all the strains. The same went for the OXA-180 gene, which was detected only in strains of ST 267. The gene responsible for encoding OXA-69 was almost exclusive to strains belonging to STs 1, 20, 81, and 195.

Other different patterns of gene distribution can be seen in Supplementary Table S2. Nonetheless, there was no significant pattern of visible distribution related to the geographic location of the isolates, except in some cases. The OXA-67 gene was exclusive to isolates (strains EC and EH) from the Czech Republic, while the ADC-81 and OXA-92 genes were observed only in the A388strain.

Otherwise, the presence of plasmid content in *A. baumannii* is already known. Among the 206 strains selected for the study, 162 were deposited with the plasmid sequence. However, there was no statistical difference regarding the number of resistance genes in strains with plasmid versus strains that did not show plasmid (p-value = 0.3081). However, qualitative differences were expected. As an example, 21 genes were found exclusively in plasmids (see Supplementary Table S3). Among the 21 exclusive plasmid genes was the MCR-4.3 gene, the only one predicted in the entire pan-resistome with action against polymyxins.

As the distribution related to the number of antibiotics was linked to each subpartition of the pan-resistome, the antibiotic with the highest amount of resistance mechanisms linked to it was cephalosporin, with about 103 resistance proteins within the formed pan-resistome (Figure 6). In contrast, antimicrobials (sulfonamide, sulfone, cephamycin, and pleuromutilin) had low amounts of resistance mechanisms related to the predicted resistome of *A. baumannii*.

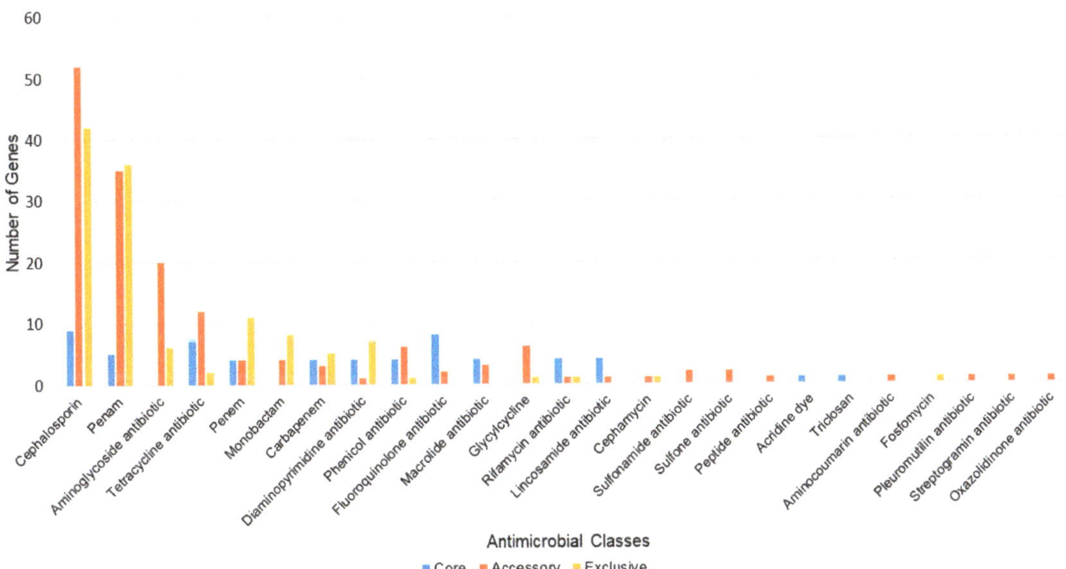

Figure 6. Distribution of the resistance mechanisms related to each antimicrobial found in the database used in the predicted total pan-resistome.

In accordance with the distribution of the types of resistance mechanisms found, 131 caused the enzymatic inactivation of the antibiotic (Figure 7). This total is equivalent to 76.6% of the predicted pan-resistome. Moreover, almost all the core resistome-related proteins are efflux pumps (8 proteins).

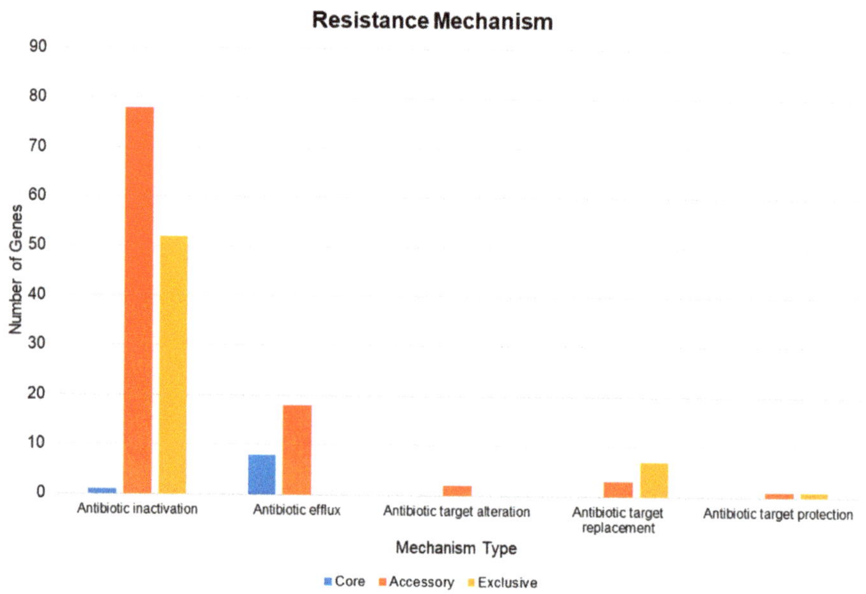

Figure 7. Distribution of the resistance mechanisms related to each type of action provided by the translated protein.

The genomics islands of resistance identified some genes, such as adeS, adeR, adeA, adeB, and adeC (within resistance island 2). Moreover, on resistance island 7 (or AbaR1 island), the following antimicrobial resistance-related genes and products were detected: sul1, qacH, AAC(3)-Ia, APH(3′)-Ia, catI, tet(A), dfrA10, ANT(3″)-IIa, OXA-10, cmlA5, arr-2, ANT(2″)-Ia, VEB-1, AAC(6′)-Ian, tet(G), floR, dfrA1, APH(6)-Id, and APH(3″)-Ib.

3. Discussion

3.1. Similarity Analysis, Geographic Distribution, and Phylogenomic Reconstruction

Recently, a more significant number of sequences of the *A. baumannii* genome provided resources for studying genomic epidemiology. Out of the 206 genomes analyzed, we identified 47 unique STs, of which STs 1, 2, 18, and 79 were distributed in significant prevalence throughout North America, Europe, and the East Asian continent, but also a diverse set of STs was indistinctly spread across the globe. The presence of few STs could be because most of the genome sequencing projects come from a single outbreak or several strains representing the same geographic location, and, in some cases, a unique ST was reported by geographic information.

According to Jeannot et al. (2014), there is a higher prevalence of strains belonging to ST 2 across the globe, but mainly in the European continent. The previous work also pointed out the polyclonality of *A. baumannii* strains within the French nation, considering the existence of different STs randomly isolated throughout the country [15].

Based on the analysis of sequence similarity of STs among the global distribution of *A. baumannii* strains, we reported several genogroups (STs) with genomic similarity less than a 98% identity in the same geographic region (Figure 1 and Supplementary Figure S1), which is opposite to the initial hypothesis that isolates from the same region exhibit high genomic similarity [23]. We also observed that the obtained result revealed a discrepancy for strains with the same ST (Figure 1 and Supplementary Figure S1). Strains from different STs in the same geographic region and the same STs isolated on other continents may maintain high similarity (>99%). In contrast, diverse genomic strains, with less than a 98% identity, are shared in nearby locations.

In previous studies, Kazmierczak and et al. (2016) reported a heterogeneous global distribution between strains of *Pseudomonas aeruginosa* and *Enterobacter* spp. based on genetic variants of lactamases. In our study, geographically close isolates may or may not have the same variant. Consequently, phylogenetic proximity is not mandatory for members of the same geographically close species [33].

The phylogenomic tree of 206 genomes displays concordance with the ST distribution, which is expected since both analyses depend on the vertical evolutionary relationship among strains. The tree phylogeny and distribution of STs, however, show a relationship with the geographical origin. This would indicate that the dissemination of *A. baumannii* does not present a population structure. Nevertheless, it is not definitive because a higher and representative number of strains is required to evaluate the population structure.

It is biologically interesting to understand the evolution and speciation of *A. baumannii* when compared to the other species of this genus. Using this analysis, it is possible to infer the origin of specific mechanisms expressed in this species and identify the closest and the most distant members within the genus to standardize comparative analyses better. Phylogenetic analysis based on a few housekeeping genes does not represent the complete evolutionary history or the final diversity between strains or members of the same genus. Nonetheless, a study based on phylogenomics shows a refined ancestry and variety caused by changes in the niche or geographic location of bacterial populations [34].

In comparison, a previous study pointed out that, for a limited number of genes, phylogenetic inference using concatenated genes is better at portraying genetic diversity and distance between different species than the use of consensus trees derived from individual genetic analyses [35]. Therefore, through the proposed method, one can evaluate a significant distancing of *A. baumannii* strains from the other species belonging to this genus, which indicates that the use of concatenated rpoB and 16 S rRNA genes is an excellent option for the inference of phylogenetic distance of strains belonging to specific genera. A previous study obtained a similar result when performing phylogenetic inference using the complete genomes of 136 strains of *Acinetobacter* within the genus [36].

3.2. Genomic Plasticity in Acinetobacter Baumannii

Previous work has reported genomic plasticity among persistent *A. baumannii* strains in Italy [14], Argentina [37], and Australia [38]. Historically, it is a species capable of receiving and donating genes to other microorganisms in the environment, a mechanism mediated by recombination events [36].

The analysis of *A. baumannii* genomes revealed the presence of 14 essential elements of resistance within genomic islands that are acquired through the horizontal transfer of genomic recombination events [39]. The largest genomic island found (RI7) has a length of 96,878 nucleotides and presents in its content a total of 25 resistance genes, characterized by an identity of more significant than 70% against the database. This same island is partially shared by strains of different ST 1 and is more similar within strains belonging to ST 2 than ST 1. Previously, this island was described as AbaR1 resistance island, and was considered to be one of the leading genomic elements responsible for the high resistance of the species members due to its size and quantity of elements. Currently, its mobile elements are known to originate from bacteria of the genera *Pseudomonas*, *Salmonella*, and *Escherichia* [1]. Based on several studies, more than 10 islands of resistance have already been identified in the genomes of *A. baumannii* [40]. In the study of the AYE strain, seven islands of resistance were detected.

Similar events of gene displacement and the presence of specific factors related to pathogenicity within genomic islands are also reported in other species with high intrinsic resistance to antimicrobials. The presence of mobile elements containing virulence or resistance factors allows for better adaptation and proliferation of *A. baumannii*. These are usually included in some phylogenetic groups, which have greater global distribution [29]. Therefore, monophyletic clades have stability in gene content, which may explain its low clonal incidence compared to other clones, which are characterized by high genomic plasticity [14]. In the literature, such mechanisms are revealed of transfer and translocation in species such as *Pseudomonas aeruginosa* [41] and *Klebsiella pneumoniae* [42], microorganisms that, like *A. baumannii*, are also considered models for understanding resistomes, virulence, and pathogenicity. This fact indicates that resistance islands are persistent in the distribution of nosocomial bacteria due to selective pressure, and they are spread and fixed in bacteria to generate adaptive fitness.

3.3. Functional Characterization through Pan-Genome Analysis

Currently, pan-genome analyses, which allow for the observation of the total genetic repertoire of a species, are incredibly relevant for determining similarity, functional characterization, and analysis of exclusive characteristics of certain strains of a microbial species. Such a report aims to assess the number of genes shared by all representatives of a taxonomic set and the genes shared by more than one, but not all, strains belonging to the group, known as accessory genomes [28]. Presently, pan-genome analyses are relevant for determining genetic variability, similarity, essential genes, functional characterization, and prediction of exclusive genes by phenotypic groups to characterize species and strains, in addition to being able to also view the discrepancies between genomes that are not perceived by conventional analyses [43]. Nowadays, there are reports of pan-genome analyses of several pathogens, such as *Streptococcus agalactiae* [44], *Legionella pneumophila* [45], *Corynebacterium pseudotuberculosis* [46], *Pasteurella multocida* [47], *Pseudomonas aeruginosa* [48], and *Treponema pallidum* [49].

In the pan-genome analysis of *A. baumannii*, a core genome containing 1999 genes was identified. Biologically, using the Kyoto Encyclopedia of Genes and Genomes database, the core genome contains all the essential genes for the survival of the bacteria in a favorable environment. Therefore, it includes pathways related to metabolism and cell division, genetic processes, and energy production [28,49]. Among the genes related to the core genome, only nine are related to resistance, and these genes represent the core resistome.

On the other hand, the accessory genome has genes related to microbial adaptation mechanisms, such as antimicrobial resistance factors, symbiosis, adaptation to the environment, and virulence, which may or may not be acquired via horizontal gene transfer [50,51]. In our study, the accessory genome revealed a total of 10,337 genes that may be related to adaptation to the host and are more represented in pathways of carbohydrate and amino acid metabolism, xenobiotic metabolism, and drug resistance. Considering the pathogenic cycle of the species, xenobiotic biosynthesis and degradation pathways are essential facilitators of bacterial adaptation, mostly when related to microbial antibiosis associated with adaptation to the host [52,53], which, in theory, provides a more prolonged microbial survival. The prevalence of carbohydrate, amino acid, and xenobiotic metabolism pathways comes in part from the pathogen's evolutionary history [54,55].

In a previous study, Hassan et al. (2016) considered 30 complete genomes of *A. baumannii* for the inference of the pan-genome, reaching values of pan- and core genomes of 7606 and 2445 genes, respectively [56]. Our work suggests a more closed pan-genome (12,336), due to an increase in the core genome (1999). This comparative result was expected, considering the increase in genomes in the dataset. This fact does not invalidate the analysis made previously by the authors but it sheds light on the development of the species' pan-genome. In contrast, Mangas et al. (2019) considered 2467 complete and draft genomes of the species for inference of the pan-genome, reaching values of the core genome and pan-genome equivalent to 2221 and 19,272 respectively [57]. The result presented in this work tends to present a more closed value in relation to the core genome, which was expected since the exclusive use of complete genomes tends to increase the accuracy of orthologs' analyses. In contrast, the number of genes found in the pan-genome was lower than that presented by the authors. This fact can be justified by the difference in methodological approaches existing between both works.

3.4. Resistome of Acinetobacter Baumannii

In the pan-resistome analysis, it was possible to ascertain the numerical presence of a variant of beta-lactamases, such as the OXA genes. Previous studies have already pointed out that this gene is widespread among the distinct geographic locations of *A. baumannii* strains, reporting that the coding gene for OXA-143 is exclusive for Brazilian strains [58]. Moreover, the same study points out that the enzyme OXA-58 is very prevalent across the globe but has a higher incidence in strains from southeastern Europe [58]. Nevertheless, using the methodology employed, the OXA-143 gene was not detected, but a similar pattern was observed for the beta-lactamase SAT-1, which is exclusive to the Brazilian strain MRSN15313. As for the OXA-58 gene, its presence was inferred for isolated strains in Italy, India, Greece, Ghana, China, and two Mexican strains, indicating and corroborating its higher prevalence in the East.

As for the efflux pumps presented, one of the most important and studied is the *ade*ABC pump, which belongs to the RND family (resistance–nodulation–division) [9]. The same *ade*IJK pump family is adequately represented in the core resistome. Previous studies indicate that efflux pumps are excellent targets for drugs, considering that their inhibition greatly amplifies the action of antimicrobials that, under normal conditions, would be eliminated by the cell [7]. Recent studies report that inhibition of the *ade*B and *ade*J portions leads to a significant reduction in microbial resistance [59]. Both are present inside the cell and anchor the pump to the membrane. As for proteins with enzymatic action, the one that stands out the most is *ampC* beta-lactamase. It has been described with high prevalence in *A. baumannii*, which is considered one of the main species responsible for the resistance to beta-lactams [31].

Interestingly, the preferred treatment for susceptible strains of *A. baumannii* is based on carbapenems [60]; however, in evaluating the pan-resistome, many mechanisms of resistance to this class have been reported, which may suggest that its presence in the genome does not indicate expression, mainly when relating to the presence of resistance mechanisms in the genome of the SDF strain. In the case of resistant strains, tigecycline treatment has been used with varying success [10,60].

Still, in this context, it is known that the number of reports of strains resistant to polymyxins within the species has grown [61,62]. However, it was possible to predict only one gene related to the resistance phenotype to that class of antimicrobials based only on the predicted proteome of the species.

4. Materials and Methods

4.1. Genomes Database, Annotation, and Data Retrieval

All the complete *A. baumannii* genomes and their plasmids were obtained through the National Center for Biotechnology Information (NCBI)/GenBank-RefSeq [63]. An in-house Python3 script was developed to extract chromosome sequences from strains with plasmid sequences, using as a criterion the extraction of the largest contig present in the fasta file. Both files, those containing only the chromosome and those containing the chromosome and plasmid, were annotated using the same parameters in the PROKKA pipeline version 1.13.7 [64], with an additional setting: the prediction of RNAs, using RNAmmer software version 1.2 [65].

4.2. Multilocus Sequence Typing and Phylogeny

The similarity analysis was performed using all the complete genomes as input to the software FastANI [21], using default parameters. The sequence type was predicted using the MLST 2.18.0 software, based on the PubMLST platform [66]. The scheme used was the *abaumannii_2*, determined and made available by the Pasteur Institute based on seven sequenced housekeeping alleles: *cpn60* (Chaperonin family protein), *fusA* (Elongation factor

G), *gltA* (Citrate synthase), *pyrG* (CTP synthase), *recA* (Protein RecA), *rplB* (50S ribosomal protein L2), and *rpoB* (a beta subunit of RNA polymerase) [67].

A customized Python3 script was used to extract the nucleotide sequences of 16S rRNA and rpoB genes, ranked by BLAST similarity (>98%) against *A. baumannii* AYE reference sequences. Subsequently, the extracted sequences were concatenated into a single file. The phylogeny was performed using the maximum likelihood method using the *rpoB* and 16S rRNA sequences. The alignment was performed using MAFFT software version 7.31.0 with default parameters [68], and the phylogenetic tree was inferred with the MEGA7 software [69], using the maximum likelihood method with statistical support of 10,000 bootstrap iterations to amplify the reliability of the formed clades. The generated tree figure was optimized using the FigTree 1.4.4 software [70].

4.3. Resistance Genes Profile

The Comprehensive Antibiotic Resistance Database (CARD) [11] was used to compare the local alignments and the determination of the presence of genes related to microbial resistance. For this purpose, the predicted proteome product of the automatic annotation of *A. baumannii* (206 strains) was used.

A customized Python 3.6 script was used to automate BLAST alignments [71] of proteomes against the CARD database. Only the results whose identity and coverage were equal to or greater than 70% and an E-value below 5×10^{-6}, respectively, were used. It was also used to generate the binary matrix of presence and absence genes, considering the previous mining files of the multiple alignments [72]. The final result was to generate the cluster map. The prediction of plasmid resistome was possible by comparing the annotation of the complete genomes (chromosome and plasmid) with the annotations of the chromosome only. The statistical difference related to the number of resistance factors between the strains that do not have a plasmid and those that do have a plasmid was made by the Wilcox-on-Mann–Whitney test.

4.4. Genomic Islands Analysis

Genomic Islands Prediction Software (GIPSy) [39] was used to perform the prediction of the genomic islands. In this analysis, the AYE strain was selected for the reference genome due to its history of resistance on the European continent and the high presence of resistance genes. As a subject, the genome of *A. baumannii* SDF was selected, which is a strain previously described as susceptible [1,39,73,74]. Subsequently, the BLAST Ring Image Generator (BRIG) software was used to visualize the genomic islands present in the genomes [75].

4.5. Pan-Genome and Pan-Resistome Analyses

The significant pan-genomic analyses were performed using the software Orthofinder [76]. It uses MCL (Markov Clustering algorithm) to determine the clusters of orthologous genes based on multiple alignments using the amino acid fasta as input. For the pan-genomic analysis, the annotation of both chromosomes and plasmids was considered. For the functional analysis of each subpartition, multiple comparisons against the Kyoto Encyclopedia of Genes and Genomes database (KEGG) [77] were considered. Obtaining the values related to the development of the pan-genome, as well as the alpha value was done through an in-house script.

The extracted core genome, resulting from the analysis using Orthofinder, was aligned with MAFFT [68] for subsequent phylogenomic inference using FastTree software with maximum likelihood methodology.

5. Conclusions

There is a wide variety of genes in the total repertoire of the species studied. Unfortunately, there is no visible clustering for the host and geographic location; however, the grouping of the strains based on ST reveals a coherent pattern, corresponding to the core genome similarity. The repertoire of the resistome was characterized in terms of the presence and similarity of genes in the total pan-genome. It demonstrated enormous plasticity when evaluating the distribution of factors throughout the groups and the analyzed phylogeny. The pan-resistome also pointed out the presence of the *adel*JK efflux pump and *amp*C enzyme in all the strains of this species, as well as the heterogeneous distribution of resistance factors across the globe. Another interesting fact is the higher amount of resistance factors to cephalosporins, aminoglycosides, and tetracycline in the studied genomes. Therefore, there is a contraindication to the use of these drugs in *A. baumannii*. These facts point mainly to the discrepancy of strains belonging to different STs within the *A. baumannii* species and its high capacity to remodel the gene repertoire to adapt to the environment or host, and, hence, can remain as an important pathogen for years. Therefore, the data collected are pertinent to better evaluate the high resistance of the species in a hospital environment and, consequently, can be used for a targeted prescription of antibiotics based on phenotyping related to a genetic presence profile. From this perspective, it is possible to use the data obtained in this work to carry out studies for new drug candidates based on the core genome and to take advantage of the assembled pan-resistome to anticipate possible escape mechanisms of *A. baumannii*.

Supplementary Materials: The following are available online at https://www.mdpi.com/article/10.3390/antibiotics10050596/s1, Figure S1: Cluster map representing the genomic similarity evaluated among all strains of *Acinetobacter baumannii* included in the study. The intensity of the color indicates a greater degree of similarity between the genomes. For each strain, the isolation site and the predicted sequence type were added respectively. The cladograms were generated using Euclidean distance. Figure S2: Phylogenetic tree based on the concatenated sequences of the 16S rRNA and rpoB genes representing the positioning of the *Acinetobacter baumannii* strains compared to the genus. Statistical support of 10,000 bootstraps was applied. The colors represent different species within the genus *Acinetobacter*. Figure S3: Cluster map representing the presence of resistance genes (*X*-axis) in all genomes (*Y*-axis) were addressed. In the case of existence, the color intensity represents the sequence similarity to the database used, with a minimum similarity of 70% versus the CARD database. The cladograms used are based on the Euclidean distance between the data. Table S1: Genomic data on the deposit made available by the National Center for Biotechnology Information (NCBI). The information is distributed respectively in Strain, BioSample, BioProject, Assembly, Size, GC%, and FTP for RefSeq access. Table S2: Matrix representing the pan-resistome of the strains under study. Numbers >0 represent the presence and similarity to the CARD database of the gene in the genome of the strain presented in the first column. The strains were grouped according to the phylogenomic proximity of the core genome. The first column represents the local isolation of each strain. The second column represents the sequence type predicted. Table S3: Resistance genes predicted exclusively in plasmids. The relative presence of genes considers only the 162 strains deposited with plasmid.

Author Contributions: Conceptualization, F.F.A., F.M.-R. and V.A.; Data curation, formal analysis and data curation, D.L.N.R., R.H. and R.G.d.S.; Writing—original draft preparation, D.L.N.R.; Writing—review and editing, F.M.-R., D.C.C., D.B., P.G., K.J.A., S.C.S., R.R., A.G.-N. and F.F.A.; Supervision, V.A., F.F.A. and F.M.-R.; Funding acquisition, V.A., F.F.A.; Visualization, D.L.N.R. All authors have read and agreed to the published version of the manuscript.

Funding: The work was financially supported by Coordenação de Aperfeiçoamento de Pessoal de Nível Superior (CAPES), Fundação de Amparo à Pesquisa do Estado de Minas Gerais (FAPEMIG), Conselho Nacional de Desenvolvimento Científico e Tecnológico (CNPq) and Pró-Reitoria de Pesquisa da Universidade Federal de Minas Gerais (PRPq) for all funding.

Institutional Review Board Statement: Not applicable.

Informed Consent Statement: Not applicable.

Data Availability Statement: The dataset used in this study is available in Supplementary Table S1.

Acknowledgments: We are very grateful to the team of the omics science network (RECOM) for all the support. KJA acknowledges a grant from Taif University Researchers Supporting Program (project number: TURSP-2020/128), Taif University, Saudi Arabia. We also thank the postgraduate program in bioinformatics and the postgraduate program in genetics for the partnership in the development of this work.

Conflicts of Interest: The authors declare no conflict of interest.

References

1. Howard, A.; O'Donoghue, M.; Feeney, A.; Sleator, R.D. Acinetobacter Baumannii-An Emerging Opportunistic Pathogen. *Virulence* **2012**, *3*, 243–250. [CrossRef]
2. Nowak, P.; Paluchowska, P. Acinetobacter Baumannii: Biology and Drug Resistance—Role of Carbapenemases. *Folia Histochem. Cytobiol.* **2016**, *54*, 61–74. [CrossRef]
3. Antunes, L.C.S.; Visca, P.; Towner, K.J. Acinetobacter Baumannii: Evolution of a Global Pathogen. *Pathog. Dis.* **2014**, *71*, 292–301. [CrossRef] [PubMed]
4. Brooks, L.E.; Ul-Hasan, S.; Chan, B.K.; Sistrom, M.J. Quantifying the Evolutionary Conservation of Genes Encoding Multidrug Efflux Pumps in the ESKAPE Pathogens to Identify Antimicrobial Drug Targets. *mSystems* **2018**, *3*. [CrossRef] [PubMed]
5. Ni, Z.; Chen, Y.; Ong, E.; He, Y. Antibiotic Resistance Determinant-Focused Acinetobacter Baumannii Vaccine Designed Using Reverse Vaccinology. *Int. J. Mol. Sci.* **2017**, *18*, 458. [CrossRef] [PubMed]
6. Towner, K. The Genus Acinetobacter. In *The Prokaryotes: A Handbook on the Biology of Bacteria Volume 6: Proteobacteria: Gamma Subclass*; Dworkin, M., Falkow, S., Rosenberg, E., Schleifer, K.-H., Stackebrandt, E., Eds.; Springer: New York, NY, USA, 2006; pp. 746–758. ISBN 978-0-387-30746-6.
7. Li, X.-Z.; Nikaido, H. Efflux-Mediated Drug Resistance in Bacteria. *Drugs* **2004**, *64*, 159–204. [CrossRef] [PubMed]
8. Peleg, A.Y.; de Breij, A.; Adams, M.D.; Cerqueira, G.M.; Mocali, S.; Galardini, M.; Nibbering, P.H.; Earl, A.M.; Ward, D.V.; Paterson, D.L.; et al. The Success of Acinetobacter Species; Genetic, Metabolic and Virulence Attributes. *PLoS ONE* **2012**, *7*, e46984. [CrossRef]
9. Peleg, A.Y.; Seifert, H.; Paterson, D.L. Acinetobacter Baumannii: Emergence of a Successful Pathogen. *Clin. Microbiol. Rev.* **2008**, *21*, 538–582. [CrossRef]
10. Gandham, P. A Review on Multidrug-Resistant Acinetobacter Baumannii. *Int. J. Curr. Microbiol. Appl. Sci.* **2014**, *3*, 5.
11. Alcock, B.P.; Raphenya, A.R.; Lau, T.T.Y.; Tsang, K.K.; Bouchard, M.; Edalatmand, A.; Huynh, W.; Nguyen, A.-L.V.; Cheng, A.A.; Liu, S.; et al. CARD 2020: Antibiotic Resistome Surveillance with the Comprehensive Antibiotic Resistance Database. *Nucleic Acids Res.* **2020**, *48*, D517–D525. [CrossRef]
12. Sultan, I.; Rahman, S.; Jan, A.T.; Siddiqui, M.T.; Mondal, A.H.; Haq, Q.M.R. Antibiotics, Resistome and Resistance Mechanisms: A Bacterial Perspective. *Front. Microbiol.* **2018**, *9*. [CrossRef] [PubMed]
13. Qin, H.; Lo, N.W.-S.; Loo, J.F.-C.; Lin, X.; Yim, A.K.-Y.; Tsui, S.K.-W.; Lau, T.C.-K.; Ip, M.; Chan, T.-F. Comparative Transcriptomics of Multidrug-Resistant Acinetobacter Baumannii in Response to Antibiotic Treatments. *Sci. Rep.* **2018**, *8*, 3515. [CrossRef] [PubMed]
14. Gaiarsa, S.; Bitar, I.; Comandatore, F.; Corbella, M.; Piazza, A.; Scaltriti, E.; Villa, L.; Postiglione, U.; Marone, P.; Nucleo, E.; et al. Can Insertion Sequences Proliferation Influence Genomic Plasticity? Comparative Analysis of Acinetobacter Baumannii Sequence Type 78, a Persistent Clone in Italian Hospitals. *Front. Microbiol.* **2019**, *10*. [CrossRef] [PubMed]
15. Jeannot, K.; Diancourt, L.; Vaux, S.; Thouverez, M.; Ribeiro, A.; Coignard, B.; Courvalin, P.; Brisse, S. Molecular Epidemiology of Carbapenem Non-Susceptible Acinetobacter Baumannii in France. *PLoS ONE* **2014**, *9*. [CrossRef]
16. Bastardo, A.; Ravelo, C.; Romalde, J.L. Phylogeography of Yersinia Ruckeri Reveals Effects of Past Evolutionary Events on the Current Strain Distribution and Explains Variations in the Global Transmission of Enteric Redmouth (ERM) Disease. *Front. Microbiol.* **2015**, *6*. [CrossRef]
17. Bastardo, A.; Balboa, S.; Romalde, J.L. *From the Gene Sequence to the Phylogeography through the Population Structure: The Cases of Yersinia Ruckeri and Vibrio Tapetis*; IntechOpen: Rijeka, Croatia, 2017; ISBN 978-953-51-2950-9.
18. Vijayakumar, S.; Mathur, P.; Kapil, A.; Das, B.K.; Ray, P.; Gautam, V.; Sistla, S.; Parija, S.C.; Walia, K.; Ohri, V.C.; et al. Molecular Characterization & Epidemiology of Carbapenem-Resistant Acinetobacter Baumannii Collected across India. *Indian J. Med. Res.* **2019**, *149*, 240–246.
19. Azevedo, F.K.S.F.; Dutra, V.; Nakazato, L.; Pepato, M.A.; Sousa, A.T.H.I.; Azevedo, C.C.S.F.; Souto, F.J.D. New Sequence Types of Acinetobacter Baumannii in Two Emergency Hospitals in the Central-West Region of Brazil. *Rev. Soc. Bras. Med. Trop.* **2019**, *52*, e20190077. [CrossRef] [PubMed]
20. Rodríguez, C.H.; Balderrama Yarhui, N.; Nastro, M.; Nuñez Quezada, T.; Castro Cañarte, G.; Magne Ventura, R.; Ugarte Cuba, T.; Valenzuela, N.; Roach, F.; Mota, M.I.; et al. Molecular Epidemiology of Carbapenem-Resistant Acinetobacter Baumannii in South America. *J. Med. Microbiol.* **2016**, *65*, 1088–1091. [CrossRef]
21. Jain, C.; Rodriguez-R, L.M.; Phillippy, A.M.; Konstantinidis, K.T.; Aluru, S. High Throughput ANI Analysis of 90K Prokaryotic Genomes Reveals Clear Species Boundaries. *Nat. Commun.* **2018**, *9*, 5114. [CrossRef]
22. Goris, J.; Konstantinidis, K.T.; Klappenbach, J.A.; Coenye, T.; Vandamme, P.; Tiedje, J.M.Y. 2007 DNA–DNA Hybridization Values and Their Relationship to Whole-Genome Sequence Similarities. *Int. J. Syst. Evol. Microbiol.* **2007**, *57*, 81–91. [CrossRef]
23. Belén, A.; Pavón, I.; Maiden, M.C.J. Multilocus Sequence Typing. *Methods Mol. Biol.* **2009**, *551*, 129–140.

24. Busse, H.-J. Review of the Taxonomy of the Genus Arthrobacter, Emendation of the Genus Arthrobacter Sensu Lato, Proposal to Reclassify Selected Species of the Genus Arthrobacter in the Novel Genera Glutamicibacter Gen. Nov., Paeniglutamicibacter Gen. Nov., Pseudoglutamicibacter Gen. Nov., Paenarthrobacter Gen. Nov. and Pseudarthrobacter Gen. Nov., and Emended Description of Arthrobacter Roseus. *Int. J. Syst. Evol. Microbiol.* **2016**, *66*, 9–37.
25. Kishi, L.T.; Fernandes, C.C.; Omori, W.P.; Campanharo, J.C.; de Macedo Lemos, E.G. Reclassification of the Taxonomic Status of SEMIA3007 Isolated in Mexico B-11A Mex as Rhizobium Leguminosarum Bv. Viceae by Bioinformatic Tools. *BMC Microbiol.* **2016**, *16*. [CrossRef]
26. Martens, T.; Heidorn, T.; Pukall, R.; Simon, M.; Tindall, B.J.; Brinkhoff, T. Reclassification of Roseobacter Gallaeciensis Ruiz-Ponte et al. 1998 as Phaeobacter Gallaeciensis Gen. Nov., Comb. Nov., Description of Phaeobacter Inhibens Sp. Nov., Reclassification of Ruegeria Algicola (Lafay et al. 1995) Uchino et al. 1999 as Marinovum Algicola Gen. Nov., Comb. Nov., and Emended Descriptions of the Genera Roseobacter, Ruegeria and Leisingera. *Int. J. Syst. Evol. Microbiol.* **2006**, *56*, 1293–1304.
27. Tindall, B.J.; Rosselló-Móra, R.; Busse, H.-J.; Ludwig, W.; Kämpfer, P. Notes on the Characterization of Prokaryote Strains for Taxonomic Purposes. *Int. J. Syst. Evol. Microbiol.* **2010**, *60*, 249–266. [CrossRef] [PubMed]
28. Carlos Guimaraes, L.; Benevides de Jesus, L.; Vinicius Canario Viana, M.; Silva, A.; Thiago Juca Ramos, R.; de Castro Soares, R.; Azevedo, V. Inside the Pan-Genome-Methods and Software Overview. *Curr. Genomics* **2015**, *16*, 245–252. [CrossRef]
29. Imperi, F.; Antunes, L.C.S.; Blom, J.; Villa, L.; Iacono, M.; Visca, P.; Carattoli, A. The Genomics of Acinetobacter Baumannii: Insights into Genome Plasticity, Antimicrobial Resistance and Pathogenicity. *IUBMB Life* **2011**, *63*, 1068–1074. [CrossRef] [PubMed]
30. He, Y.; Zhou, X.; Chen, Z.; Deng, X.; Gehring, A.; Ou, H.; Zhang, L.; Shi, X. PRAP: Pan Resistome Analysis Pipeline. *BMC Bioinform.* **2020**, *21*, 20. [CrossRef]
31. LIU, Y.; LIU, X. Detection of AmpC β-Lactamases in Acinetobacter Baumannii in the Xuzhou Region and Analysis of Drug Resistance. *Exp. Ther. Med.* **2015**, *10*, 933–936. [CrossRef]
32. Corvec, S.; Caroff, N.; Espaze, E.; Giraudeau, C.; Drugeon, H.; Reynaud, A. AmpC Cephalosporinase Hyperproduction in Acinetobacter Baumannii Clinical Strains. *J. Antimicrob. Chemother.* **2003**, *52*, 629–635. [CrossRef]
33. Kazmierczak, K.M.; Rabine, S.; Hackel, M.; McLaughlin, R.E.; Biedenbach, D.J.; Bouchillon, S.K.; Sahm, D.F.; Bradford, P.A. Multiyear, Multinational Survey of the Incidence and Global Distribution of Metallo-β-Lactamase-Producing Enterobacteriaceae and Pseudomonas Aeruginosa. *Antimicrob. Agents Chemother.* **2016**, *60*, 1067–1078. [CrossRef] [PubMed]
34. Shapiro, B.J.; Polz, M.F. Microbial Speciation. *Cold Spring Harb Perspect Biol.* **2015**, *7*. [CrossRef] [PubMed]
35. Gadagkar, S.R.; Rosenberg, M.S.; Kumar, S. Inferring Species Phylogenies from Multiple Genes: Concatenated Sequence Tree versus Consensus Gene Tree. *J. Exp. Zool. Part B Mol. Dev. Evol.* **2005**, *304B*, 64–74. [CrossRef] [PubMed]
36. Sahl, J.W.; Gillece, J.D.; Schupp, J.M.; Waddell, V.G.; Driebe, E.M.; Engelthaler, D.M.; Keim, P. Evolution of a Pathogen: A Comparative Genomics Analysis Identifies a Genetic Pathway to Pathogenesis in Acinetobacter. *PLoS ONE* **2013**, *8*, e54287. [CrossRef]
37. Mussi, M.A.; Limansky, A.S.; Relling, V.; Ravasi, P.; Arakaki, A.; Actis, L.A.; Viale, A.M. Horizontal Gene Transfer and Assortative Recombination within the Acinetobacter Baumannii Clinical Population Provide Genetic Diversity at the Single CarO Gene, Encoding a Major Outer Membrane Protein Channel. *J. Bacteriol.* **2011**, *193*, 4736–4748. [CrossRef]
38. Valenzuela, J.K.; Thomas, L.; Partridge, S.R.; van der Reijden, T.; Dijkshoorn, L.; Iredell, J. Horizontal Gene Transfer in a Polyclonal Outbreak of Carbapenem-Resistant Acinetobacter Baumannii. *J. Clin. Microbiol.* **2007**, *45*, 453–460. [CrossRef]
39. Soares, S.C.; Geyik, H.; Ramos, R.T.J.; de Sá, P.H.C.G.; Barbosa, E.G.V.; Baumbach, J.; Figueiredo, H.C.P.; Miyoshi, A.; Tauch, A.; Silva, A.; et al. GIPSy: Genomic Island Prediction Software. *J. Biotechnol.* **2016**, *232*, 2–11. [CrossRef]
40. Wang, H.; Wang, J.; Yu, P.; Ge, P.; Jiang, Y.; Xu, R.; Chen, R.; Liu, X. Identification of Antibiotic Resistance Genes in the Multidrug-Resistant Acinetobacter Baumannii Strain, MDR-SHH02, Using Whole-Genome Sequencing. *Int. J. Mol. Med.* **2017**, *39*, 364–372. [CrossRef]
41. Jani, M.; Mathee, K.; Azad, R.K. Identification of Novel Genomic Islands in Liverpool Epidemic Strain of Pseudomonas Aeruginosa Using Segmentation and Clustering. *Front. Microbiol.* **2016**, *7*. [CrossRef]
42. Lery, L.M.; Frangeul, L.; Tomas, A.; Passet, V.; Almeida, A.S.; Bialek-Davenet, S.; Barbe, V.; Bengoechea, J.A.; Sansonetti, P.; Brisse, S.; et al. Comparative Analysis of Klebsiella Pneumoniae Genomes Identifies a Phospholipase D Family Protein as a Novel Virulence Factor. *BMC Biol.* **2014**, *12*, 41. [CrossRef]
43. Medini, D.; Donati, C.; Tettelin, H.; Masignani, V.; Rappuoli, R. The Microbial Pan-Genome. *Curr. Opin. Genet. Dev.* **2005**, *15*, 589–594. [CrossRef]
44. Tettelin, H.; Masignani, V.; Cieslewicz, M.J.; Donati, C.; Medini, D.; Ward, N.L.; Angiuoli, S.V.; Crabtree, J.; Jones, A.L.; Durkin, A.S.; et al. Genome Analysis of Multiple Pathogenic Isolates of Streptococcus Agalactiae: Implications for the Microbial Pan-Genome. *Proc Natl. Acad. Sci. USA* **2005**, *102*, 13950–13955. [CrossRef] [PubMed]
45. D'Auria, G.; Jiménez-Hernández, N.; Peris-Bondia, F.; Moya, A.; Latorre, A. Legionella Pneumophila Pangenome Reveals Strain-Specific Virulence Factors. *BMC Genom.* **2010**, *11*, 181. [CrossRef] [PubMed]
46. Soares, S.C.; Silva, A.; Trost, E.; Blom, J.; Ramos, R.; Carneiro, A.; Ali, A.; Santos, A.R.; Pinto, A.C.; Diniz, C.; et al. The Pan-Genome of the Animal Pathogen Corynebacterium Pseudotuberculosis Reveals Differences in Genome Plasticity between the Biovar Ovis and Equi Strains. *PLoS ONE* **2013**, *8*. [CrossRef]

47. Hurtado, R.; Carhuaricra, D.; Soares, S.; Viana, M.V.C.; Azevedo, V.; Maturrano, L.; Aburjaile, F. Pan-Genomic Approach Shows Insight of Genetic Divergence and Pathogenic-Adaptation of Pasteurella Multocida. *Gene* **2018**, *670*, 193–206. [CrossRef] [PubMed]

48. Freschi, L.; Vincent, A.T.; Jeukens, J.; Emond-Rheault, J.-G.; Kukavica-Ibrulj, I.; Dupont, M.-J.; Charette, S.J.; Boyle, B.; Levesque, R.C. The Pseudomonas Aeruginosa Pan-Genome Provides New Insights on Its Population Structure, Horizontal Gene Transfer, and Pathogenicity. *Genome Biol. Evol.* **2019**, *11*, 109–120. [CrossRef] [PubMed]

49. Jaiswal, A.K.; Tiwari, S.; Jamal, S.B.; de Castro Oliveira, L.; Alves, L.G.; Azevedo, V.; Ghosh, P.; Oliveira, C.J.F.; Soares, S.C. The Pan-Genome of Treponema Pallidum Reveals Differences in Genome Plasticity between Subspecies Related to Venereal and Non-Venereal Syphilis. *BMC Genom.* **2020**, *21*, 33. [CrossRef]

50. Blaustein, R.A.; McFarland, A.G.; Ben Maamar, S.; Lopez, A.; Castro-Wallace, S.; Hartmann, E.M. Pangenomic Approach to Understanding Microbial Adaptations within a Model Built Environment, the International Space Station, Relative to Human Hosts and Soil. *mSystems* **2019**, *4*. [CrossRef] [PubMed]

51. Rocha, E.P. Evolutionary Patterns in Prokaryotic Genomes. *Curr. Opin. Microbiol.* **2008**, *11*, 454–460. [CrossRef] [PubMed]

52. Klotz, L.-O.; Steinbrenner, H. Cellular Adaptation to Xenobiotics: Interplay between Xenosensors, Reactive Oxygen Species and FOXO Transcription Factors. *Redox Biol.* **2017**, *13*, 646–654. [CrossRef]

53. Patterson, A.D.; Gonzalez, F.J.; Idle, J.R. Xenobiotic Metabolism–A View through the metabolometer. *Chem. Res. Toxicol.* **2010**, *23*, 851–860. [CrossRef] [PubMed]

54. Bergogne-Bérézin, E. Acinetobacter Spp., Saprophytic Organisms of Increasing Pathogenic Importance. *Zentralbl. Bakteriol.* **1994**, *281*, 389–405. [CrossRef]

55. Doughari, H.J.; Ndakidemi, P.A.; Human, I.S.; Benade, S. The Ecology, Biology and Pathogenesis of Acinetobacter Spp.: An Overview. *Microbes Environ.* **2011**, *26*, 101–112. [CrossRef] [PubMed]

56. Hassan, A.; Naz, A.; Obaid, A.; Paracha, R.Z.; Naz, K.; Awan, F.M.; Muhmmad, S.A.; Janjua, H.A.; Ahmad, J.; Ali, A. Pangenome and Immuno-Proteomics Analysis of Acinetobacter Baumannii Strains Revealed the Core Peptide Vaccine Targets. *BMC Genomics* **2016**, *17*. [CrossRef] [PubMed]

57. Mangas, E.L.; Rubio, A.; Álvarez-Marín, R.; Labrador-Herrera, G.; Pachón, J.; Pachón-Ibáñez, M.E.; Divina, F.; Pérez-Pulido, A.J. Pangenome of Acinetobacter Baumannii Uncovers Two Groups of Genomes, One of Them with Genes Involved in CRISPR/Cas Defence Systems Associated with the Absence of Plasmids and Exclusive Genes for Biofilm Formation. *Microb. Genom.* **2019**, *5*. [CrossRef]

58. Evans, B.A.; Amyes, S.G.B. OXA β-Lactamases. *Clin. Microbiol. Rev.* **2014**, *27*, 241–263. [CrossRef] [PubMed]

59. Abdi, S.N.; Ghotaslou, R.; Ganbarov, K.; Mobed, A.; Tanomand, A.; Yousefi, M.; Asgharzadeh, M.; Kafil, H.S. Acinetobacter Baumannii Efflux Pumps and Antibiotic Resistance. *Infect. Drug Resist.* **2020**, *13*, 423–434. [CrossRef]

60. Moubareck, C.A.; Halat, D.H. Insights into Acinetobacter Baumannii: A Review of Microbiological, Virulence, and Resistance Traits in a Threatening Nosocomial Pathogen. *Antibiotics* **2020**, *9*, 119. [CrossRef]

61. Elham, B.; Fawzia, A. Colistin Resistance in Acinetobacter Baumannii Isolated from Critically Ill Patients: Clinical Characteristics, Antimicrobial Susceptibility and Outcome. *Afr. Health Sci.* **2019**, *19*, 2400–2406. [CrossRef]

62. Genteluci, G.L.; Gomes, D.B.C.; Souza, M.J.; de Carvalho, K.R.; Villas-Bôas, M.H.S.; Genteluci, G.L.; Gomes, D.B.C.; Souza, M.J.; de Carvalho, K.R.; Villas-Bôas, M.H.S. Emergence of Polymyxin B-Resistant Acinetobacter Baumannii in Hospitals in Rio de Janeiro. *J. Bras. Patol. Med. Lab.* **2016**, *52*, 91–95. [CrossRef]

63. Pruitt, K.D.; Tatusova, T.; Maglott, D.R. NCBI Reference Sequences (RefSeq): A Curated Non-Redundant Sequence Database of Genomes, Transcripts and Proteins. *Nucleic Acids Res.* **2007**, *35*, D61–D65. [CrossRef]

64. Seemann, T. Prokka: Rapid Prokaryotic Genome Annotation. *Bioinformatics* **2014**, *30*, 2068–2069. [CrossRef]

65. Lagesen, K.; Hallin, P.; Rødland, E.A.; Stærfeldt, H.-H.; Rognes, T.; Ussery, D.W. RNAmmer: Consistent and Rapid Annotation of Ribosomal RNA Genes. *Nucleic Acids Res.* **2007**, *35*, 3100–3108. [CrossRef] [PubMed]

66. Jolley, K.A.; Maiden, M.C. BIGSdb: Scalable Analysis of Bacterial Genome Variation at the Population Level. *BMC Bioinform.* **2010**, *11*, 595. [CrossRef] [PubMed]

67. Diancourt, L.; Passet, V.; Nemec, A.; Dijkshoorn, L.; Brisse, S. The Population Structure of Acinetobacter Baumannii: Expanding Multiresistant Clones from an Ancestral Susceptible Genetic Pool. *PLoS ONE* **2010**, *5*. [CrossRef] [PubMed]

68. Katoh, K.; Standley, D.M. MAFFT Multiple Sequence Alignment Software Version 7: Improvements in Performance and Usability. *Mol. Biol. Evol.* **2013**, *30*, 772–780. [CrossRef] [PubMed]

69. Kumar, S.; Stecher, G.; Tamura, K. MEGA7: Molecular Evolutionary Genetics Analysis Version 7.0 for Bigger Datasets. *Mol. Biol. Evol.* **2016**, *33*, 1870–1874. [CrossRef]

70. Rambaut, A. Rambaut/Figtree; GitHub Repository. 2020. Available online: https://github.com/rambaut/figtree.git (accessed on 18 May 2021).

71. Mount, D.W. Using the Basic Local Alignment Search Tool (BLAST). *Cold Spring Harb. Protoc.* **2007**, *2007*. [CrossRef]

72. Rodrigues, D.L.N. PanViTa Tool; GitHub Repository. 2020. Available online: https://github.com/dlnrodrigues/panvita (accessed on 18 May 2021).

73. Adams, M.D.; Goglin, K.; Molyneaux, N.; Hujer, K.M.; Lavender, H.; Jamison, J.J.; MacDonald, I.J.; Martin, K.M.; Russo, T.; Campagnari, A.A.; et al. Comparative Genome Sequence Analysis of Multidrug-Resistant Acinetobacter Baumannii. *J. Bacteriol.* **2008**, *190*, 8053–8064. [CrossRef] [PubMed]

74. Vallenet, D.; Nordmann, P.; Barbe, V.; Poirel, L.; Mangenot, S.; Bataille, E.; Dossat, C.; Gas, S.; Kreimeyer, A.; Lenoble, P.; et al. Comparative Analysis of Acinetobacters: Three Genomes for Three Lifestyles. *PLoS ONE* **2008**, *3*, e1805. [CrossRef]
75. Alikhan, N.-F.; Petty, N.K.; Ben Zakour, N.L.; Beatson, S.A. BLAST Ring Image Generator (BRIG): Simple Prokaryote Genome Comparisons. *BMC Genomics* **2011**, *12*, 402. [CrossRef] [PubMed]
76. Emms, D.M.; Kelly, S. OrthoFinder: Phylogenetic Orthology Inference for Comparative Genomics. *Genome Biol.* **2019**, *20*, 238. [CrossRef] [PubMed]
77. Kanehisa, M.; Furumichi, M.; Tanabe, M.; Sato, Y.; Morishima, K. KEGG: New Perspectives on Genomes, Pathways, Diseases and Drugs. *Nucleic Acids Res.* **2017**, *45*, D353–D361. [CrossRef] [PubMed]

MDPI

St. Alban-Anlage 66

4052 Basel

Switzerland

Tel. +41 61 683 77 34

Fax +41 61 302 89 18

www.mdpi.com

Antibiotics Editorial Office

E-mail: antibiotics@mdpi.com

www.mdpi.com/journal/antibiotics